Laser Cleaning II

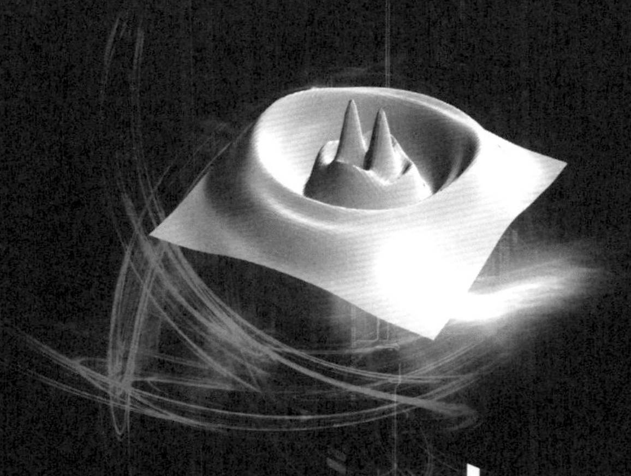

Laser Cleaning II

D M Kane

Macquarie University, Sydney, Australia

World Scientific

NEW JERSEY • LONDON • SINGAPORE • BEIJING • SHANGHAI • HONG KONG • TAIPEI • CHENNAI

Published by

World Scientific Publishing Co. Pte. Ltd.
5 Toh Tuck Link, Singapore 596224
USA office: 27 Warren Street, Suite 401-402, Hackensack, NJ 07601
UK office: 57 Shelton Street, Covent Garden, London WC2H 9HE

British Library Cataloguing-in-Publication Data
A catalogue record for this book is available from the British Library.

LASER CLEANING II

ISBN-13 978-981-270-372-9
ISBN-10 981-270-372-1

Printed in Singapore by World Scientific Printers (S) Pte Ltd

LIST OF CONTRIBUTORS

N. Arnold
Institut für Angewandte Physik, Johannes-Kepler-Universität Linz, A-4040 Linz, Austria

K. G. H. Baldwin
Atomic and Molecular Physics Laboratories, Research School of Physical Sciences and Engineering, The Australian National University, Canberra, ACT 0200, Australia

P. Balling
Department of Physics And Astronomy, University of Aarhus, Ny Munkegade, Dk-8000 Aarhus C, Denmark

S. Barcikowski
Laser Zentrum Hannover E.V., Hollerithallee 8, 30419 Hannover, Germany

N. Bärsch
Laser Zentrum Hannover E.V., Hollerithallee 8, 30419 Hannover, Germany

D. Bäuerle
Institut für Angewandte Physik, Johannes-Kepler-Universität Linz, A-4040 Linz, Austria

D. Brodoceanu
Institut für Angewandte Physik, Johannes-Kepler-Universität Linz, A-4040 Linz, Austria

T. Burmester
Laser Zentrum Hannover E.V., Hollerithallee 8, 30419 Hannover, Germany

J. Bunte
Laser Zentrum Hannover E.V., Hollerithallee 8, 30419 Hannover, Germany

R. J. Carman
Department of Physics, Macquarie University, Sydney, NSW 2109, Australia

R. J. Chater
Department Of Materials, Exhibition Road, Imperial College, London SW7 2AZ, UK

A. J. Fernandes
Department Of Physics, Macquarie University, Sydney, Nsw 2109, Australia

D. Freeman
Laser Physics Centre, The Australian National University, Canberra, ACTt 0200, Australia

E. G. Gamaly
Laser Physics Centre, Research School of Physical Sciences and Engineering, The Australian National University, Canberra, ACT 0200, Australia

A. Gervais
Norddeutsches Zentrum Für Materialkunde Von Kulturgut E.V. (North German Centre for the Material Science of Cultural Assets), Scharnhorststr. 1, 30175 Hannover, Germany

B. Gong
Surface Science and Technology, University Of New South Wales, Sydney, NSW 2052, Australia

T. Gumpenberger
Institut für Angewandte Physik, Johannes-Kepler-Universität Linz, A-4040 Linz, Austria

D. Hallam
National Museum of Australia, Acton Peninsula, Canberra, ACT 2600, Australia

J. Heitz
Institut für Angewandte Physik, Johannes-Kepler-Universität Linz, A-4040 Linz, Austria

D. Hirschausen
Department of Physics, Macquarie University, Sydney, NSW 2109, Australia

J. Hughes
The University of Canberra, Bruce, ACT 2617, Australia

D. Jang
Department of Mechanical Engineering, POSTECH, Pohang, 790-784, Korea

A. M. Joyce
Department of Physics, Macquarie University, Sydney, Nsw 2109, Australia

D. M. Kane
Department of Physics, Macquarie University, Sydney, Nsw 2109, Australia

D. Kim
Department of Mechanical Engineering, POSTECH, Pohang, 790-784, Korea

J. Kofler
Institut für Angewandte Physik, Johannes-Kepler-Universität Linz, A-4040 Linz, Austria

V. Z. Kolev
Laser Physics Centre, The Australian National University, Canberra, ACT 0200, Australia

R. N. Lamb
Surface Science and Technology, University of New South Wales, Sydney, NSW 2052, Australia

G. Langer
Institut für Angewandte Physik, Johannes-Kepler-Universität Linz, A-4040 Linz, Austria

B. Luther-Davies
Laser Physics Centre, Research School of Physical Sciences and Engineering, The Australian National University, Canberra, ACT 0200, Australia

N. R. Madsen
Laser Physics Centre, Research School of Physical Sciences and Engineering, The Australian National University, Canberra, ACT 0200, Australia

J. Marczak
Institute Of Optoelectronics, Military University of Technology, 2 Kaliskiego Street, 00-908 Warsaw, Poland

M. Meier
Niedersächsisches Landesamt Für Denkmalpflege (Lower Saxony Department of Preservation of Ancient Monuments), Scharnhorststr. 1, 30175 Hannover, Germany

R. P. Mildren
Department of Physics, Macquarie University, Sydney, NSW 2109, Australia

B. Oh
Department of Mechanical Engineering, POSTECH, Pohang, 790-784, Korea

A. Ostendorf
Laser Zentrum Hannover E.V., Hollerithallee 8, 30419 Hannover, Germany

R. Ostrowski
Institute Of Optoelectronics, Military University of Technology, 2 Kaliskiego Street, 00-908 Warsaw, Poland

K. Piglmayer
Institut für Angewandte Physik, Johannes-Kepler-Universität Linz, A-4040 Linz, Austria

S. Pleasants
Department of Physics, Macquarie University, Sydney, NSW 2109, Australia (but now at Dept. of Information and Image Sciences, Chiba University, Chiba, Japan)

A. V. Rode
Laser Physics Centre, Research School of Physical Sciences and Engineering, The Australian National University, Canberra, ACT 0200, Australia

M. Strzelec
Institute of Optoelectronics, Military University of Technology, 2 Kaliskiego Street, 00-908 Warsaw, Poland

J. Ulrich
Laser Zentrum Hannover E.V., Hollerithallee 8, 30419 Hannover, Germany

A. Wain
Australian War Memorial, ANZAC Parade, Campbell ACT, 2612, Australia

J. Walter
Laser Zentrum Hannover E.V., Hollerithallee 8, 30419 Hannover, Germany

B. K. Ward
Department of Physics, Macquarie University, Sydney, NSW 2109, Australia

PREFACE

Laser Cleaning II is the second book to update research on the use of lasers to remove contamination from surfaces, with a primary, but non-exclusive, focus on contaminant particle removal. It is sub-micron and nano-scale sized particles, in particular, that can adhere very strongly to surfaces and present a key cleaning challenge, in the broad context of cleaning science and technology. Some of these challenges, relevant to the semiconductor fabrication industry, are documented in the International Technology Roadmap for Semiconductors*. Other significant application areas are optics and photonics, biotechnology, nuclear decontamination and the automotive industry. Discoveries related to the near-field focussing of the laser beam by sub-micron and nano-scale particles on surfaces have led to a new branch of investigation using this as a method of nano-patterning and nano-structuring surfaces described in this book. Laser cleaning for Art and Cultural Heritage Conservation is a related, mature field of research and there are also chapters here-in that are cross-linked to this research area. The first book *Laser Cleaning*, edited by Boris Luk'yanchuk, was published in 2002. This second book is dedicated to him as an appropriate way to mark the occasion of his 60th birthday. It includes chapters that are original, peer reviewed, research papers, presented at the 4th International Workshop on Laser Cleaning/ New Trends in Laser Cleaning III, (IWLC4/NeToLAC III) held in Sydney in December 2004, and research review chapters.

As the knowledge and understanding of laser cleaning science and processes increases, some key insights into the interplay of ideal theory and physical reality are emerging. If nano-scale particles are adhered to a surface by van der Waals-force-adhesion, then dry laser cleaning of these particles from the surface will not be achieved without laser-induced damage to the underlying surface. However, it emerges that in many "real" physical cases, the particles, either introduced "dry" or from a suspension, do not approach the surface closely enough (due to electrostatic, capillary and/or moisture film effects) and so can have an adhesion to the surface orders of magnitude smaller than the

*International Technology Roadmap for Semiconductors 2003 Edition, 2004 tables Updates, http://public.itrs.net/, accessed 14/2/05, in particular "Front End Processes" section.

idealised “0.4 nm spaced” van der Waals force. Thus, the theoretical predictions of laser pulse energy required for laser cleaning, based on van der Waals adhesion, are a “worst case” scenario. The adhesion in real cleaning applications can be significantly less and therefore successful dry laser cleaning can be achieved in many cases. If the adhesion is consistent with van der Waal’s; which will be the case in ultraclean, dry, electrically-neutral environments; then steam laser cleaning and hybrid laser cleaning (that combine laser cleaning with other cleaning methods such as ultrasonic/megasonic cleaning, dry CO_2 (snow) cleaning, shockwave cleaning, brushing etc.) may be preferred. Such techniques may be preferred for process/manufacturing-related reasons in any case. Progress in laser cleaning research, as described in Laser Cleaning and Laser Cleaning II (this book), has been facilitated by the strong interaction between top theoretical physicists and experimentalists in the field. In my opinion, the “story” of laser cleaning is of interest to a broader readership than those in the field for this reason. A significant part of this story is the work of Professor Boris Luk’yanchuk and it is a pleasure to celebrate his career — a lifetime of dedication to high-level theoretical physics — and to mark the occasion of his 60th birthday by dedicating this book to him.

Deb Kane
Department of Physics
Macquarie University, Sydney
Australia

Professor Boris Luk'yanchuk

TRIBUTE TO PROFESSOR BORIS LUK'YANCHUK: TO MARK HIS CONTRIBUTIONS TO PHYSICS ON THE OCCASION OF HIS 60TH BIRTHDAY

Professor Boris Luk'yanchuk was born 22 March 1944 and graduated from M. V. Lomonosov Moscow State University, Russia, with Specialty: Physics, in 1967. He then graduated from High Postgraduate Courses at Moscow Institute for Physics and Technology in 1970 (Cathedra at L. D. Landau Theoretical Physics Institute, Academy of Sciences of USSR), with Specialty: Theoretical Physics. He joined the Scientific Research Institute for Optical and Physical Measuring, Moscow as a Senior Engineer in 1970 and then was Principal Designer at the Scientific Research Institute "Altair", Moscow from 1973-1980. His PhD in Physics and Mathematics was awarded in 1979 from the P. N. Lebedev Physical Institute, Academy of Sciences of USSR. This was followed by the Doctor of Sciences (Physics and Mathematics) — Second Doctorate from the General Physics Institute, Academy of Sciences of USSR, in 1991; and the State Professor's Degree (Physics and Mathematics) from the State Highest Certifying Committee of the Russian Federation in 1992. From 1980–1983 he was a Senior Researcher at the P. N. Lebedev Physical Institute, Academy of Sciences of USSR, Moscow. From 1983–2000 he was Head of Laboratory at the General Physics Institute, Russian Academy of Sciences, Moscow. In parallel with these positions he has held visiting and invited Professor positions at Johannes Kepler University, Linz, Austria (1989–1999); University of Lecce, Italy (1993–1999); University of Marseilles, France, (1993–1998); Tokyo Institute of Technology, Japan, (1999); and Data Storage Institute, Singapore (1999-2000). He is an Honorary Professor at Johannes Kepler University, Austria. Since 2000 he has been Professor and Senior Scientist at the Data Storage Institute, Agency for Science, Technology and Research, Singapore. He has published six monographs and more than 230 research papers. It is a hallmark of Boris that he proudly and regularly acknowledges the privilege and benefit of being taught by physics greats. One of his postgraduate teachers was the famous teacher, Prof. A. Abrikosov, at the Landau Institute of Theoretical Physics. Boris' research has been in many fields of physics. He started by making important contributions in the theory of quantum liquids, and then in the theory of shock waves in solids. But most of his life's work to date is in solving many and varied, challenging theoretical problems of laser matter interactions.

Boris has supervised and contributed to the supervision of more than 30 PhD students, in addition to being an excellent lecturer of both undergraduate and postgraduate physics. Such a list of qualifications, positions and contributions, associated with such prestigious institutions, is impressive and establishes a high benchmark of science credibility but it is the bare bones of a scientific career, albeit a very strong skeleton. It tells nothing of the specific knowledge created, the growth in understanding of the physical world and the significance of these for society; and the impact the scientist has on the lives of others – this is the flesh on the bones. And in the case of a generous spirited scientist like Boris Luk'yanchuk it is the warmest flesh. For his 60th birthday Boris received, on the day, the warm wishes from colleagues and collaborators from around the world, all wanting to acknowledge the positive impact of his life and work on their own. What better recommendation could you want, to show the impact of an amazingly talented individual, playing a key role in weaving quality science (an interwoven, interconnected fabric) through leading in, and sharing with a network of colleagues.

One of the things colleagues and friends enjoy about conversations with Boris is gaining his insights into the world of Nobel prizes in Physics. His knowledge of the winners, and the physics honoured by the prize, is prodigious. It extends to having gained a reputation for being a canny predictor of Nobel Laureates. Boris likes to recount that in 1999 he suggested a person as an invited speaker to an International Conference only to be told by the organizers that they did not know this person and they rejected the nomination. Boris exclaimed: "But this person is very close to winning the Nobel Prize!" The organizers just smiled wryly and were unmoved. In 2000 this person received Nobel Prize in Physics. It was Prof. Zhores Alferov. Later in 2002 Boris suggested his teacher, Prof. Alexei Abrikosov, as an invited lecturer to Singapore, saying that in his view he was very close to receiving the Nobel Prize. Once again people smiled and rejected the suggestion. Prof. Abrikosov received his Nobel Prize in Physics in 2004. Another person Boris predicted to be an expected Nobel Prize winner was Prof. Vitaly Ginzburg. News of the lost opportunity to "bet" on Boris' predictions reached his Director, at DSI. It is reported the Director exclaimed "Next time Boris suggests anybody for a public lecture in Singapore invite this person immediately!" The next suggested invitee was Prof. Vladimir Zakharov who was invited and did give a lecture in

Singapore. Within six months Prof. Zakharov was awarded Gold Medal of Dirac exactly for the topic of his lecture.

For new colleagues and friends, a first meeting with Boris will likely involve a question of whether you can write down Maxwell's equations. He will not be too harsh on those who cannot. If visiting you in your office, he will comment on your book collection approving one that has all of the volumes of Landau and Lifshitz and question any that do not. A developing link will likely involve Boris demonstrating his phenomenal knowledge of current research, and deep understanding of the underpinning physics, over the broadest range, through discussions and presentations. His powerpoint tutorial presentations are much admired and catalogued on colleagues computers around the world. His strong preference to link with experimental scientists; his ability to talk, listen, and interact productively and creatively with all physicists; and his passion for the subject make him a physicist's physicist and everybody's physicist. He is always looking to make a ground breaking contribution in fertile new areas of research. Some examples of his recent contributions (2000–2004) are discussed briefly below.

Determination of the basic parameters related to light scattering by a particle on a surface

Boris met this problem in 1999, solving problems related to laser cleaning of micro- and nano-sized particles on surfaces using short laser pulses [1]. At this time the generally accepted model of dry laser cleaning considered that a small particle was not influenced by the thermal expansion of the substrate, which produced the necessary acceleration for the dry laser cleaning effect. When Boris examined this problem precisely using Mie theory and the more complicated theory of "particle on surface" which he developed (this theory takes into account effects of secondary scattering of radiation reflected by substrate) [2], he found that the small transparent particle functions as an efficient lens in the near-field region, producing enhancement of the energy flux, on a nano-scale, at the surface, and near to the surface. The enhancement of the field strength varied from ten to one hundred times for different conditions (even for particles with sizes of the order of the radiation wavelength and smaller). This enhancement was sufficient for use in many technological applications, e.g. for new technology of processing on the nano-scale region, because the efficiency of conversion of light energy under the particle was condensed to many orders of amplitude higher than achieved with the scanning near field optical microscopy (SNOM) technique, where the efficiency of light transmission is extremely small (10^{-5}–10^{-6}). The DSI group confirmed that a spherical particle of micrometer size can achieve surface modification on the scale of about 100 nm experimentally in 2000 [3]. This discovery was confirmed by many experimental groups within one year of his theoretical prediction and has become the basis for the development of many applications. Since 2001 there have been a lot of publications devoted to different applications and designs using the technique.

The precise analysis of the particle on the surface physics problem was a very complicated one and it took some time to fully treat and clarify the many aspects of the interactions [1, 4–14]. This has resulted in the ability to fully determine the basic parameters related to light scattering for a particle on a surface, which has been applied in technologies by the DSI group (and others). For example, they created a three-dimensional theory of dry laser cleaning, which precisely describes experimental results [14], while conventional 1D-theory disagrees with experiments by 1–2 orders of magnitude.

Application of near field effects for material processing

It was clear that near-field processing lends itself to multiplication using an array of particles. The group of Prof. P. Leiderer (Konstanz University, Germany) suggested, in 2001, the use of colloidal particles to produce a self-assembled 2D matrix array of particles for such processing. The DSI, Singapore, group also worked with this technique applying nanosecond and femtosecond pulses [1, 6, 9, 12, 13, 15–22]. Among the DSI achievements was the processing of metallic films. This is rather complicated for nanoscale processing due to the high reflectivity and high thermal conductivity of the metal. At the same time the theory of a particle on a surface predicts that the surface works as a mirror coupled with a spherical resonator and this increase in the reflectivity leads to an enhancement of surface intensity and sharpening in the spatial intensity distribution. Both of these effects permit efficient processing of metallic films. It was experimentally demonstrated that an excimer laser, with nanosecond pulselength, produced a lattice of holes with a size below 50 nm (using an array of 140 nm PS particles) [1, 9, 12, 16]. Another application was related to use of the backscattered radiation from metallic nanoparticles, situated on the backside of a transparent material, e.g. silica in the IR region. With the use of this technique the DSI group produced holes with diameter of about 40 nm, using a CO_2 laser and gold colloid particles [13]. The feature sizes achieved (below $\lambda/260$) were a world record for optical processing which does not use

SNOM, at the time, or a technique where laser is combined with atomic force microscope.

Optical and magneto-optical recording

The DSI group applied the technique with an array of particles for phase change in a GeSbTe film to see the effect of bit point localization [20]. Coupling of the particle with the surface demonstrates a non-trivial effect in the distribution of intensity with the angle of incidence for the laser beam ([19, 22]). At some angles and intensities one can see a chain of small holes around the border of the geometrical shadow of the particle. This effect is also successfully explained within the theory Boris created for light scattering by a particle on a surface.

Formation of nanoclusters during laser ablation

Formation of nanoclusters is, by itself, a rather complicated process, which depends on many experimental parameters. Boris analysed the formation of nanoclusters by the laser ablation method ([23–27]). Different stages of cluster formation were analysed in the frame of the Zeldovich–Raizer theory. The most complicated part related to the dynamics of the plume expansion into the background gas. This was analysed on the basis of a numerical solution of the two-dimensional gas-dynamic problem. The predicted size distribution function for the nanoclusters was close to that found in experiments. The theory also successfully explained the non-monotonous experimental dependence of the mean size of clusters versus the background gas pressure [28]. The modelling permitted a method of controlled synthesis of nanoclusters and nanostructures by conventional pulsed laser ablation to be developed [23].

Highly multiplexed scanning nanoscopic imaging

Boris put forward a new proposal for the use of an array of particles, in a particular trapped array of microspheres for highly multiplexed nanoscopic imaging [29, 30]. This technique potentially can be used for rapid analysis of DNA sequencing. The first experimental studies were done in collaboration with the Physics Department, University of Wisconsin-Milwaukee, USA. A novel-imaging instrument based on multifoci optical tweezers and nanoscale light concentration is under development. A 7×7 2D assembly of 1-μm diameter polystyrene microspheres has been achieved (multiplexing about 50) [30], but there are no basic physical limitations for further increase in multiplexing.

Plasmonics

Scattering of light on nanoparticles with a surface plasmon resonance has peculiarities, which have not been previously analyzed. Since the time of Rayleigh and Debye it has been generally accepted that the behavior of a small particle is close to that of a point radiating dipole (PRD). A new analysis [31] shows that for the particle with a plasmon resonance distribution of fields the Poynting vector deviates significantly from that of a PRD. Analysis beyond the dipole approximation demonstrates a highly complex structure to the near-field Poynting vector, e.g. formation of "energy vortices" for some parameters. Another theoretically predicted effect is a reversal hierarchy for extinction cross-sections of small particles with plasmon resonance [32]. Understanding the peculiarities of the energy flux in the near-field region is a necessary part of design of optical elements to be used in technologies harnessing surface plasmon resonance effects.

Laser ablation and plume dynamics

Boris has worked a lot on these problems, in collaboration with many research groups. Reviews of the research have been published in [33–34].

Laser thermochemistry

These problems were the focus of Boris' attention for many years. For example he was invited by Prof. Ilya Prigozhin in 1988 to give a lecture on these topics in Brussels University. The recent book [35] is an English translation of Boris' and his co-workers book, which was previously published in Russian, in Russia (two editions).

Friend and poet

In addition to being a top class physicist Boris enjoys appreciation as a poet. Thus, it seems appropriate that part of this tribute should be in a poetic form. His former student and colleague Dr Nikita Arnold develops this below. One other (who prefers to remain anonymous) has also submitted an "Alphabet Poem to Boris". By virtue of his life as a theoretical physicist this has to be the Greek alphabet, of course, but with some poetic license.

As one of Boris' pupils and long time collaborator, I know him not only from the scientific side. In addition to his numerous talents, he writes poetry on a semi-professional basis and is included in the anthology of the poets that graduated from Moscow University. (http://poesis.ru/poeti-poezia/lukjanchuk/biograph.htm). Poetry is a national pastime in Russia and was even more so in former times, so I am not foreign to it and Boris often asked for my opinion of his verses. My comments were mostly critical, probably even more so than in our scientific discussions. This does not mean that I did not like his poems — it is just easier to make critical comments than constructive ones. To give Boris the opportunity for easy revenge and to give the whole (not only Russian-speaking) laser cleaning community the taste of how poetry can interlace with the everyday and scientific life, I have made my congratulation rhymed or at least rhythmed. But to avoid too harsh a criticism, I hide behind a well known bard.

Epigraph:

"... congratulate you as I would my father
on similar occasion eulogize ..."
B. Pasternak to V. Bryusov

To clean or not to clean: that is the question:
Whether it's better with the brush to clean
The dirt and dust of outrageous smallness,
Or take the soap and wash them all away?
Or use the ultrasound? Chemical solution?
Not possible — it will dissolve the chip,
Contaminate remaining wafer surface;
So, how to clean – in megasonic bath?
With snowflakes of carbon dioxide?
Or ultrasonic shock, Or — laser?
Laser… The shining beam of purest light there is
The pulse that hits like lightning, only faster;
Invisible… or visible — depends
Upon wavelength, which we are free to choose,
And pulse duration, that one has to vary,
In order to achieve the best results…
Therefore the laser — this is our way,
For who would tinker with the brush or soap,
Or humid mess of ultrasonic bath,
When laser gives it all — its action brings along
Assortment of effects, predominantly thermal
Expansion, sound, focusing of light,
Ablation, small tsunamis, near field,
And focusing again (in near field this time),
Adhesion, condensation, all the forces,
All formulas on earth and all Landau
Lifshitz… O, Boris Luk'yanchuk — if not for you
would Laser Cleaning never be so sumptuous.

Nikita Arnold

Greek Alphabet Poem for Boris

Α α Ablation action; applied
Β β brilliance;
Γ γ gracious,
Δ δ delightful, delver;
Ε ε expanding, elastic, excellence; enthusiastic
Ζ ζ zeal;
Η η insightful
Θ θ theorist;
Ι ι inspirational interaction; intellectual,
Κ κ creative, collaboration;
Λ λ longtime linkage;
Μ μ Mie matter(s)
Ν ν near-field, nano-network;
Ξ ξ xenogamous xenophile;
Ο ο oscillating
Π π potential; productive, pulsing, power;
Ρ ρ Russian romantic;
Σ σ silicon/silica scientist;
Τ τ tolerant traveler;
Υ υ vigorous vanguard;
Φ φ physics personified;
Χ χ cheery chair;
Ψ ψ psychologising psalmist;
Ω ω ongoing, open-hearted odyssey.

Anon.

References

1. Luk'yanchuk B. S., Mosbacher M., Zheng Y. W., Münzer H.- J., Huang S. M., Bertsch M., Song W. D., Wang Z. B., Lu Y. F., Dubbers O., Boneberg J., Leiderer P., Hong M. H., and Chong T. C. , Chapter 3. Optical resonance and near-field effects in dry laser cleaning, In: Laser Cleaning, Ed. By B. S. Luk`yanchuk (World Scientific, New Jersey, London, Singapore, Hong Kong, 2002), pp. 103-178.

2. Luk'yanchuk B. S., Zheng Y. W., and Lu Y. F., Laser Cleaning of the surface: Optical resonance and near-field effects, Proc. SPIE, vol. **4065**, pp. 576-587 (2000).

3. Lu Y. F., Zhang L., Song W. D., Zheng Y. W., and Luk'yanchuk B. S**.,** Laser writing of sub-wavelength structure on silicon (100) surfaces with particle enhanced optical irradiation, JETP Letters, vol. **72**, No. 9, pp. 457-459 (2000).

4. Zheng Y. W., Luk'yanchuk B. S., Lu Y. F., Song W. D., and Mai Z. H., Dry laser cleaning of Particles from Solid Substrates: Experiments and Theory, J.. Appl. Phys., vol. **90**, No. 5, pp. 2135-2142 (2001).

5. Huang S. M., Hong M. H., Luk'yanchuk B. S., Lu Y. F., Song W. D., and Chong T. C., Sub-50 nm nanopatterning of metallic layers by green pulsed laser combined with atomic force microscopy, Journal of Vacuum & Science Technology B, vol. **20** (3), pp. 1118-1125 (2002).

6. Huang S.M., Hong M.H., Lu Y.F., Luk'yanchuk B. S., Song W.D., and Chong T. C. Pulsed laser-assisted surface structuring with optical near-field enhanced effects. Appl. Phys., vol. **92** (5), pp. 2495-2500 (2002).

7. Luk'yanchuk B. S., Zheng Y. W., and Lu Y. F., Particle on the surface: Basic physical problems related to laser cleaning, Proc. SPIE, vol. **4426**, pp. 284-289 (2002) – invited.

8. Lu Y. F., Zhang L**.,** Song W. D., Zheng Y.W., and Luk'yanchuk B. S. Particle-Enhanced Near-Field Optical Effect and Laser Writing for Nanostructure Fabrication Proc. SPIE, vol. **4426**, pp. 143-145 (2002).

9. Huang S. M., Hong M. H., Luk'yanchuk B. , and Lu Y. F. Laser assisted nanofabrications on metal surfaces with optical near field effects, Proc. SPIE, vol. **4760**, pp. 185-191 (2002).

10. Luk'yanchuk B. S., Huang S. M., and Hong M. H., D effects in dry laser cleaning, Proc. SPIE, vol. **4760**, pp. 204-210 (2002) – invited.

11. Luk'yanchuk B. S., Arnold N., Huang S. M., Wang Z. B., and Hong M. H., Three-dimensional effects in dry laser cleaning Appl. Phys. **A 77**, issue 2, pp. 209-215 (2003).

12. Huang S. M., Hong M. H., Luk'yanchuk B., and Wang Z. B., Nanostructures fabricated on metal surfaces assisted by laser with optical near-field effects, Appl. Phys. **A 77**, issue 2, pp. 293-296 (2003).

13. Huang S. M., Luk'yanchuk B., Hong M. H., and Chong T. C., Direct and sub-diffraction limit laser nanofabrication in silicon, Appl. Phys. Letters, vol. **82**, pp. 4809-4811 (2003).

14. Luk'yanchuk B. S., Wang Z. B., Song W. D., and Hong M. H., Particle on surface: 3D - effects in dry laser cleaning Appl. Phys. A **79**, pp. 747–751 (2004).

15. Hong M. H., Huang S. M., Luk'yanchuk B. S., Lu Y. F., and Chong T. C., Laser assisted nanostructuring, In: Pacific RIM Workshop on Transducers and Micro/Nano Technologies, Ed. by Chuan-Hong Chen (Xiamen University Press, 2002), pp. 243-246.

16. Hong M.H., Huang S.M., Luk'yanchuk B. S., and Chong T.C., Laser assisted surface nanopatterning, Sensors and Actuators A-Physical, vol. **108** (1-3), pp. 69-74 (2003).

17. Hong M. H., Huang S. M., Luk'yanchuk B., Wang Z. B., Lu Y. F., and Chong T.C., Laser assisted nanofabrication, Proc. SPIE, vol. **4977**, pp. 142-155 (2003).

18. Hong M. H., Huang S. M., Wang W. J., Tiaw K. S., Teoh S. H., Luk'yanchuk B., and Chong T. C., Unique functional micro/nano-structures created by femtosecond laser irradiation, Mat. Res. Soc. Symp. Proc., vol. **780**, pp. Y2.2.1-Y2.2.11 (2003).

19. Luk'yanchuk B. S., Wang Z.B., Hong M.H., Chong T. C., and Arnold N., Optical Resonance and Near-Field Effects: Applications for Nanopatterning, Proc. SPIE, vol. 5448, pp. 37-44 (2004) (Plenary Lecture.)

20. Wang Z. B., Hong M. H., Luk'yanchuk B. S., Huang S. M., Wang O.F., Shi L.P., and Chong T. C., Parallel nanostructuring of GeSbTe films with particle-mask, Appl. Phys. A **79**, pp. 1603–1606 (2004).

21. Hong M. H., Luk'yanchuk B. S., Huang S. M., Ong T. S., Van L. H., and Chong T. C.,Femtosecond laser application for high capacity optical data storage, Appl. Phys. A **79**, pp. 791–794 (2004).

22. Wang Z. B., Hong M. H. , Luk'yanchuk B. S., Lin Y., Wang Q.F., and Chong T. C., Angle effect in laser nanopatterning with particle-mask, J. Appl. Phys., vol. **96**, No. 7 (2004).

23. Marine W., Patrone L., Luk'yanchuk B. S., and Sentis M., Strategy of Nanocluster and Nanostructure Synthesis by Conventional Pulsed Laser Ablatio, Appl. Surf. Sci., vol. **154-155**, pp. 345-352 (2000).

24. Luk'yanchuk B. S., and Marine W. , On the Delay Time in Photoluminescence of Si- Nanoclusters, Produced by Laser Ablation, Appl. Surf. Sci., vol. **154-155**, pp. 314-319 (2000).

25. Luk'yanchuk B. S., and Marine W., Detection of Si-Nanoclusters, Produced by Laser Ablation: Delay Time in Photoluminescence and Rayleigh Scattering, Proc. SPIE, vol. **3885**, pp. 182-192 (2000) – invited.

26. Kuwata M., Luk'yanchuk B., and Yabe T., Nanoclusters formation within the vapor plume, produced by ns-laser ablation: Effect of the initial density and pressure distributions, Jpn. J. Appl. Phys., vol. **40**, Part 1, No. 6A, pp. 4262-4268 (2001).

27. Ohkubo T., Kuwata M., Luk'yanchuk B. , and Yabe T., Two-Dimensional Simulation of Nanocluster Formation and Comparison with Experiments, Proc. SPIE, vol. **4760**, pp. 156-163 (2002) – invited.

28. Ohkubo T., Kuwata M., Luk'yanchuk B., and Yabe T., Numerical analysis of nanocluster formation within ns-laser ablation plume, Appl. Phys. **A 77**, issue 2, pp. 271-275 (2003).

29. Yakovlev V. V., and Luk'yanchuk B., Multiplexed nanoscopic imaging, Laser Physics, vol. **14**, No. 8, pp. 1065-1071 (2004).

30. Faustov A., Scheslavskiy V., Petrov G. I., Luk'yanchuk B., and Yakovlev V.V., Highly multiplexing scanning nanoscopic imaging, Proc. SPIE, vol. 5331, pp. 21-28 (2004).

31. Wang Z. B., Luk'yanchuk B. S., Hong M. H., Lin Y., and Chong T. C., Energy flows around a small particle investigated by classical Mie theory, Phys. Rev. B. **70**, issue 3, 032427 (2004).

32. Tribelsky M. I., and Luk'yanchuk B. S., Anomalous Light Scattering by Small Particles, Abstract for EM-NANO 2004 International Symposium on Organic and Inorganic Electronic Materials and Related Nanotechnologies June 7-10, 2004, Toki Messe, Niigata, Japan.

33. Anisimov S. I., and Luk'yanchuk B. S., Selected Problems of Laser Ablation Theory, Physics - Uspekhi, vol. **45**, No. 3, pp. 293-324 (2002).

34. Bityurin N., Luk'yanchuk B. S., Hong M. H., and Chong T. C., Models for laser ablation of polymers, Chemical Reviews, vol. **103**, No. 2, pp. 519-52 (2003)

35. Karlov N. V., Kirichenko N. A., and Luk'yanchuk B. S., Laser Thermochemistry. Fundamentals and Applications. Cambridge International Science Publishing, (Cambridge, UK, 2000) - 380 p.

Deb Kane
Department of Physics
Macquarie University, Sydney
Australia

ACKNOWLEDGEMENTS

It is a pleasure to acknowledge the various research funding agencies around the world (Australia, Austria, Denmark, European Union, Germany, Korea, Poland, Singapore and the United Kingdom) that have supported the research reported here-in. DMK would like to extend her appreciation to all the contributors to this research monograph and to all those at World Scientific Publishing who have been involved with the publication of the book, in particular Rhaimie Wahap. The editorship of this book was completed at a time when the editor assumed a senior management role in her university that was unexpected when the book was first planned. DMK is grateful for the understanding of colleagues involved in the project in this changed circumstance throughout the book production.

CONTENTS

Chapter 1

LASER CLEANING AND SURFACE MODIFICATIONS: APPLICATIONS IN NANO- AND BIOTECHNOLOGY

DIETER BÄUERLE, THOMAS GUMPENBERGER, DANIEL BRODOCEANU, GREGOR LANGER, JOHANNES KOFLER, JOHANNES HEITZ, KLAUS PIGLMAYER

Institut für Angewandte Physik, Johannes-Kepler-Universität Linz, A-4040 Linz, Austria, Tel. ++43 2468-9244,-9243, Fax. ++43 2468-9242; E-mail: dieter.baeuerle@jku.at

This article gives an overview of the fundamentals of laser-matter interactions with regard to the cleaning of solid surfaces from either tiny particulates or extended contamination layers. Various different applications and limitations of the technique are discussed together with surface modifications that open up completely new possibilities in nano- and biotechnology.

1. Introduction

The term laser cleaning denotes two quite different fields: the removal of *particulates* and the removal of extended *contamination layers* from solid surfaces [Luk'yanchuk 2002, Bäuerle 2000]. The latter application includes such different areas as cleaning stainless steel of scale or organic impurities, the removal of paint from metal surfaces, the cleaning of semiconductor surfaces and microelectronic or micromechanical devices, the removal of contamination layers from mechanical or electrical contacts, switches, metallic photocathodes, silicon field emitter arrays [Takai *et al.* 2002], the pretreatment of LCD-glass substrates or adhesion surface areas, etc. The application of *la*sers in the *con*servation of *a*rtworks (LACONA) has also become a rapidly growing field. Among the examples are the restoration of ancient metal artwork [Drakaki *et al.* 2004], medieval stained glass [Troll *et al.* 1999], paper [Rudolph *et al.* 2004, Kollia *et al.* 2004], paintings, frescos and stone [Zafiropulos 2002, Teule 2001, Klein *et al.* 1999] etc. With many of these applications, the removal of contamination layers can be understood in terms of laser ablation. On a wider scope, one can also include processes where a thin surface layer is "cleaned" by *outdiffusion of impurities* under the action of laser light. With systems that are strongly inhomogeneous with respect to their optical, thermal, and elastic properties, thermo- and photomechanical mechanisms may become important. With very high fluences laser-induced shock waves and fragmentation processes are observed [Bäuerle 2000]. Clearly, these latter regimes are inadequate for well-defined cleaning of sensitive substrates and devices.

Completely *new* aspects arise with the cleaning of particulates from surfaces Particle sizes (diameters) in the range between some 10 nm and several microns have been laser cleaned. Efficient techniques to clean off such small particulates from solid surfaces, heat-sensitive coatings, devices, etc. become increasingly important in semiconductor device fabrication, micromechanics, optics, telecommunication, etc. Here, conventional techniques such as ultrasonic and megasonic cleaning, wiping and scrubbing, high-pressure jet spraying, CO_2 snow cleaning, conventional etching, plasma cleaning etc. are often inadequate for the removal of particulate contaminations. The reason is that small particulates adhere on substrates or devices with relatively strong forces that are difficult to overcome with these traditional cleaning techniques. As a consequence, these techniques are often ineffective, result in the addition of other contaminations, the damaging of prefabricated parts, etc. It has been demonstrated that, under certain conditions and with certain systems, laser cleaning enables efficient removal of micron and submicron particles from solid surfaces.

In the present chapter, we give an overview on the applications of laser techniques to the cleaning of surfaces from both thin extended contamination layers and particulates. We start with the latter field and also discuss the submicron- and nanopatterning of surfaces by means of two-dimensional lattices of microspheres. This application can be considered as a spin-off from investigations on dry-laser cleaning of particulates from surfaces. In the fourth part we discuss the removal of contamination layers within both inert and reactive ambient media. In both cases, laser cleaning is often correlated with physical and/or chemical changes of surface properties. Among those are surface smoothening (polishing), glazing, passivation, depletion and/or substitution of particular species, the formation of stochastic or periodic structures, etc. The last part deals with laser cleaning and light-induced surface modifications of polytetrafluoroethylene (PTFE) in NH_3 atmosphere with special emphasis on applications in biotechnology.

Further details on various different topics that are only briefly discussed in this overview can be found in other chapters of this book.

2. Removal of Particulates

For cleaning-off particulates from solid surfaces, the action of the incident laser radiation must overcome the strong adhesion forces between the tiny particulates and the surface. Among these adhesion forces are mainly Van der Waals forces K_{VdW}, capillary forces K_{c}, and electrostatic forces K_{e} [Bäuerle 2000, Bowling 1995, Israelachvili 1992]. The size of these forces and their relative importance in laser cleaning depends on the size of particulates, the physical and chemical properties of particulate and substrate materials, the ambient medium etc. In fact, laser cleaning is often classified according to the ambient medium that is employed during the experiments.

Dry laser cleaning (DLC) is performed in vacuum or a dry atmosphere. If, on the other hand, the removal of particulates is assisted by the evaporation of a thin liquid film, the technique is denoted as *steam laser cleaning* (SLC). If laser cleaning is performed in an atmosphere of high relative humidity, e.g. in water vapor, we use the term *wet laser cleaning* (WLC).

In all of these different techniques, the cleaning efficiency for single-pulse laser irradiation is defined by $\eta = 1 - N_{\mathrm{f}} / N_{\mathrm{i}}$ where N_{i} and N_{f} are the respective surface densities of particulates before and after laser-light irradiation. η depends on the size of the adhesion forces, the laser parameters, the optical and thermal properties of both the substrate and particulate material, etc. Cleaning efficiencies achieved with N_ℓ laser pulses can be described by $\eta(N_\ell) \approx 1-(1-\eta)^{N_\ell}$. The number of laser pulses typically employed in laser cleaning is between 1 and 10.

Because of the various different dependences of the cleaning efficiency on the material and laser parameters, the ambient medium, the size and shape of particulates etc., investigations of fundamental interaction mechanisms require well-defined cleaning experiments with well-defined model contaminants. Well-defined contamination of the substrate surface is achieved by spin coating of colloidal solutions. Such solutions are commercially available for different materials and for particle sizes between some ten nanometers and a few microns. The size distribution of such particles is relatively narrow, typically below ± 10%. The shape of particles is often *spherical*. To avoid the coalescence of particles and the formation of aggregates during spin coating, one has to adjust the concentration of particles in the suspension, the rotation speed of the spinner, etc.

An overview on the various different systems that have been investigated can be found in [Kane *et al*. 2002].

2.1. *Dry Laser Cleaning (DLC)*

Dry laser cleaning can be classified into cases where particle removal relies on strong absorption of the laser light by either the substrate or the particulate, or both of them.

2.1.1. *Substrate absorption*

Let us first consider non-absorbing rigid (non-elastic) particulates on an absorbing semi-infinite substrate. If we ignore the influence of the particulates on the radiation field, uniform pulsed-laser irradiation will cause uniform (one-dimensional) thermal expansion of the substrate. Let us ignore also capillary and electrostatic forces between the particulates and the substrate, and set $K_c = K_e = 0$. Then, we find from Newton's law and an acceleration of particulates normal to the surface [Bäuerle 2000] the threshold fluence for cleaning

$$\phi_{cl} \approx \xi \frac{c_p}{\beta_T A} \frac{\tau_\ell^2}{r_p^q} \propto \frac{\tau_\ell^2}{r_p^q} \tag{2.1}$$

where ξ slightly depends on various different parameters. c_p is the specific heat at constant pressure, β_T the coefficient of (linear) thermal expansion, A the absorptivity, τ_ℓ the laser pulse length and r_p the radius of the particulate. q is an exponent which, for adhesion forces proportional to r_p or r_p^2 is within the range $1 \leq q \leq 2$. Within this one-dimensional (1D) model the particulates are "kicked off" from the surface due to rapid thermal expansion and subsequent contraction of the substrate. While this model is much too simple, it already shows that surface cleaning becomes increasingly difficult with decreasing particle size. An upper limit of the fluence that can be employed is set by material damage due to fracture, melting and ablation. More sophisticated 1D models take into account elastic deformations, the influence of pulse shapes, etc.

While such 1D models *qualitatively* describe some trends in laser cleaning, the calculated threshold fluence ϕ_{cl}, wavelength dependences, etc. often disagree from the experimental results by one to two orders of magnitude. This has several reasons. Among those are enhancements of the electromagnetic field underneath the particulates. These field enhancements can be related to the focusing of the incident radiation by the particle ($r_p >> \lambda$) or to Mie-scattering ($r_p . \lambda$). In any case, local field enhancements cause a *local* temperature rise in addition to the uniform temperature rise related to uniform laser-light irradiation. The local temperature rise causes local 3D-thermal expansion of the substrate material.

Calculations of this type have been performed by Luk'yanchuk *et al.* (2004), Arnold *et al.* (2004), Pleasants *et al.* (2004), and others.

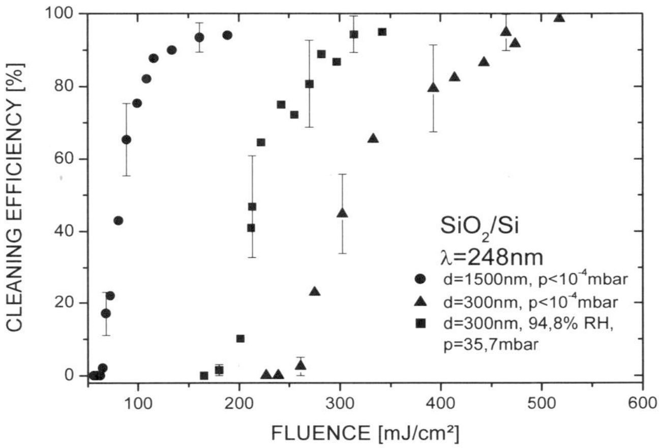

Figure 2.1: Cleaning efficiencies η achieved with single-pulse KrF-laser radiation on (100) Si substrates contaminated with spherical SiO_2 particles of diameter $d = 2r_p$. Experiments were performed in vacuum ($d = 1500$ nm, $d = 300$ nm) and water atmosphere of relative humidity $RH \approx 95$ % ($T \approx 27^oC$) ($d = 300$ nm) [adapted from Schrems 2003].

Detailed *experimental* investigations have been performed on the laser cleaning of Al_2O_3, SiO_2, Si_3N_4, polystyrene (PS) particulates from Si [Mosbacher *et al.* 2003, Schrems *et al.* 2003, Wu *et al.* 2000], NiP [She *et al.* 1999], glass [Pleasants and Kane 2003], and polyimide (PI) [Fourrier *et al.* 2001] substrates. Cleaning efficiencies were evaluated by either counting the particles before and after cleaning by using an optical microscope in combination with special computer software, or by measuring the change in scattered light from a HeNe probe laser illuminating the area under consideration [Chaoui *et al.* 2003]. Figure 2.1 shows the typical behavior of the cleaning efficiency of SiO_2 particles on Si as a function of fluence for KrF-laser radiation and for two different particle sizes and atmospheres. The cleaning efficiency η strongly increases above a certain threshold fluence and then remains almost constant. In *qualitative* agreement with (2.1), ϕ_{cl} increases with decreasing particle size. However, the situation may be even more complicated. With SiO_2 spheres on Si substrates strong deformation of the spheres during sample storage has been found. Depending on the size of particulates, ϕ_{cl} can significantly increase with storage time [Schrems *et al.* 2003].

A detailed analysis of the threshold fluences ϕ_{cl} measured as a function of particle size cannot be explained theoretically even on the basis of 3D thermal expansion models. This is quite understandable from experimental observations. With many systems investigated so far, local substrate damage and the formation of holes underneath the particulates is observed – even with fluences ϕ . ϕ_{th}. Here, ϕ_{th} is the threshold fluence for *uniform* substrate damage/ablation. Thus, for such systems, DLC is apparently *not* based on the acceleration of particulates due to thermal expansion, but rather on local substrate ablation. Clearly, due to local field enhancement, the fluence underneath the particle, ϕ(local), exceeds the incident uniform fluence ϕ(uniform). This phenomenon can be used for well-defined patterning of material surfaces (see part 3).

In the present case of non-absorbing spherical particles with $r_p >> \lambda$ the *average* intensity on the substrate surface underneath the particle can be approximated by

$$I_s \approx \begin{cases} \dfrac{27\,n^4}{(4-n^2)^3}\, I_0 & \text{with } \rho < \rho_c \\ 0 & \text{with } \rho > \rho_c \end{cases} \tag{2.2}$$

where n is the index of refraction of the microsphere and I_0 the (uniform) incident intensity. With $n(SiO_2) = 1.42$ we obtain $I_S \approx 14\, I_0$. ρ is the distance from the axis of symmetry on the substrate surface and ρ_c the caustic radius. The latter is given by

$$\rho_c = \frac{(4-n^2)^{3/2}}{3\sqrt{3}\,n^2}\, r_p \tag{2.3}$$

With $\rho_c\,(SiO_2) \approx 0.27\, r_p$. For transparent particles with $r_p << \lambda$ we can use the dipole approximation [Arnold 2003]

$$I_s = \left(1 + \frac{n^2-1}{n^2+2}\, k^2 r_p^{\,2}\right)^2 I_0 \tag{2.4}$$

where $k\, r_p$ is the Mie parameter with $k = 2\pi/\lambda$. While the *overall* intensity enhancement by microspheres is qualitatively well described by the approximations (2.2) and (2.4), the real situation is much more complex. For example, for transparent particles and $r_p >> \lambda$, detailed calculations of the intensity $EE^* \equiv |E|^2$ where E is the electric field, yield a pronounced double-peak structure. This is shown in Fig. 2.2. This "fine" structure may become relevant with strongly nonlinear processes and short laser pulses. In fact, experiments using ultrashort laser pulses give clear evidence for such a distribution. Among the examples are double-hole structures observed after irradiation of PS particulates on Si substrates with 100 fs Ti:Sapphire-laser pulses [Münzer *et al.* 2002] and of SiO_2 particulates on Ni foils with 500 fs KrF-laser pulses [Bäuerle *et al.* 2003]. With longer pulses this structure is "smeared out" due to spatial dissipation of the excitation energy. Integration of $|E|^2$, $|H|^2$ or the Poynting vector $|S|$ over the caustic area $\pi \rho_c^2$ in the plane $z \approx r_p$ yields, with Mie parameters / 30, and within ± 20 %, the same result as (2.2). Further details on the imaging properties of microspheres have been discussed in [Kofler 2004, Luk'yanchuk 2002]. It should be noted, however, that the scattering of the radiation reflected from the substrate yields an additional contribution to the electromagnetic field and further increases the overall intensity below the particulate [Luk'yanchuk *et al.* 2004, 2002].

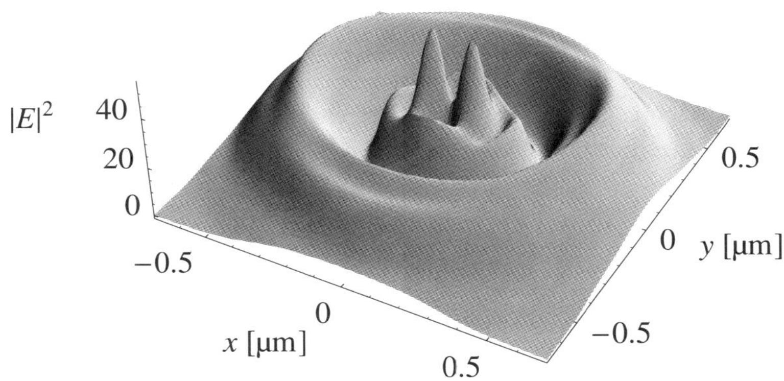

Figure 2.2: Intensity $|E|^2$ underneath a focusing microsphere at $z \approx r_p$ in the x,y-plane. The origin of the coordinate system is located in the center of a sphere with radius r_p = 3.1 μm. The incident electric field is a plane wave (wavelength λ = 248 nm), propagating in z- and polarized in x-direction with unity strength. The refractive index of the sphere is n = 1.42 and the Mie parameter is $k\, r_p = 2\, \pi\, r_p / \lambda \approx 78.5$. The picture was produced with the method of Bessoid matching and reveals a pronounced double-peak structure in the direction of initial polarization [after Kofler 2004].

In any case, the amplification of the intensity underneath the particulates will result in *local* ablation at fluences ϕ (uniform) < ϕ_{th}. Threshold fluences for "cleaning" calculated on the basis of particle removal by *local ablation* of the substrate still do not describe the experimental data quantitatively but, at least, yield more consistent slopes of the dependence $\phi_{cl} = \phi_{cl}(r_p)$ [Arnold *et al.* 2004]. A quantitative description of such a process seems to be almost impossible. Near threshold, the ablation rate and the related pressure caused by the ablated species changes drastically within a small range of fluences (local laser-induced temperature rise). Simple estimations yield pressures up to several 10^2 bar. Furthermore, with the onset of local substrate ablation, adhesion of the particulates becomes meaningless.

2.1.2. *Particulate absorption*

Experimental investigations have been performed, e.g., for SiO_2 substrates contaminated with Cu and Al particles [Lu *et al.* 1998] and for PMMA contaminated with PS particles [Fourrier *et al.* 2001]. In both cases KrF-laser radiation has been employed.

In many cases, the situation may be quite similar to that described before – except that thermal expansion of the particulate becomes important. However, with strong absorption and particulates that can be easily decomposed in a photochemical, photophysical, or thermal process, cleaning may become dominated by the *ablation* of the particles. This mechanism differs significantly from that described above. For a purely thermally activated process the fluence that causes vaporization of the particulate can be estimated from the energy balance. If we ignore thermal losses we obtain

$$\phi_{cl} \approx \frac{\rho_p}{A} \left[c_p T_{\mathrm{v}} + \Delta H_{\mathrm{v}} \right] r_p \, . \tag{2.5}$$

The physical parameters for the particulates, mainly A, make this process very selective. T_{v} and ΔH_{v} are the temperature and the enthalpy of evaporation of the particles, respectively, and ρ_p the mass density . Clearly, these quantities, and in particular the mass and the absorptivity of the particulate change during evaporation. Thus, (2.5) is only a very crude approximation.

In all cases of dry cleaning, recondensation of particulates on the cleaned substrate can be drastically reduced by employing vacuum conditions or a laminar flow of H_2 or He and an appropriate geometry of the setup with $\hat{\boldsymbol{n}} \downarrow\downarrow \boldsymbol{g}$ or $\hat{\boldsymbol{n}} \perp \boldsymbol{g}$ where $\hat{\boldsymbol{n}}$ is the surface normal and $\boldsymbol{g}$ the acceleration due to gravity.

2.1.3. *Particulate and substrate absorption*

The removal of metallic particles such as Au, Cu, and W from Si [Curran *et al.* 2002] and metal surfaces has been studied by means of the fundamental and harmonics of Nd:YAG-laser radiation. Here, it has been found that substrate damage can be often diminished by using instead of perpendicular laser-light incidence glancing angles [Lee *et al.* 2000].

Surface cleaning was also demonstrated by exciting surface acoustic waves (SAW) by means of a *focused* laser beam [Kolomenskii *et al.* 1998]. In this technique, *non-local* cleaning within the region of the propagating SAW can be achieved. This region may cover an area of several millimeters around the laser-irradiated spot.

2.2. *Steam Laser Cleaning (SLC)*

With many systems, the removal of very small particulates from solid surfaces can be significantly enhanced by a liquid film that is deposited, e.g. via a nozzle, onto the particle-contaminated surface prior to pulsed-laser irradiation. High cleaning efficiencies have been achieved with strongly *absorbing substrates* and liquid films that are transparent at the incident laser wavelength. For example, with KrF- and Nd:YAG-laser radiation, particulates of Au, Cu, Al_2O_3, Fe_2O_3, Si, PS, etc. have been efficiently cleaned off from Si wafers and membranes [Lang *et al.* 2003; Neves *et al.* 2002, Wu *et al.* 2000; Allen *et al.* 1997; Zapka 2002]. In most experiments, the liquid film consists of water mixed with 10 to 20 % alcohol and its thickness is, typically, between about 100 nm and several μm. The alcohol improves substrate wetting.

The dependence of the cleaning efficiency on laser fluence for particles of *different* sizes (60 nm – 1300 nm) and different materials (Al_2O_3 , SiO_2 , PS) and geometries (spherical SiO_2 and PS and arbitrarily shaped Al_2O_3) has been investigated in [Lang *et al.* 2003, Mosbacher *et al.* 2003]. In contrast to the situation in DLC, almost the *same* cleaning threshold has been found for *all* types and sizes of particles. For other laser wavelengths and other types of substrate materials and liquid films, this threshold fluence may change significantly. In any case, the threshold fluences for cleaning required in SLC are, in general, much lower than those required for very tiny particulates in DLC. This is particularly important in connection with substrate damage or/and in situations where particles melt or/and react or form an alloy with the substrate material. Nevertheless, systematic investigations on substrate damage in SLC are still lacking.

The microscopic mechanisms that dominate SLC are not completely clear. The reason that water films are employed in most cases of SLC may be related to the high transient pressure caused by superheated water. With superheating to the critical temperature $T_{cr}(H_2O) \approx 375^{\circ}$ C peak pressures up to $p_{cr}(H_2O) \approx 220$ bar can be achieved. However, in a number of investigations it has been demonstrated that with the moderate laser-light intensities employed in most cases of SLC, the detachment of particulates is not related to this pressure, but to the pressure wave which results from fast-growing bubbles near and above the boiling temperature. Depending on the degree of superheating and the surface properties of the substrate, the typical pressures were measured to be . 30 bar [Kim and Lee 2003; Leiderer *et al.* 2002; Grigoropoulos and Kim 2002]. It should be noted, however, that superheating and bubble nucleation is not only determined by the laser and material parameters. Of strong influence is also the roughness of the substrate surface [Leiderer *et al.* 2002]. In any case, for micron- and submicron-sized particles, even pressures of 30 bar can cause accelerations of up to 10^{11} cm/s^2. Thus, although the adhesion forces for the different types of particles investigated in [Lang *et al.* 2003, Mosbacher *et al.* 2003] vary by more than an order of magnitude, the cleaning forces are so high that they exceed by far the adhesion forces for *all* of these different particulates. This may explain the "universal" cleaning threshold observed during these investigations. Clearly, the real situation may be much more complex. Cavitation effects, changes in adhesion forces, e.g. via the decrease of the Hamaker constant in water, the temperature jump at the solid-liquid interface etc. may play an important role as well.

2.2.1. *Absorbing liquid films*

Laser cleaning of polymer surfaces has been demonstrated by using KrF-laser radiation and either a strongly *absorbing* film of acetone [transmittivity D(248nm) $\approx$ 0] or *transparent* isopropanol (IPA) [transmittivity D(248nm) $\approx$ 0.97]. Here, efficient cleaning was achieved only with IPA and it was significantly enhanced in comparison to dry cleaning [Lee *et al.* 1998].

Steam cleaning of Si substrates from different types of particles with sizes ranging from about 0.1 to 10 μm has also been demonstrated for CO_2-laser radiation in combination with (absorbing) water films [Allen *et al.* 1997, Boughaba *et al.* 1999].

2.3. *Wet Laser Cleaning (WLC)*

A technique which combines some of the advantages of DLC and SLC and which avoids some of their disadvantages is wet laser cleaning. In WLC, the substrate is immersed in an atmosphere of high relative humidity (*RH*) of water. Thus, there is no need of any thickness control of the liquid film or synchronization of the laser pulse with the water jet. WLC avoids an overall contamination of the surface related to the liquid film. Here, the liquid condenses only in the interstices between the particulates and the substrate, where it forms a stable well-defined meniscus. For complete wetting, the volume of capillary condensed water is $V_\ell = 4\pi R_K^2 r_p$ where $R_K \approx 0.52 / \ln^{-1}(RH^{-1})$ nm is the Kelvin radius. With $RH = 95$ % we find $R_K = 10$ nm. Thus, the technique may be advantageous for very small particulates where the volume of the superheated water that is condensed in the interstice between the particulate and the substrate causes a high enough pressure for particle removal. Experimentally, we have found that for SiO_2 particles on Si substrates the threshold for cleaning is well below that for DLC for particle sizes $d = 2\, r_p$. 500 nm [Schrems 2003]. Figure 2.1 shows this influence for 300 nm particles. Further investigations in this field, in particular for particle sizes d . 100 nm and humidities $RH > 95$ %, seem to be quite promising. A larger water meniscus would certainly favor bubble formation.

3. Submicron- and Nanopatterning

The local field enhancement observed below and near particulates in dry laser cleaning may result in local material ablation and/or in physical and chemical surface modifications. This phenomenon can be employed for single-step surface patterning. Besides of stochastic patterns, well-defined periodic structures can be fabricated. In the latter case we employ regular two-dimensional (2D) lattices of microspheres which form by well-known self-assembly processes from the colloidal suspensions already mentioned before. Because of their thermal stability and optical properties, we mainly employ microspheres of a-SiO_2 and a transparent support. The microspheres act like microlenses and focus the incident laser radiation onto the substrate, albeit with significant (spherical) aberration. Due to aberration the maximum intensity is shifted from the geometrical focus

$$f = \frac{n}{n-1}\frac{r_p}{2} \tag{3.1}$$

to the diffraction focus

$$f_d \approx f \left(1 - \sqrt{\frac{3\pi}{4\,k\,r_p} \; \frac{n(3-n)-1}{n\,(n-1)}} \right) . \qquad (3.2)$$

This expression approximates the position of the maximum intensity within an error of < 5% for kr_p / 100 and values of the refractive index 1.4 . n . 1.6. Besides of primary patterns, subpatterns (secondary structures) have been observed for substrate distances that exceed the focal length of the microspheres. These secondary structures originate from interferences of beams propagating from single microspheres.

The 2D lattices of microlenses permit one to produce on a substrate surface thousands or millions of single submicron features with a single or a few laser shots. By using single-shot-laser radiation and a fused quartz support with a monolayer of a-SiO_2 spheres of various diameters, we have demonstrated various different types of surface patterning. Among the examples are 2D ablation patterns on polymers [Denk *et al.* 2002; Piglmayer *et al.* 2002], arrays of circular cones on (100) Si wafers [Wysocki *et al.* 2003], and holes in W films that have been etched in WF_6 atmosphere. Arrays of W and Pd dots have been deposited from gaseous mixtures of WF_6 and H_2 [Denk *et al.* 2003] and from aqueous solutions of $PdCl_2$ in NH_3 [Bäuerle *et al.* 2004, 2003], respectively.

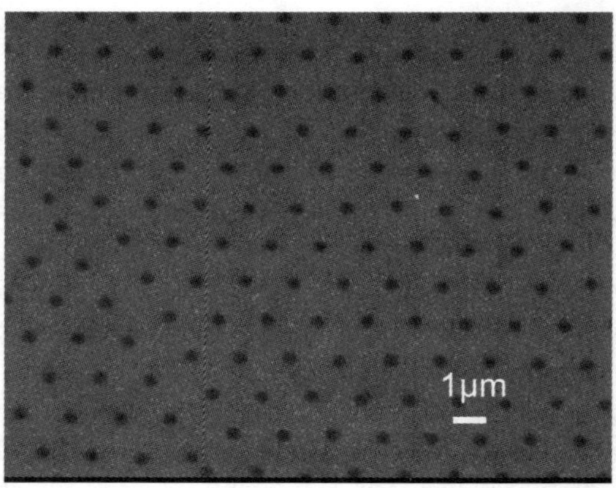

Figure 3.1: $YBa_2Cu_3O_{7-\delta}$ film on (100) MgO patterned by means of 248 nm KrF laser radiation ($\tau_P \approx 24$ ns) and a 2D lattice of a-SiO_2 microspheres ($d = 1.5$ μm) [after Brodoceanu *et al.* 2005].

Figure 3.1 shows a typical hexagonal pattern generated by means of a 2D lattice of microlenses and a single KrF-laser pulse on a thin film of $YBa_2Cu_3O_{7-\delta}$

on a MgO substrate. In such a primary pattern, the distance between neighboring spots is equal to the size of the microspheres, $d = 2r_p$. It has already been demonstrated in [Liberts *et al.* 1988] that laser-light irradiation of high-temperature superconductors causes a depletion of oxygen and thereby changes the material properties from superconducting to semiconducting (δ / 0.45). The dark dots in Fig. 3.1 consist of small humps with a maximum height of 5 to 8 nm with, probably, local depletion of oxygen. Such a structure may yield additional well-defined periodic pinning centers for vortex "lines" [Brodoceanu *et al.* 2005].

Well-defined surface "roughening" based on secondary intensity maxima has been demonstrated for polyethylenenaphthalate (PEN). The dimensions of the structures are significantly smaller than those of the corresponding primary structures [Bäuerle *et al.* 2003].

Laser-induced forward transfer (LIFT) has been performed by using thin metal foils in close contact between the microspheres and the substrate. By this means, hexagonal patterns of metal dots on arbitrary substrate materials together with the corresponding holes in the metal foils have been produced by single-shot KrF-laser irradiation. Similar patterns have been generated from metal films that were directly evaporated onto the surface of the microspheres. Scanning electron microscope (SEM) pictures of the 2D lattice of microspheres *after* the LIFT process reveal that the evaporated film is removed from the surfaces of spheres only within a certain region of diameter ρ. Similar to the size of the metal dots generated on a substrate, ρ depends on the size and index of refraction of the microspheres, the type and thickness of the evaporated film, and the laser parameters [Landström *et al.* 2004, Bäuerle *et al.* 2003]. In any case, this technique permits one to generate a 2D lattice of microlenses with well-defined apertures. Such a system can be employed, e.g., as a highly efficient contact mask. Due to the focusing nature of the microspheres, the incident intensity can be almost totally used for surface patterning. The situation is quite different to standard shadow masks where a significant amount of the incident intensity is lost. The amplification of the local intensity related to the focusing microspheres is of the order of $(r_p/\lambda)^q$ with 1 . q . 2. Thus, with the present conditions we obtain an amplification factor of 10 to 100 in comparison to a shadow mask.

Figure 3.2 shows a 2D lattice of microspheres with apertures $\rho < \rho_c \approx 0.53$ µm. Apertures with $\rho < \lambda$ are particularly well suited for investigating near field effects.

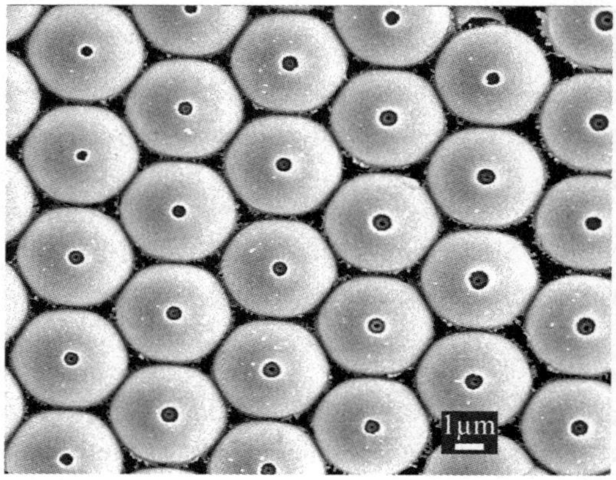

Figure 3.2: Apertures fabricated on a 2D lattice of SiO_2 microspheres (d = 4 μm) covered with 75 nm Au by single-pulse Ti:Sapphire-laser irradiation ($\lambda \approx 800$ nm, $\tau_P \approx 120$ fs) [after Langer *et al.* 2005].

4. Removal of Contamination Layers

The removal of contamination layers from technical or medical devices or from art work is mainly based on pulsed-laser ablation (PLA). Pulsed-laser ablation permits one to widely suppress the dissipation of the excitation energy beyond the volume that is ablated during the pulse. This is fulfilled if the thickness of the layer ablated per pulse, Δh, is of the order of the heat penetration depth, $l_T \approx 2 (D\tau_\ell)^{1/2}$ or the optical penetration depth $l_\alpha = \alpha^{-1}$, depending on which is larger

$$\Delta h \approx \max (l_T , l_\alpha) \qquad (4.1)$$

Here, D is the heat diffusivity and α the absorption coefficient. This (simplified) condition is, in fact, the basic requirement for applications of the technique in laser cleaning. With fluences around and above the threshold fluence for ablation, i.e. with ϕ / ϕ_{th} we can employ the approximation $\Delta h \propto \phi - \phi_{th}$. This relation holds for thermal, photophysical, and photochemical ablation mechanisms which have been discussed in detail in [Bäuerle 2000].

With laser radiation that is strongly absorbed within the contamination layer, pulsed-laser ablation permits one to remove such layers with a single or multiple pulses and submicrometer thickness control. This is a very important property of the technique and a prerequisite for many applications such as the cleaning of prefabricated devices, the restoration of paintings etc. Furthermore, contrary to mechanical tools or chemical techniques, laser light avoids any contamination of the material being processed. With many systems, the threshold fluence for substrate ablation exceeds the fluence required for the removal of the contamination layer. In such cases, cleaning becomes self-terminated. This process can be optimized via the laser parameters, in particular the laser wavelength and pulse duration. With medical and biological applications it is also important that laser beams are absolutely sterile tools.

With strongly inhomogeneous systems where the physical (optical, thermal, elastic, etc.) properties of the contamination layer differ significantly from those at the underlying material, thermo- and photomechanical ablation mechanisms may become important [Bäuerle 2000]. Stresses (pressures) related to such mechanisms can cause crack formation, exfoliation (removal of macroscopic fragments or flakes), instabilities etc. Thus, cleaning becomes uncontrolled and may cause substrate damage. With sensitive substrate materials and devices, it is important to suppress such mechanisms by choosing adequate laser parameters. Here, it is often sufficient to use short enough laser pulses and a wavelength that is strongly absorbed by the contamination layer.

With very high fluences, laser-induced shock waves become increasingly important in material ablation and fragmentation processes. Clearly, this regime is inadequate for the cleaning of sensitive substrates and devices.

With many systems, the removal of a contamination layer causes also physical and/or chemical modifications of the surface. Such surface modifications may be advantageous or disadvantageous, depending on the particular system and application. Finally, a reactive ambient medium may significantly enhance laser cleaning rates and cause well-defined surface modifications.

4.1. *Surface polishing and glazing*

With some applications it is essential that the removal of the contamination layer is correlated with a polishing (smoothening) or glazing of the surface. The mechanisms of these processes have been discussed in [Bäuerle 2000]. Clearly, surface smoothening decreases the effective surface area and thereby suppresses rapid recontamination based on physical, chemical or biological processes. Additionally, surface smoothening may decrease tribological effects, increase the optical transparency of the material, etc. A recent example for the latter application is the ablation and polishing of polytetrafluoroethylene (PTFE) foils for various different applications, as e.g. for liquid crystal displays (LCD). Figure 4.1 shows scanning electron microscope (SEM) pictures of a rough PTFE foil before and after F_2-laser irradiation in N_2 atmosphere. The untreated surface shows a fibrous structure which is typical for this type of dense PTFE. The average surface roughness measured by means of an atomic force microscope (AFM) is $R_A = 127$ nm. The aspect ratio of this roughness defined by the steepest slope found on any typical traverse scan by the AFM-cantilever was at least 20E. This value is determined by the resolution of the AFM cantilever which cannot recess in between the fibers. Thus, the real aspect ratio is much higher.

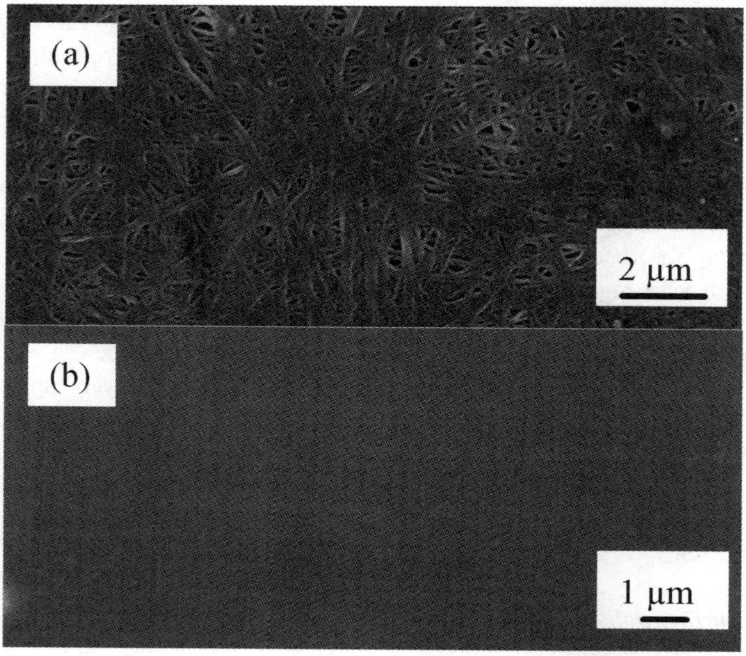

Figure 4.1: SEM pictures of a 40 µm thick PTFE foil. (a) Original surface showing fibrous structure of dense PTFE. (b) After irradiation with 157 nm F_2-laser light ($\phi = 20$ mJ/cm^2, pulse repetition rate $\nu_r = 20$ Hz, number of pulses $N_\ell = 1003$) in N_2 atmosphere at $p(N_2) \approx 1.1$ bar [after Gumpenberger *et al.* 2005b].

Figure 4.1b shows the same foil after 157 nm F_2-laser irradiation in N_2 atmosphere. The laser-treated surface is smooth and featureless. The surface roughness is $R_A \approx 64$ nm and the aspect ratio about 3E. While the total optical transmission of treated and untreated foils is almost the same within the wavelength region 250 nm $\leq \lambda \leq$ 800 nm, the collimated transmission – in particular within the region between 250 nm and 600 nm – is significantly increased in laser-treated samples. The decrease in light scattering observed in such polished samples can also be seen from the reflectivity spectra in Fig. 4.2. The laser-induced temperature rise can be estimated from Eq. (7.5.2) in [Bäuerle 2000]. For a fluence of $\phi \approx 5$ mJ/cm^2 and a pulse length of about 20 ns we obtain $\Delta T \approx$ 350EC. This temperature rise agrees quite well with the decomposition temperature of PTFE. Clearly, this estimation does not account for the influence of the surface roughness and the related dimensionality of heat diffusion.

Surface smoothening together with glazing is an important "side effect" in some cases of laser cleaning of stone reliefs, sculptures, and buildings. Glazing increases the surface hardness and decreases its porosity.

4.2. *The influence of an ambient medium*

A *non-reactive* ambient medium may influence the cleaning process in various different ways. Among those are the transport of the material removed from the surface, the attenuation of the incident laser light, and the promotion of surface instabilities. For gaseous media, all of these effects become more pronounced with increasing pressure.

In many cases it is advantageous to clean off contamination layers by irradiating the surface in the presence of a *reactive* ambient medium. Such a medium may enhance the cleaning efficiency and/or suppress physical or chemical changes of the surface and/or passivate or modify the surface simultaneously with cleaning. The enhancement of ablation or etch rates in gaseous and liquid ambient media has been discussed in detail for various different systems in [Bäuerle 2000].

5. Surface Modifications of PTFE, Applications in Biotechnology

Fluorocarbon polymers are widely used in various different fields of physical and chemical technology, in medicine and biotechnology. The wide range of applications of these materials is related to their high chemical resistivity and thermal stability, to their low adhesion properties, their low electrical conductivity and wettability for various liquids, etc. For applications that require a coating or hydrophilic/lipophilic properties of the surface, either the whole surface or restricted areas thereon must be modified. To achieve this, radiation treatment has been extensively investigated [Gumpenberger *et al.* 2005a, Benson 2002, Bäuerle 2000]. In this paragraph we discuss our present activities in this field, in particular with respect to applications in biotechnology.

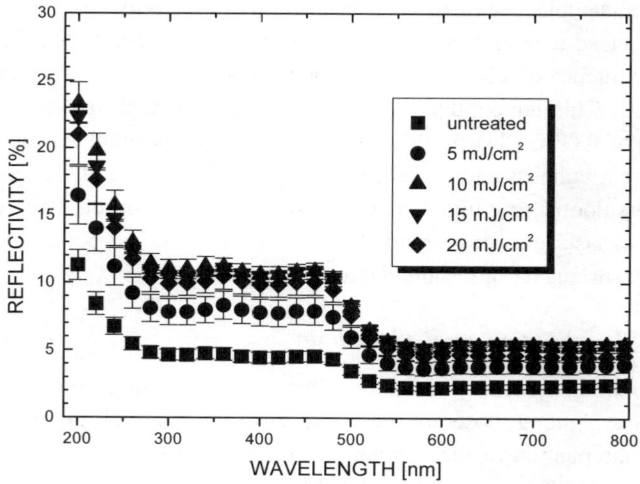

Figure 4.2: Reflectivity spectra of PTFE foil (40 μm) treated with 157 nm F_2-laser radiation at different fluences in N_2 atmosphere (pulse repetition rate ν_r = 20 Hz, scanning velocity of the irradiated area on the sample surface v_s = 40 μm/s, $p(N_2) \approx$ 1.1 bar). Measurements were performed with an Au mirror as reference [after Gumpenberger *et al.* 2005b].

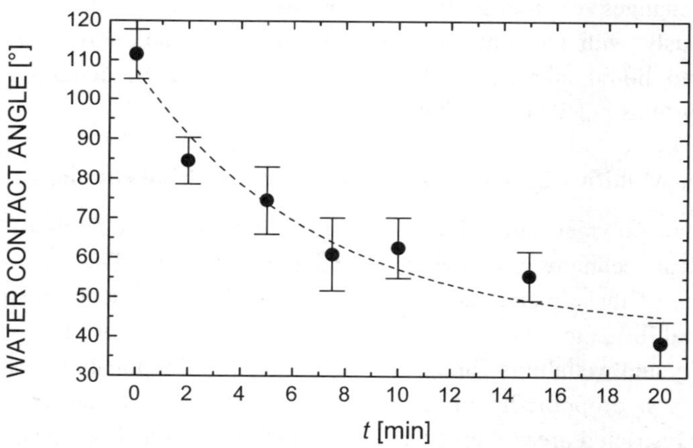

Figure 5.1: Water contact angle on dense PTFE versus time of irradiation with 172 nm Xe_2^*-light in 5 mbar NH_3 [Gumpenberger *et al.* 2005b].

Figure 5.1 shows the water contact angle of PTFE surfaces for 172 nm Xe_2*-lamp irradiation in NH_3 atmosphere. The biocompatibitily of surfaces treated in this way was evaluated by seeding biological cells onto them. Figure

5.2 shows the population density of human umbilical vein endothelial cells (HUVEC) on untreated and treated PTFE for both different irradiation times and times of cultivation. For comparison, results achieved with standard polystyrene (PS) Petri dishes have been included as a reference. First of all, good adhesion of cells is observed only on irradiated PTFE surfaces. Here, the cells start to proliferate and then grow to dense confluent multiple layers within about 8 days. The population densities are comparable to those achieved in PS Petri dishes. On untreated PTFE, the number of cells does not change significantly even after 8 days. For *all* samples, the population density of cells immediately after seeding is much lower than the density of seeded cells (76 000 cells/cm^2; full line in Fig. 5.2).

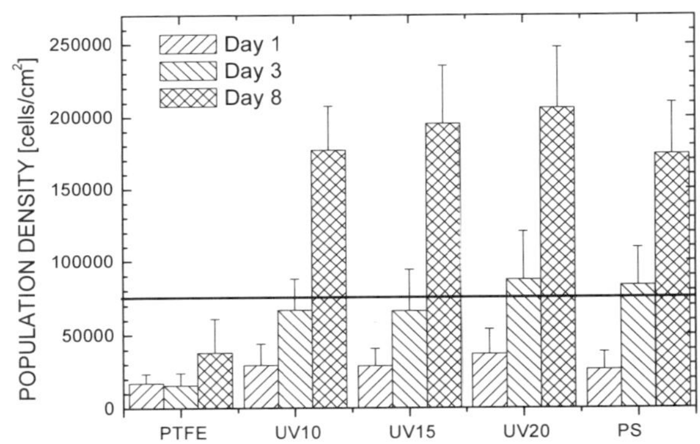

Figure 5.2: Population densities of human umbilical vein endothelial cells (HUVECs) on various samples after 1, 3 and 8 days of cultivation. PTFE: untreated material; UV10: PTFE surface exposed to 172 nm Xe_2^*-light in 5 mbar NH_3 for 10 min; UV15: 15 min; UV20: 20 min; PS: standard polystyrene Petri dish [adapted from Gumpenberger *et al.* 2003].

The high wettability and biocompatibility of treated surfaces is ascribed to polar groups introduced into the PTFE surface by substitution of F atoms. The subtraction of F is mediated via atomic H which is generated in the dissociation reaction $NH_3 + h\nu\,(172\text{ nm}, 193\text{ nm}) \rightarrow NH_2 + H$. Here, 172 nm radiation is more efficient with respect to the wettability achieved, but it also causes more structural damage of the material surface than 193 nm ArF-laser radiation. The high wettability is correlated with an enhanced biocompatibility of the surface. This is either due to the adsorption of adhesion-mediating proteins from the cell culture medium or due to direct interactions of the cell membrane, or specific receptors thereon, with the new species that substitute F in the modified PTFE surface.

Endothelial cells as, e.g., HUVECs, form the inner surface of blood vessels in direct contact with the blood stream. They play an important role in the avoidance of thrombosis, the immuno-response after injuries or the vascularization of tissues [Alberts *et al.* 2002]. Therefore, they are regarded as the optimal coating for artificial blood vessels, which are currently fabricated either from expanded PTFE or from knitted polyethyleneterephthalate (PET) fibers. The biocompatibility of both polymers can be improved by UV-irradiation in a reactive atmosphere [Heitz *et al.* 2004]. This offers the possibility to coat artificial blood vessels before implantation with the own (autologous) endothelial cells of the patient.

Modified PTFE surfaces show a high degree of biocompatibility with good cell adhesion and proliferation, which is confined to the irradiated areas. However, the UV-treatment results also in a loss of mechanical stability due to the scission of polymer chains, especially for light-sources with wavelengths below 193 nm [Gumpenberger *et al.* 2005a]. Figure 5.3a shows the decrease in tear strength of thin PTFE foils with increasing time of 172 nm Xe_2*-lamp irradiation. This decrease in mechanical stability can be diminished by creating only "pinning centers" for cell adhesion instead of uniform modification of the surface. Cell adhesion to extra-cellular proteins as, e.g., collagen, fibronectin, or fibrin, is based on interactions of receptors in the cell membrane with specific ligands in the proteins. The adhesion receptors with dimensions of, typically, 10 nm, are not homogeneously spread in the cell membrane, but form local adhesion points with lateral dimension of about 1 μm [Alberts *et al.* 2002]. Therefore, by diminishing photochemical surface modifications to spots with a diameter of about 1 μm and with a suitable spacing, cell adhesion similar to that obtained with homogeneously modified surfaces should be achieved. In preliminary experiments, we have demonstrated this by using again a 2D lattice of SiO_2 microspheres (d = 6.8 μm). The result is shown in Fig. 5.3b. The PTFE sample was first irradiated for 30 min through the 2D lattice of microspheres with 172 nm Xe_2*-light in 5 mbar NH_3. Subsequently, the microspheres were removed and HUVECs were seeded onto the surface after sterilization. Adhesion and proliferation of the HUVECs on such locally modified surfaces was similar to that on homogeneously irradiated surfaces. Again, cells did not adhere on non-irradiated areas. The distance between the dark points in Fig. 5.3b is equal to the diameter of the SiO_2 microspheres. We assume that these points are formed by local photochemical modification of the PTFE surface due to focussing of the light by the microspheres.

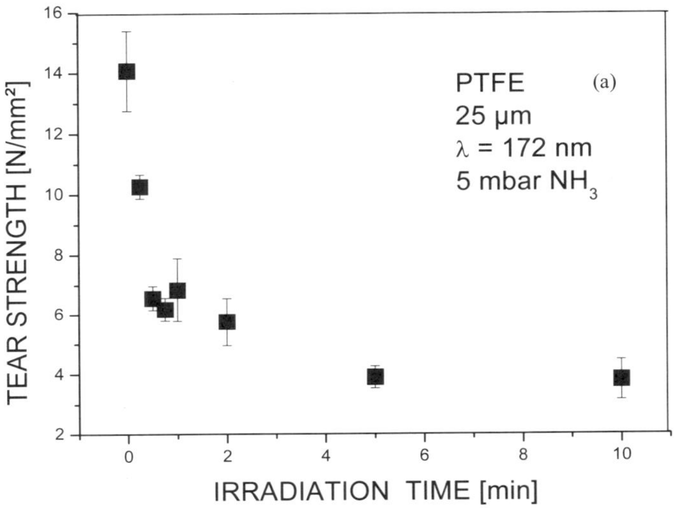

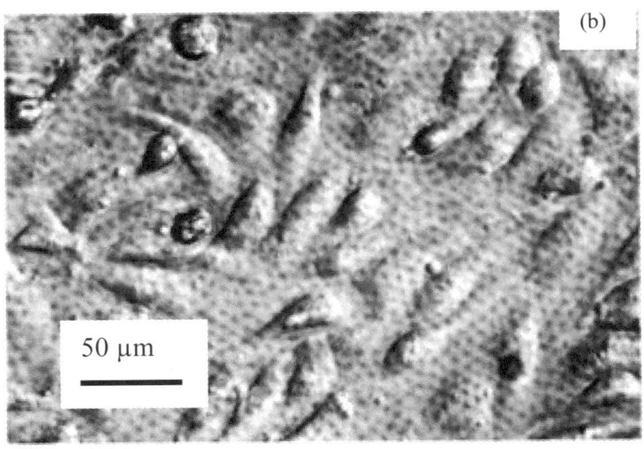

Figure 5.3: Modification of PTFE foil by 172 nm Xe_2^* light in 5 mbar NH_3. a) Change in tear strength with irradiation time [adapted from Gumpenberger *et al.* 2005b]. b) Phase contrast microscope picture of HUVECs grown on a PTFE surface irradiated for 30 min through a 2D lattice of SiO_2 microspheres (d = 6.8 μm). The image was taken 4 days after seeding with 64 000 cells/cm^2 [after Bäuerle *et al.* 2005].

The selectivity of adhesion and growth of cells can be tested by irradiating the surface with 172 nm Xe_2^*-light through a contact mask. Phase contrast microscope pictures of such surfaces taken 3 days after seeding with HUVECs and for different seeding densities are shown in Figs. 5.4a,b. If we define the selectivity of adhesion by the ratio of the number density of cells on the irradiated spots and the total number density of cells $N^* = N_{spots} / N_{total}$, we find that N^* is between 70 % and 90 %. This is a very high value, in particular if we take into account that only about 8.7 % of the total surface was irradiated. The high selectivity and resolution achieved in these experiments are very promising with respect to applications of this technique in the fabrication of micro-cell arrays. Such micro-cell arrays permit high-throughput analysis of gene functions, pharmacological testings, etc. in living cells [Ziauddin and Sabatini 2001, Wu *et al.* 2002, Silva *et al.* 2004]. Miniaturization also allows efficient use of potentially rare cells or biological samples. Furthermore, such cell arrays are very useful for cell multiplexing, because hundreds of arrays can be produced from a single set of source plates.

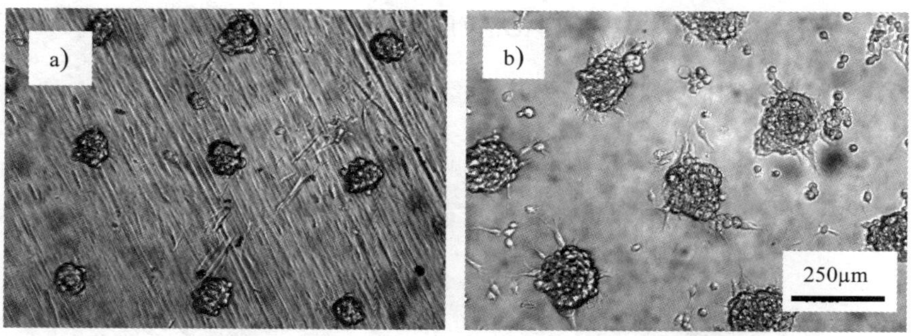

Figure 5.4: Phase contrast microscope pictures of HUVECs grown on a PTFE surface irradiated through a contact mask for 20 min with 172 nm Xe_2^*-light in 5 mbar NH_3. The images were taken 3 days after seeding with (a) 21 000 cells/cm^2 (b) 32 000 cells/cm^2 [adapted from Mikulikova *et al.* 2005].

Cell arrays, maybe even single-cell arrays with one cell per spot can also be fabricated by generating pinning centers by means of the microsphere technique. Clearly, the spacing between such centers must exceed at least the dimensions of single cells. Such a technique would have a wide range of novel applications.

6. Conclusions

Investigations on laser cleaning of solid surfaces from submicron- and nano-particulates have led to a deeper understanding of fundamental laser-matter interaction processes. Local substrate damage observed in many cases of dry laser cleaning (DLC) is unacceptable for most applications under consideration. Among the exceptions may be systems where, e.g., strongly absorbing "dust" particles can be easily evaporated/ablated without substrate damage. The situation is more promising with steam laser cleaning (SLC). The "universal threshold" observed in the experiments permits one to clean with essentially the same laser fluence substrates that are contaminated with particulates of different sizes and shapes and which consist of different materials. For submicron- and nano-particulates the threshold fluences are much lower than those required in DLC. However, whether SLC really becomes an adequate tool for large-scale applications is not yet clear. Here, further investigations with respect to possible surface modifications, the optimization of process parameters and efficiencies, the improvement of the reliability of the technique, etc. are certainly required. Wet laser cleaning (WLC) may combine advantages of DLC and SLC. At present, however, the few investigations performed in this area do not permit any conclusions. Local field enhancements observed in the vicinity of particulates in DLC can be employed for various different kinds of surface patterning. The technique permits one to generate both stochastic and periodic structures with variable surface densities and feature sizes below 100 nm in a single processing step.

Laser cleaning of solid surfaces from extended contamination layers is a well-established field. Cleaning can be performed in an inert or in a reactive ambient medium. Applications range from the cleaning of small devices up to extended areas including different types of artwork and even whole buildings. The removal of contamination layers often causes physical and/or chemical modifications of the surface. Among those are surface smoothening, densification and glazing, changes of the optical and wetting properties of the surface, etc. Laser cleaning together with such changes in surface properties open up completely new applications in various different fields of device- and biotechnology.

Acknowledgements

We wish to thank Irmengard Haslinger, Heidi Piglmayer-Brezina, Alois Mühlbachler, and Alfred Nimmervoll for expert technical assistance and the

Austrian Research Fund FWF (Fonds zur Förderung der wissenschaftlichen Forschung) under contract no. P16133-N08 for financial support.

References

Alberts B., D. Bray and J. Watson: Molecular Biology of the Cell, 4th edition, Garland Publications 2002.

Allen S.D., A.S. Miller and S.J. Lee: Laser assisted particle removal 'dry' cleaning of critical surfaces, Materials Science and Engineering **B49**, 85 (1997).

Arnold N., G. Schrems and D. Bäuerle: Ablative thresholds in laser cleaning of substrates from particulates, Appl. Phys. A **79**, 729 (2004).

Arnold N.: Theoretical description of dry laser cleaning, Appl. Surf. Sci. **208–209**, 15 (2003).

Bäuerle D., L. Landström, J. Kofler, N. Arnold and K. Piglmayer: Laser-processing with colloid monolayers; Proc. SPIE 5339, 20 (2004).

Bäuerle D., G. Wysocki, L. Landström, J. Klimstein, K. Piglmayer and J. Heitz: Laser-induced single step micro/nanopatterning, Proc. SPIE 5063, 8 (2003).

Bäuerle D.: *Laser Processing and Chemistry*, 3rd edition, Springer Verlag 2000.

Bäuerle D., T. Gumpenberger, J. Heitz, D. Brodoceanu, S. Moritz and C. Romanin: (2005).

Benson R.S.: Use of radiation in biomaterials science, Nucl Instr Meth B **191,** 752 (2002).

Boughaba S., J.B. Héroux, M. Curcio, E. Sacher and M. Meunier: Removal of particulate contamination from silicon surfaces with laser-based cleaning technology. Particles on Surfaces **5 & 6**: Detection, Adhesion and Removal, 297 (1999).

Bowling R.A.: A theoretical review of particle adhesion. In *Particles on Surfaces: Detection, Adhesion, and Removal,* Ed. by K.L. Mittal, Marcel Dekker Inc. (New York, Basel, Hong Kong 1995).

Brodoceanu D., I. Puica, W. Lang, J.D. Pedarnig and D. Bäuerle: (2005).

Chaoui N., J. Solis, C.N. Afonso, T. Fourrier, T. Mühlberger, G. Schrems, M. Mosbacher, D. Bäuerle, M. Bertsch and P. Leiderer: A high sensitivity in-situ optical diagnostic technique for laser cleaning of transparent substrates, Appl. Phys. A **76**, 767 (2003).

Curran D., J.M. Lee and K.G. Watkins: UV laser removal of small metallic particles from silicon wafers, Optics and Lasers in Engineering **38**, 405 (2002).

Denk R., K. Piglmayer and D. Bäuerle: Laser-induced etching and deposition of W using a-SiO2 microspheres, Appl. Phys. A **76**, 549 (2003).

Denk R., K. Piglmayer and D. Bäuerle: Laser-induced nanopatterning of PET using a-SiO2 microspheres, Appl. Phys. A **74**, 825 (2002).

Drakaki E., A.G. Karydas, B. Klinkenberg, M. Kokkoris, A.A. Serafetinides, E. Stavrou, R. Vlastou and C. Zarkadas: Laser cleaning on Roman coins, Appl. Phys. A **79**, 1111 (2004).

Fourrier T., G. Schrems, T. Mühlberger, J. Heitz, N. Arnold, D. Bäuerle, M. Mosbacher, J. Boneberg and P. Leiderer: Laser cleaning of polymer surfaces, Appl. Phys. A **72**, 1 (2001).

Grigoropoulos C.P. and D. Kim: Liquid-assisted pulsed laser cleaning with near infrared and UV-pulsed lasers, Chapter 5 in Luk'yanchuk (2002), p. 229.

Gumpenberger T., J. Heitz, D. Bäuerle and T. Rosenmayer: Modification of expanded polytetrafluoroethylene by UV irradiation in reactive and inert atmosphere, Appl. Phys. A **80**, 27 (2005a).

Gumpenberger T., J. Heitz, D. Bäuerle and T. Rosenmayer: Laser polishing of Polytetrafluoroethylene (2005b).

Gumpenberger T., J. Heitz, D. Bäuerle, H. Kahr, I. Graz, C. Romanin, V. Svorcik and F. Leisch: Adhesion and Proliferation of Human Endothelial Cells on Photochemically Modified Polytetrafluoroethylene, Biomaterials **24**, 5139 (2003).

Heitz J., T. Gumpenberger, H. Kahr and C. Romanin: Adhesion and proliferation of human vascular cells on UV-light modified polymers, Biotechnol. Appl. Biochem. **39**, 59 (2004).

Israelachvili J.N.: Intermolecular and Surface Forces, 2nd edition, Academic Press (San Diego San Francisco New York) 1992.

Kane D.M., A.J. Fernandes and D.R. Halfpenny: Pulsed Laser Cleaning of Particles from Surfaces & Optical Materials, Chapter 4 in Luk'yanchuk (2002), p. 181

Kim D. and J. Lee: On the physical mechanisms of liquid-assisted laser cleaning, J. Appl. Phys. **93**, 762 (2003).

Klein S., T. Stratoudaki, V. Zafiropulos, J. Hildenhagen, K. Dickmann and Th. Lehmkuhl: Laser-induced breakdown spectroscopy for on-line control of laser cleaning of sandstone and stained glass, Appl. Phys. A **69**, 441 (1999).

Kofler J.: Focusing of light in axially symmetric systems within the wave optics approximation. Master Thesis, Johannes-Kepler-University of Linz, Austria, September 2004.

Kollia, Z., E. Sarantopoulou, A.C. Cefalas, S. Kobe and Z. Samardzija: Nanometric size control and treatment of historic paper manuscript and prints with laser light at 157 nm, Appl. Phys. A **79**, 379 (2004).

Kolomenskii A.A., H.A. Schuessler, V.G. Mikhalevich and A.A. Maznev: Interaction of laser-generated surface acoustic pulses with fine particles: surface cleaning and adhesion studies, J. Appl. Phys. **84**, 2404 (1998).

Landström L., J. Klimstein, G. Schrems, K. Piglmayer and D. Bäuerle: Single-step patterning and the fabrication of contact masks by laser-induced forward transfer, Appl. Phys. A **78**, 537 (2004).

Lang F., M. Mosbacher and P. Leiderer: Near field induced defects and influence of the liquid layer thickness in steam laser cleaning of silicon wafers, Appl. Phys. A **77**, 117 (2003).

Langer G., D. Brodoceanu, K. Piglmayer and D. Bäuerle: (2005).

Lee J.M., K.G. Watkins and W.M. Steen: Angular laser cleaning for effective removal of particles from a solid surface, Appl. Phys. A **71**, 671 (2000).

Lee Y.P., Y.F. Lu, D.S.H. Chan, T.S. Low and M.S. Zhou: Steam laser cleaning of plasma-etch-induced polymers from via holes, Jpn.J. Appl. Phys. **37**, 2524 (1998).

Leiderer P., M. Mosbacher, V. Dobler, A. Schilling, O. Yavas, B.S. Luk'yanchuk and J. Boneberg: Steam laser cleaning of silicon wafers: Laser-induced bubble nucleation and efficiency measurements, Chapter 6 in Luk'yanchuk (2002), p. 255.

Liberts G., M. Eyett and D. Bäuerle: Laser-induced Surface Reduction of the High T_c Superconductor $YBa_2Cu_3O_{7-x}$, Appl. Phys. A **45**, 313 (1988).

Lu Y.F., W.D. Song, C.K. Tee, D.S.H. Chan and T.S. Low: Wavelength effects in the laser cleaning process, Jpn.J. Appl. Phys. **37**, 840 (1998).

Luk'yanchuk B.S., Z.B. Wang, W.D. Song and M.H. Hong: Particle on surface: 3D-effects in dry laser cleaning, Appl. Phys. A **79**, 747 (2004).

Luk'yanchuk B.S. (Ed.): *Laser Cleaning*, Series: Optical Physics, Applied Physics and Materials Science, World Scientific (New Jersey, London, Singapore, Hong Kong) 2002.

Mikulikova R., S. Moritz, T. Gumpenberger, M. Olbrich, J. Heitz, C. Romanin, L. Bacakova and V. Svorcik: Cell microarrays on photochemically modified polytetrafluoroethylene, to be published (2005).

Mosbacher M., H.-J. Münzer, M. Bertsch, V. Dobler, N. Chaoi, J. Siegel, R. Oltra, D. Bäuerle, J. Boneberg and P. Leiderer: Laser assisted particle removal from silicon wafers, In *Particles on Surfaces 7: Detection, Adhesion and Removal*, ed. by K.L. Mittal (VSP, Zeist 2003), p. 291.

Neves P., M. Arronte, R. Vilar and A.M. Botelho do Rego: KrF excimer laser dry and steam cleaning of silicon surfaces with metallic particulate contaminants, Appl. Phys. A **74**, 191 (2002).

Piglmayer K., R. Denk and D. Bäuerle: Laser-induced surface patterning by means of microspheres, Appl. Phys. Lett. **80**, 4693 (2002).

Pleasants S., B.S. Luk'yanchuk and D.M. Kane: Modelling laser cleaning of low-absorbing substrates: the effect of near-field focussing, Appl. Phys. A **79,** 1595 (2004).

Pleasants S. and D.M. Kane: Laser cleaning of alumina particles on glass and silica substrates – Experiment and quasistatic model, J. Appl. Phys. **93** (2003).

Rudolph P., F.J. Ligterink, J.L. Pedersoli, M.van Bommel, J. Bos, H.A. Aziz, J.B.G.A. Havermans, H. Scholten, D. Schipper and W. Kautek: Characterization of laser-treated paper, Appl. Phys. A **79**, 181 (2004).

Schrems G.: Cleaning and Patterning of Various Surfaces by Pulsed Laser Irradiation, PhD Thesis, Johannes-Kepler-University Linz, Austria, September 2003.

Schrems G., M.P. Delamare, N. Arnold, P. Leiderer and D. Bäuerle: Influence of storage time on laser cleaning of SiO_2 on Si, Appl. Phys. A **76**, 847 (2003).

She M., D. Kim and C.P. Grigoropoulos: Liquid-assisted pulsed laser cleaning using near-infrared and ultraviolet radiation, J. Appl. Phys. **86**, 6519 (1999).

Silva J.M., H. Mizuno, A. Brady, R. Lucito and G.J. Hannon: RNA interference microarrays: High-throughput loss-of-function genetics in mammalian cells, Proc Natl Acad Sci USA **10**, 6548 (2004).

Takai M., N. Suzuki and O. Yavas: Cleaning for Field Emitter Arrays, Chapter 10 in Luk'yanchuk (2002), p. 417.

Teule R.: Surface cleaning of artwork using 248 nm laser radiation, Lambda Highlights May 2001, No. 58, p.1.

Troll C., H. Römich, K. Dickmann and J. Hildenhagen: Cleaning of corrosion crusts on stained glass windows with excimer lasers, ICOM Committee for Conservation, 12th Triennial Meeting Lyon, 29.8.-3.9.1999.

Wu R.Z., S.N. Bailey and D.M. Sabatini: Cell-biological applications of transfected-cell microarrays, Trends Cell Biol **12**, 485 (2002).

Wu X., E. Sacher and M. Meunier: The modeling of excimer laser particle removal from hydrophilic silicon surfaces, J. Appl. Phys. **87**, 3618 (2000).

Wysocki G., R. Denk, K. Piglmayer, N. Arnold and D. Bäuerle: Single-step fabrication of silicon-cone arrays, Appl. Phys. Lett. **82**, 692 (2003).

Zafiropulos V.: Laser ablation in cleaning of artworks, Chapter 8 in Luk'yanchuk (2002), p. 343.

Zapka W.: The road to "Steam Laser Cleaning", Chapter 1 in Luk'yanchuk (2002), p. 23.

Ziauddin J. and D.M. Sabatini: Microarrays of cells expressing defined cDNAs. Nature **411**, 107 (2001).

Chapter 2

AN OVERVIEW OF EXPERIMENTAL RESEARCH INTO THE LASER CLEANING OF CONTAMINANTS FROM SURFACES

A J FERNANDES AND D M KANE

Department of Physics, Macquarie University, Sydney, NSW 2109, Australia

An overview of the literature on laser cleaning experiments is provided in a tabulated form. The details of a large cross-section of experimental studies on the removal of particles and contaminants from various material surfaces by pulsed laser cleaning is provided and discussed. The tables aim to provide a useful, clear and quick comparison of the numerous different sets of experimental parameters used in laser cleaning studies, and to enable a better understanding of the removal of particulates from surfaces.

1. An Overview of the Information in the Tables

The field of laser cleaning has grown enormously in the last fifteen years, and has led to the development of new fields, such as the use of UV irradiation to remove surface hydroxyls (which is detailed in the chapter entitled 'UV Laser-induced Dehydroxylation of UV Fused Silica Surfaces' in this book), as well as the nano-structuring and patterning of various optical and semiconductor surfaces. In *Laser Cleaning*[1], we presented an overview of the literature to date, detailing the experimental studies performed to remove particles and contaminants from various pure material surfaces, using pulsed laser cleaning[2]. The focus is on industrial applications and does not include the prodigious and excellent work in complex material systems motivated by using laser cleaning in art and cultural heritage conservation.

The aim of this work is to provide a useful comparison of the effect on laser cleaning of the different parameters and techniques involved, thus enabling a better and clearer understanding of the many experimental factors which affect particle removal. Also, the many measurement and analysis techniques used are introduced. The tables have been updated to include published research done in the last few years, to the time of writing. The range of materials studied has increased, as have the range of process methods and analysis and measurement techniques, compared to those mentioned in Kane et al.[2]. The tables do not represent an exhaustive summary of all laser cleaning studies on pure materials. We have kept abreast of research into pulsed laser cleaning of particulates from surfaces. We apologise in advance for any work which has not been included in the summary.

The summary of the laser cleaning work has been tabulated in Tables 1.1-5, and Table 2. Tables 1.1-5 are keys to some of the abbreviations used to describe various aspects of the reported research. New additions to the main summary table and the key tables since Kane et al.[2] are marked with an asterisk (*). Table 2 is the main summary table detailing experimental research published at the time of writing. The major focus of this summary is on pulsed laser cleaning of particulates from surfaces, using pulsed laser cleaning, but some other types of contamination are also included. Some examples of laser cleaning using ablation are also provided, to enable comparison with dry, wet, and steam laser cleaning which are non-ablative. This comparison between the different types of laser cleaning highlights the different mechanisms and parameters that can contribute to contaminant removal.

2. Categories of Laser Cleaning

The information in Table 2 lists the type of laser cleaning (ablation, dry, steam or wet) which was used in the research summarised therein. These methods employ different removal mechanisms, which may affect how easily contaminants are removed when using a given laser fluence. A brief description of these techniques is given here.

Ablative laser cleaning is normally used to remove layers of contamination from surfaces such as pollutants from artwork[11], encrustations from marble[45], and oxides from copper[18]. This uses absorption of the laser energy by the material to cause ablation to a certain thickness/depth in the contaminant layer, and thus the material characteristics, laser fluence and the wavelength used are important parameters. The laser fluence used has to be above the ablation threshold fluence which is material and wavelength dependent. For example, removal of medical powders from metal pharmaceutical punches[16] was most efficient at a wavelength of 1.06μm and not efficient at 532nm. However, removing water repellent treatments from limestone required less laser fluence at UV wavelengths compared to using visible and infra-red wavelengths[40]. By increasing the number of pulses used to remove the contamination, lower fluences could be used. Despite taking a longer time for processing, this minimises the risk of damage and allows greater control over the depth of removal. For example, 100% removal of photoresist and metallic polymer from storage nodes[17] was obtained using a KrF excimer laser, with a pulse fluence of 130mJ/cm^2 using three pulses, whereas by using six pulses, the pulse fluence required for 100% removal was 90mJ/cm^2.

Pulsed laser cleaning can be classified as either wet, steam or dry laser cleaning, and is used primarily for removing particulate contamination. The mechanisms used here are different to those in ablative laser cleaning as short pulsed laser cleaning removes the particle as a whole. In dry laser cleaning, the most important mechanism is where the short pulse length causes a rapid thermal expansion of the particle and/or the substrate, and the resulting accelerations produced can overcome the particle-substrate adhesion forces leading to particle removal. This requires either the substrate and/or the particle material to be absorbing at the laser wavelength.

Wet laser cleaning uses a liquid layer which is strongly absorbing at the laser wavelength being used. The liquid is used to cover the substrate and contaminants, and can surround and penetrate below the particles. The laser irradiation can rapidly heat and evaporate the liquid, and this can force the particles off the surface[100]. Steam laser cleaning uses a very thin liquid layer, which is condensed onto the substrate surface. This is superheated (and evaporated), and causes the nucleation and growth of bubbles at the liquid-solid interface. This results in high pressures below the contaminant which can be large enough to remove the particle from the surface[101]. In general, steam and wet laser cleaning require lower fluences for particle removal than dry laser cleaning, although application of liquids to the substrate may not always be practical. More detailed explanations of dry, wet, steam, and ablative laser cleaning can be found in *Laser Cleaning*[1] as well as in many of the references provided in this chapter.

Further information on the laser cleaning processes used is listed in Table 1.1. This includes details such as the type of liquid used (for example, steam cleaning with water[98] or wet cleaning with an isopropanol film[51]). The cleaning process used can be further enhanced using a number of techniques which are also listed in Table 1.1. This includes exposing the samples to a humid environment when using dry laser cleaning[90], as this causes capillary condensation on the substrate and aids removal in a similar way that a liquid layer does in steam or wet laser cleaning. Work has also been done changing the way the irradiation is applied, such as irradiating from the back surface which has been shown to remove carbon particles from paper without affecting the ink on it[59], and focussing the laser beam into a line which is scanned across the surface[76] (although the mechanism for removal using this technique was not clear).

Other mechanisms for laser cleaning which are neither ablative nor primarily thermo-mechanical are surface acoustic wave cleaning, and shock wave laser cleaning. The laser beam has been used to generate surface acoustic waves in the substrate, where the mechanical forces associated with the wave were sufficient to remove particles[91]. However the large fluences required to generate high amplitude waves for particle removal meant that laser cleaning took place in the ablation regime, that is, unwanted damage to the substrate occurred. An investigation into the substrate surface displacement during pulsed laser irradiation was performed[92]. This showed that while the maximum displacement depended on the total laser energy deposited onto the substrate, the maximum acceleration produced (which needs to be sufficiently high to overcome the particle-substrate adhesion forces) has a strong dependence on the temporal laser pulse shape. That is, the cleaning process can be enhanced if the laser pulse shape can be optimised.

Table 1.1: Abbreviations used in Table 2 to describe the different types of laser cleaning. New additions since Kane et al.[2] are marked with an asterisk (*).

Key	Type of Laser Cleaning
AF	Acetone film
AlF	Water/alcohol 300nm layer
EF	Ethanol film
EtAb	Etching and laser ablation*
IF	Isopropanol film
LAS	Laser-assisted shock wave
LB	Surface scanned with line beam
LFE	Liquid film enhanced
Lyte	Sample immersed in electrolyte to fix electrochemical potential of the surface
Rear	Rear surface irradiated*
RH/HH	Raised humidity, High humidity
RH100	Relative humidity 100%
RH40	Relative humidity 40%
SAW	Non-local laser cleaning using laser generated surface acoustic waves
SuD	Surface displacement investigated
U	Ultrasonics
WA/WAV	Water/alcohol, Water/alcohol vapour
WF/WFt	Water layer*, Water layer 0.1mm thick*
WI	Water/Isopropanol*
WV/WVC	Water vapour, Water vapour condensed layer

In shock wave laser cleaning, the laser beam has been sharply focussed to cause a breakdown in air which produces a shock wave. The forces associated with this wave are also able to remove particles, including tungsten particles which were very difficult to remove using 'conventional' laser cleaning techniques[77].

Laser cleaning has also been found to be most effective in some applications, when used in conjunction with other cleaning methods. This includes com combining laser ablation and etching[72], and also with organic solvents[39]. Furthermore, the effectiveness of steam and dry laser cleaning has been directly evaluated with ultrasonics[97].

Wet laser cleaning has been used in conjunction with electrolytes[37]. In this case, oxidised iron samples were immersed in an electrolyte under potentiostatic control in order to fix the electrochemical potential of the surface. It was found that when a potential was applied to the sample, the laser cleaning efficiency increased. It was proposed that a diffusion of hydrogen was the underlying mechanism for the enhanced cleaning.

3. Deposition Techniques

Table 1.2 lists the different methods by which particles and other contaminants are deposited on the various substrates being cleaned in scientific studies testing the technique. The number of techniques has grown greatly due to the use and acceptance of laser cleaning as a contaminant removal tool in a greater number of applications. This is evident from the large range of both contaminant materials and substrates listed in Table 2. These methods can be classified into the following general categories: natural contamination; particle suspension coating; and direct deposition.

Many experiments have used 'naturally occurring' contamination from a variety of sources. This includes natural (environmental) contamination such as volcanic dust[9], or pollution, and nuclear contamination[4]. Contamination from the fabrication method or processing[17] can occur, and includes such materials as organic glue[44], epoxy resin films[118], and residues after etching[119]. Metal pharmaceutical punches were contaminated from their use in tablet forming processing[16], while other metals may also oxidise on exposure to air during the fabrication process[25], or may have layers of oil and grease contaminants[6]. Exposure in a workshop environment[23] or during the packaging process[35] also contributes to contamination.

A number of experiments use a mix of particles and a liquid to form a suspension or solution, which is then applied to the substrate either directly or using a spin-coating machine. This is to provide a more uniform distribution of particles on the substrate than direct deposition. It also helps avoid particle aggregation, which is a problem in many direct deposition techniques. Spin-coating also provides lower and even more uniform distributions than when applying suspensions directly. The liquid used to form the suspension has been included in the summary table for those references which have specifically mentioned it. In many cases, the liquid is allowed to evaporate before laser cleaning commences, either by drying in air, or with liquified CCl_2F_2 spray can drying or spin drying.

Direct deposition of the contamination can take place in a variety of ways. Particles can be dusted on[12] or scattered intentionally[73], or sprayed onto the substrate[38], in some instances with an atomiser[34]. The Particle Measuring System particle generator consists of a nebuliser, a drying tube, and an output nozzle, and deposits particles in a circular pattern[81]. The laser-assisted deposition technique uses a laser pulse to hit a target of particles, which are then ejected onto a substrate. This aids control over particle concentrations, which vary systematically across the surface as a function of distance from the target[33]. The dip and tap technique involves covering glass slides with particles. Substrates can then be tapped to remove any excess particles which are not adhered to the surface[27].

Particles were deposited using electrostatic attraction, where a potential was applied to the substrate, and the particles were held on a spatula a few centimetres away from the surface[7].

Other types of non-particulate contamination are included in Table 2. Samples were baked[18] or exposed to high temperatures[116], which produced oxidation, while some were drawn on using pens and markers, to investigate the removal of inks[39].

It is important to note that the technique used to deposit the particles can have an effect on their adhesion to the surface. For example, suspension residues or humidity condensation can form capillary bridges between the particle and the substrate, which contributes to the total particle-substrate adhesion force, but this effect can also be used to assist removal if the capillary bridges are evaporated by the laser irradiation[90]. The particle shape may also change or deform depending on how it is applied, and therefore affect the total adhesion force. Silica spheres were found to deform over time, and increase the contact area and total adhesion force, making them harder to remove[88].

Table 1.2: Abbreviations used in Table 2 to describe the particle/contaminant deposition technique used to prepare samples for laser cleaning experiments. (* New additions since Kane et al.[2]).

Key	Deposition Technique
AJ	Air jet drying
Alsus	Coated with alumina/acetone suspension*
Asus	Acetone suspension
Bake	Copper baked to produce oxidized samples*
Dr	Surface drawn on
DT	Substrate dipped in particles and tapped to remove excess particles
Du	Dusted on
EA	Electrostatic attraction
Esus	Alcohol/ethanol suspension
EWsus	Ethanol/water suspension*
Fab	During fabrication/exposure to air
Film	Epoxy resin films*
Glue	Organic glue
Isus	Isopropyl alcohol suspension
LAD	Laser assisted deposition
Lay	Layer of oil and grease
LiqC	Liquified CCl_2F_2 spray can drying
LPar	Liquid paraffin sprayed on by atomizer*
Msus	Methanol suspension
Nat	Natural/pollution
Nucl	Nuclear contamination*
Pack	During packaging process
PMS	Particle Measuring System particle generator
Proc	Deposited/contaminated during processing*
Res	Residue after etching
Scat	Scattered intentionally*
SCCS	Spin-coating of colloidal solution/suspension
Sol	Solution
Spin	Spin drying*
Spra	Spraying*
Sus	Suspension*
Tabl	Contaminated from use in tablet forming process*
Temp	Sample exposed to high temperature*
WExp	Exposure in workshop*
WS	Water suspension

4. Laser Cleaning Efficiency

The ability of the laser cleaning technique used to remove contamination easily and without damage to the substrate is clearly an important issue. Early studies in laser cleaning looked at the possibility of using laser irradiation to remove the contamination for a specific application, compared to other cleaning methods (for example, see [97], where steam and dry laser cleaning were compared with ultrasonic cleaning). As the field has grown, ways of improving the laser cleaning process have been investigated. Standard analysis techniques have been introduced which have allowed for quantification of contaminant removal. This enables different techniques and parameters of the laser cleaning process to be evaluated and demonstrated clearly. As a result, it is possible to determine the mechanisms causing removal for a given application, and therefore the best method for achieving optimal cleaning without unwanted surface damage.

When cleaning artwork, or stones and metals, where layers or films of contamination are removed, the cleaning process is normally said to be effective or non-effective in removing the entire layer of contamination, and the laser fluence (energy per unit area) at which this occurs is specified. However for particle removal there is a threshold fluence at which removal first commences, although complete removal of the contamination (usually) requires higher fluences. A given region of interest can either be examined both before and after the laser irradiation is applied, or compared to a different non-irradiated area of the contaminated substrate, to determine the efficiency of the cleaning process.

Table 2 lists the reported cleaning efficiency achieved in many of the various studies summarised. This is often given in terms of a percentage of either the number of particles removed, or the area cleaned in a given region (area) of interest on the substrate. In addition, Table 1.3 lists further comments on the effectiveness of a given laser cleaning system and procedure used. *Optimal* indicates the parameters used for which the highest removal efficiency was obtained with a given type of laser cleaning method and system. For example, it was found that dust from an aluminium mirror was best removed with a fluence of 160±30mJ/cm^2 using a KrF laser with a pulse length of 30ns and with 5 pulses, compared to other lasers with pulse lengths between 50-200ns and when using 1-1000 pulses[10].

In some cases the cleaning efficiency is dependent on the incident angle of the laser beam onto the substrate, and also whether the substrate is irradiated from the front or the rear surface. For example, complete removal of 2.5μm silica spheres from silicon was obtained using dry laser cleaning at normal incidence[28]. The cleaning efficiency dropped to zero when angles greater than 15°

from the normal incidence were used. This effect was attributed to the spherical particles focussing the laser irradiation on the substrate surface. At normal incidence, the irradiation was focussed directly beneath the particle, leading to an enhanced temperature rise and improved cleaning efficiencies. When normal incidence was not used, the focussing beneath the particle was reduced, leading to less or no cleaning.

In contrast, a different study found that when removing 0.3μm silica spheres from silicon, there was no effect on the dry laser cleaning efficiency when the angle of incidence was changed[85]. However, this cleaning efficiency was approximately 10%. This same study found that irregularly shaped 0.3μm silicon nitride (Si_3N_4) particles were easier to remove from silicon when glancing angles of incidence close to the surface were used, as higher cleaning efficiencies were obtained[85]. The different results between the 0.3μm silica spheres and silicon nitride particles were attributed to the influence of the particle shape on the van der Waals adhesion forces, and the lower reflectivity of the silica. It was proposed that as the adhesion force of the irregularly shaped silicon nitride particles is comparable to the horizontal component of the laser beam radiation pressure force, this would cause the particles to roll off the surface, whereas the adhesion force of the spherical silica would be higher than the horizontal component of the radiation pressure force by a factor of 100, and therefore cleaning at angles would have no effect[85].

Although no systematic investigation into the cleaning of different sized particles at varying irradiating angles of incidence has been done, it is suggested that the reason that 2.5μm silica spheres seem to have a varying cleaning efficiency which is dependent on angle is due to the relative ease in removing larger sized silica particles. The difficulty in completely removing the smaller 0.3μm silica spheres may mean that the effect of different irradiation angles may not be evident.

Further work on angular laser cleaning investigated the removal of copper particles from a copper substrate[20]. It was found that using glancing angles of incidence led to a great increase in the cleaning efficiency and reduction in the threshold fluence for removal, compared to that obtained at normal incidence. This was attributed to the direct irradiation of the particle-substrate interface when using glancing angles, allowing for improved laser absorption at this point. This means that less fluence will be required to overcome the particle-substrate adhesion forces, making particle removal easier.

The cleaning efficiency dependence on front or rear surface irradiation is affected by the material properties of the contaminant and the substrate. The

Table 1.3: Abbreviations used in Table 2 to describe the effectiveness of the laser cleaning procedure and parameters used. New additions since Kane et al.[2] are marked with an asterisk (*).

Key	Comment on Laser Cleaning Effectiveness
Acc	Accelerations of 10^{10}m/s^2 remove 0.05μm particles
BGI	Beam at glancing angle cleans area larger than beam spot
CleDam	Significant cleaning, but damage also occurs*
Comb	Most effective by combining 532nm irradiation with organic solvents*
Eff	Efficient
FIEff	Front irradiation more effective than reverse irradiation*
IE	Improved efficiency
IN	Increases with number of pulses
LA	Increased at glancing beam incidence (large) angle
LT	Particles >0.5μm removed with some efficiency
NAGSW	Not as good as shorter wavelength
NC	No improvement with changing beam incidence angle
NE/IR	Not effective/Incomplete removal
NegD	Negligible damage*
NI	Normal Incidence
Opt	Optimal
Part	Coating is partially or completely eliminated*
PhRem	Photoresist removed, but not polymer*
RIEff	Reverse irradiation more effective than front irradiation*

removal of glass particles and plate-like silicon particles from glass was investigated using dry laser cleaning, with a frequency-doubled Nd:YAG laser[69]. The glass particles were easier to remove using front irradiation, as the particles acted as a lens and caused near-field focussing of the irradiation onto the particle-substrate interface. The higher laser intensity at this interface (and thermal expansion) allowed for high cleaning efficiencies. However for rear surface irradiation the focussing occurred on the top of the glass particle, rather than at the particle-substrate interface, so the obtained cleaning efficiencies were relatively lower. In contrast, the plate-like silicon particles were easier to remove from the rear surface due to the physical properties of silicon and quartz. The temperature in the particle decreases with distance from the irradiated surface. This meant that the peak temperature rise of the particle-substrate interface was higher for rear irradiation than for front irradiation. Again, as the temperature increase is related to the cleaning efficiency, in this case higher removal rates were obtained with rear irradiation.

In some instances, rear irradiation can be used to avoid unwanted substrate damage. Carbon particles were removed from paper with ink lines on it[59]. Front and rear irradiation both produced the same cleaning efficiency, although ink damage occurred when using front irradiation. It was found that the temperature increase when using rear irradiation was an order of magnitude lower than that when using front irradiation. The ink damage was due to the high temperatures, and could be avoided by using rear irradiation.

The laser wavelength chosen affects the cleaning efficiency, as this determines the amount of thermal energy absorbed by the contaminant and/or the substrate, or by the liquid if wet or steam laser cleaning is used. In general, lower threshold fluences and higher cleaning efficiencies are obtained at shorter wavelengths. For example, 1μm copper particles were removed from silicon wafers using dry laser cleaning using 10 pulses from a Nd:YAG laser[112]. If the laser was frequency-quadrupled, to a wavelength of 266nm, a fluence of 0.18J/cm^2 was required for complete removal. At a longer wavelength of 532nm (frequency-doubled), a higher fluence of 0.46J/cm^2 was required. Using the fundamental wavelength of 1.06μm, a still larger fluence of 0.60J/cm^2 was required for particle removal, and complete removal did not occur, due to the material absorption characteristics.

Ablative cleaning of epoxy resins was investigated using a Nd:YAG laser at 355nm, 532nm and 1.06μm, and with an Er:YAG at 2.94μm. The Er:YAG was optimal as the epoxy was highly absorbing at this wavelength, especially in relation to the substrate which was also absorbing. The other wavelengths discoloured the area surrounding the irradiated region[118].

The laser pulse length is a parameter which has started to be investigated more recently. The majority of laser cleaning studies use nanosecond pulse lengths, to achieve the surface accelerations required for particle removal[100]. Some work has been done using shorter pulse lengths. Polystyrene (PS) spheres were removed from silicon using steam laser cleaning, where the pulse length of the dye laser used could be changed[95]. With a pulse length of 2.5ns, the threshold fluence for cleaning was 50mJ/cm^2. At a shorter pulse length of 30ps, this was reduced to 20mJ/cm^2, meaning that removal was easier. This is due to a shorter heat diffusion length in silicon when using a picosecond laser, which means that a lower fluence is require to reach a peak temperature at the surface, than when using nanosecond lasers. As steam laser cleaning is used, the shorter pulse length also means that higher superheating of the liquid layer can be achieved. These effects contribute to easier particle removal at shorter pulse lengths.

The number of pulses used to irradiate the substrate can affect the cleaning efficiency. When a single pulse at a fluence of 25.5mJ/cm^2 was used to dry laser clean 400nm silica (SiO2) particles from polymethylmethacrylate (PMMA), 35% were removed. When multiple (20) pulses at the same fluence were used, more particles could be removed, and a higher cleaning efficiency of 80% was achieved[61]. By increasing the number of pulses used, it is also possible to reduce the laser fluence used and still retain high cleaning efficiencies, while minimising the risk of surface damage when using high fluences. The results reported in [61] also show that removal of silica particles from polyimide was ~92% efficient with one laser pulse at a fluence of 17.5mJ/cm^2. By increasing the number of pulses to 5, and using a lower fluence of 15mJ/cm^2, a similar cleaning efficiency was obtained. Note that the energy applied must still be enough to cause particle removal. In the case of dry laser cleaning of fingerprints from microscope cover slides, 2 laser pulses at a fluence of 450mJ/cm^2 were sufficient for complete removal, but 18000 pulses at a lower fluence of 70mJ/cm^2 resulted in incomplete removal[48].

The effect of particle shape was mentioned above, but it is also important to consider the size of particles and the properties of all the materials involved. In general, smaller particles are harder to remove, due to the increase in the van der Waals adhesion force with decreasing size[75]. This is also evident in the results presented in [61], where the threshold fluence for removing 1700nm PS spheres from PMMA with dry laser cleaning is 35mJ/cm^2. This rises to 60mJ/cm^2 for the smaller 800nm PS spheres, and even higher to 110mJ/cm^2 for the even smaller 320nm PS spheres. In contrast, a universal threshold fluence of 100mJ/cm^2 was required for removing different sized (140-1300nm) PS spheres from silicon by steam laser cleaning[98]. This was because the laser-induced evaporation of the liquid layer was the major contributing mechanism to particle removal. Note that when using water, a more pronounced near-field focussing effect caused higher laser intensities and surface damage. By using an isopropanol film instead, the particles could be removed without causing damage.

The material properties of the particle are an important factor. Tungsten particles are notoriously difficult to remove using conventional laser cleaning, and required very high fluences (~2J/cm^2) using a Nd:YAG laser for effective cleaning from silicon[12]. Removal at shorter wavelengths proved difficult due to the small thermoelastic force induced at the particle-substrate interface when using tungsten, (compared to gold which was easier to remove for example)[78].

An alternative is to use shock wave laser cleaning, where the laser beam was sharply focussed at a certain distance from the substrate, and caused a breakdown in air[77]. This resulted in airborne plasma shock waves where the shock wave front pressure was able to overcome the particle-substrate adhesion force and remove the tungsten particles without causing surface damage.

The different conventional laser cleaning techniques also affect the laser cleaning efficiency. Wet or steam laser cleaning is usually more effective than dry laser cleaning, due to the use of the absorbing liquid layer which can be chosen specifically for the application (for example, the improved efficiency of oxide removal from iron using wet laser cleaning, compared to dry laser cleaning[37]). Exposure of the samples to a higher humidity can also increase the dry laser cleaning efficiency[90], while the use of a line beam[76] and surface acoustic waves[91] are more recent avenues of research to enhance particle removal with dry laser cleaning. It should also be noted that the efficiency of contaminant removal can also be improved by combining a laser cleaning technique with conventional cleaning methods such as organic solvents[39].

5. Laser Characteristics

There are a number of different lasers used in the cleaning experiments, many of which are listed in Table 1.4. The effectiveness of the laser is dependent on the pulse length and the wavelength, due to the material absorption of the laser irradiation, as outlined above in §4. A short pulsed and short wavelength laser is generally more effective at overcoming the particle-substrate adhesion forces, but other longer wavelengths may be highly absorbed by the materials (such as the CO_2 lasers by water[38]), and may also be highly effective in removing contamination. It is therefore important to consider not only the contaminant and substrate involved, but also the absorption of any applied liquid. If a material is too absorbing, damage may occur, as was the case when water was used instead of isopropanol[98]. However, the very high absorption of epoxy compared to the substrate allowed it to be removed easily at 2.94μm without damage[118].

6. Analysis Methods

The cleaning efficiency can be evaluated using a number of different techniques, as listed in Table 1.5. The analysis methods can be classified into the following groups: Microscopy and image analysis; Light scattering, reflection, and deflection techniques; Surface analysis; and Miscellaneous, which covers a range of specific new techniques which have been developed to analyse the cleaning efficiency.

Table 1.4 Abbreviations used in Table 2 to denote various lasers used in laser cleaning studies, and their specified wavelengths. New additions since Kane et al.[2] are marked with an asterisk (*).

Key	**Laser**	**Wavelength**
Excimer	Excimer	Unspecified
ArF	ArF excimer	193nm
KrF	KrF excimer	248nm
UV CVL	Frequency doubled copper vapour laser	255.3nm
Nd:YAG^4	Frequency quadrupled Nd:YAG	266nm
Dye^2	Frequency doubled dye laser	292nm*
XeCl	XeCl excimer	308nm
Nitrogen	Nitrogen	337nm
Nd:YLF^3	Frequency tripled Nd:YLF	349nm
XeF	XeF excimer	351nm*
Nd:YAG^3	Frequency tripled Nd:YAG	355nm
Ti:Sapph^{400}	Ti:Sapphire	400nm*
WTO	Wavelength tunable OPO, pulsed	400-800nm*
Nd:YLF^2	Frequency doubled Nd:YLF	523nm
Nd:YAG^2	Frequency doubled Nd:YAG	532nm
CWD	CW diode pumping	532nm*
Dye	Dye laser	583nm
Nd:YAG^O	Nd:YAG/OPO	638nm
Nd:YAG^{OPO}	Nd:YAG pumped OPO	800nm*
Ti:Sapph^{800}	Ti:Sapphire	800nm*
Nd:YLF	Nd:YLF	1.047μm
Nd:YAG	Nd:YAG	1.06μm
Er:YAG	Er:YAG	2.94μm*
$CO_2^{9.2}$	TEA CO_2	9.220μm*
$CO_2^{9.3}$	TEA CO_2	9.317μm*
$CO_2^{9.6}$	TEA CO_2	9.6μm
$CO_2^{10.6}$	TEA CO_2	10.6μm

The majority of studies measure the particle density, or the number of particles in a given region of interest, before and after irradiation. This can be done using microscopy techniques and image processing and is easily incorporated into laser cleaning systems, but may be resolution limited. This includes visual observation of whether cleaning has occurred or not[59], and may use a video monitor[110], Polaroid[12] or CCD[30] camera. Optical microscopy (OM)[6], optical micrographs (OMG)[36], dark field optical microscopy (DFM)[82], and polarised light microscopy (PLM)[39], can also be used to examine surfaces visually. Image analysis software (IP)[27] or a video-based system (VB)[9] can be used to count particles, or compare particle densities (PD) before and after irradiation[105]. In some instances, the cleaning efficiency is defined as the ratio of the cleaned area to the laser spot area (AR)[20]. An automated optical surface analyser (OSA) can also measure particle distributions before and after laser cleaning[91]. This can classify particles by their size.

Other studies use Particle Counters[68] such as the PMS[7] or CANS[85] where scattered light is also used to classify particles on a substrate according to their size. A photomultiplier (PM)[98] can also be used to detect scattered light from a sample, which is proportional to the number of particles on the surface. This can be used to determine the cleaning efficiency. A photodiode was also used to detect the light scattering, where the spectral reflection of the probe HeNe laser was blocked (SR)[95].

A Diffractive Optical Element based sensor (DOE) was used to obtain information about the contamination and surface roughness of metal pharmaceutical punches[16]. Scattered light from the punch surface is incident on the sensor, and a light spot matrix image is detected using a CCD. The intensity projections can be compared to observe the effect of laser cleaning. As contamination is removed, the intensity increases. However, if the intensity subsequently starts to decrease, this is an indication of damage occurring and causing a rough surface.

The laser cleaning efficiency of cleaning mirror surfaces can be monitored using reflectivity measurements (LS). This can be done online using a HeNe laser, or offline with a spectrometer[10]. Due to the shadow cast on the mirror by the contaminants, cleaning causes an increase in the observed reflectivity.

Optical reflectance (OR) can also be used to observe the onset of steam (or wet) laser cleaning[55]. A HeNe laser probe was directed onto the surface of NiP, and the reflected beam was collected by a photodetector. At the laser cleaning threshold, bubble nucleation of the liquid layer commences. The HeNe light is scattered by the bubbles, droplets, and vapour leaving the surface, and causes a drop in the reflectance signal. This allows for determination of when cleaning occurs.

Voltage readings (VR) using a probe HeNe laser can also provide information about changes to the surface[62]. The HeNe beam is divided using a beamsplitter. The reflected beam is then sent to a reference photodiode detector, and the transmitted beam is focussed onto the sample (PS spheres on polyimide) where the intensity transmitted through is measured by another photodiode. The difference between the reference voltage and the transmitted voltage is affected by particle removal, and/or modification to the surface (such as damage).

Table 1.5 Abbreviations used in Table 2 to denote the analysis techniques used. New additions since Kane et al.[2] are marked with an asterisk (*).

Key	Analysis Technique
AES	Surface analysed by Auger Electron Spectroscopy
AFM	Atomic Force Microscopy
AR	Efficiency = clean area / laser spot area
AWM	Acoustic wave monitoring
CCD	CCD camera
DEKTAK3	DEKTAK3 surface profilometer
DFM	Dark field microscopy
DOE	Diffractive optical element (intensity projection)*
ECM	Emission current monitoring
EDS	Energy dispersion spectroscopy*
EDXA	Energy dispersive x-ray analyser*
EPM	Surface analysed before and after by electron probe microanalysis
Fluor	Photodiode monitoring of laser induced fluorescence*
FTIRS	Fourier transform infrared spectroscopy*
Fuzzy	Fuzzy rule system*
Gamma	Gamma spectrometry (measure sample activity)
HI	Heterodyne interferometer (surface displacement)
IP	Before and after images, image analysis software
LEED	Low-Energy Diffraction Spectroscopy*
LIPS	Laser induced plasma spectroscopy*
LS	Light scattering with HeNe laser and spectrometer
OCM	Optical/chromatic monitoring*
OM/OMG	Optical microscope / Optical micrographs*
OR	Optical reflectance*
OSA	Automated optical surface analyser (limit of resolution 1μm)
PC/CANS	Particle counter / Censor ANS100 Light scattering particle counter
PD	Comparing particle densities before and after
PDT	Probe deflection technique*
PLM	Polarised light microscopy*
PM	Photomultiplier
PMS	Particle Measuring System Inc, SAS 3600 particle counting system
Pol	Polaroid camera

Table 1.5 (*Continued*)

Key	Analysis Technique
Profile	Profilometer*
SEM	Scanning electron microscopy
SIMS	Secondary Ion Mass Spectrometry*
Spect	Spectrocolorimeter*
SPV	Surface photovoltage monitored
SR	Si photodiode monitors specular reflection of HeNe beam from substrate
Surf	Surface roughness*
VB	Video-based, PC driven system counts particles >~75μm diameter
Visual	Visual observation of contaminant removal
VM	Video monitor
VR	Voltage readings
XPS	X-ray Photoelectron Spectroscopy

A probe deflection technique (PDT) was also used to monitor the shock wave generation during the dry laser cleaning of particles[31]. The HeNe probe laser was parallel to and above the substrate surface, and incident on a photo-diode connected to an oscilloscope. An optoacoustic wave was generated during the laser cleaning process, and could be observed by deflection of the probe beam. During the first few pulses when cleaning occurred, strong shock waves were detected. The following pulses produced weaker waves/signals, until contaminant removal was complete, at which point the measured signal was constant.

Further studies were done on audible acoustic wave real-time monitoring (AWM). This showed that the amplitude of the acoustic wave was dependent on the laser fluence, the number of laser pulses, and also the substrate material characteristics[3]. It was also possible to determine the nature of the laser-material interactions using this technique. Below the ablation threshold of the materials, as particles are removed from the substrate surface, the amplitude of the signal reduces. This reaches a constant level once cleaning is complete.

An optical/chromatic monitoring system (OCM) has been used to observe the surface during laser cleaning of paper and stone[45]. This produces measurements which are dependent on the spectral signature of the reflected light on the surface (and is independent of light intensity). It provides information of the surface cleanliness and damage, and also chromatic information about the surface.

A spectrocolorimeter (SPECT) can also be used to measure the chromatic properties of samples, and monitor changes induced by laser cleaning and other treatment methods. The removal of water repellent treatments from limestone was analysed using this technique[40]. The application of the repellent treatment causes a chromatic change in the stone surface, and the laser cleaning of an untreated surface also causes discolouration. The repellent treatment also changes the roughness of the surface. The changes in the sample surface roughness profile and depth as a result of laser cleaning could be assessed using a surface roughness instrument (SURF)[40].

Surface analysis techniques can provide a great deal of information about the surface being cleaned. Scanning Electron Microscopy (SEM) can be used to directly image the substrate surface[6], while Atomic Force Microscopy (AFM) can be used to profile the removal of contaminant layers from a surface or holes[119]. Low-Energy Diffraction spectroscopy (LEED) can also be used to monitor surface modifications caused by the laser irradiation[72]. A surface profilometer (PROFILE)[111] such as the DEKTAK3[13] can also be used to produce a cross-sectional profile of the ablated area on a target, to assess depth of removal and the cleaning efficiency.

Auger Electron Spectroscopy (AES) measurements can be done for samples both before and after cleaning[6]. The spectrum before cleaning shows weak peaks for the substrate material (Cu), and large peaks for the contamination (C, O). After cleaning, the substrate material appears as a large peak, and the contamination peaks are relatively weak. Analysis with Secondary Ion Mass Spectrometry (SIMS) is similar, however the ratios of appropriate species peaks need to be monitored. The removal of oxides from copper required monitoring of the ratio of the copper oxide peak to the copper peak. This ratio decreased when the oxides were removed with laser cleaning[18].

An Energy Dispersive X-ray Analyser (EDXA) was used to monitor the removal of paint residues from surfaces[38]. The spectrum showed signals of the paint and the bulk surface, where the intensity signal ratio of the paint to the bulk surface was monitored (similarly to SIMS). As cleaning occurred, this ratio decreased.

Energy Dispersion Spectroscopy (EDS) was used to monitor the removal of an oxide layer from steels[116]. The spectra showed the contamination peaks, and were monitored for changes in the spectrum and peak intensity.

Fourier Transform Infra-red Spectroscopy (FTIRS) was used to examine and monitor the removal of a photoresist coating from silicon[111]. The spectrum of the contaminated sample showed the characteristic peaks of the photoresist, compared to the spectrum of the silicon without the coating.

X-Ray Photoelectron Spectroscopy (XPS) can be used to analyse the laser cleaning of resin from circuit boards[63]. Although high fluences lead to high cleaning efficiencies, this can also induce damage to the boards in terms of oxidation. This oxidation can be detected in the XPS spectrum obtained.

Electron Probe Analysis (EPM) was used to examine the removal of contamination from glass[48]. A spectrum for the clean (non-contaminated) glass surface, a contaminated glass surface, and a laser cleaned glass surface was obtained. The contaminated spectrum showed the presence of the inorganic and organic contamination, while the laser cleaned spectrum had these peaks removed (and resembled the spectrum of the clean glass).

Laser-Induced Plasma Spectroscopy (LIPS) can be used to monitor the point at which ablative irradiation/cleaning should stop. The plasma emission spectrum has enough structure to discriminate between the removal of carbonaceous dirt, pigments, and parchment[60]. However, it is hard to monitor the 'stop point' if the contamination is not uniformly distributed across the substrate.

Photodiode monitoring of the laser-induced fluorescence (FLUOR) allows both real-time observation of the removal rate and when cleaning of oil-based contaminants from microcavities is complete[47]. This is possible as the oil-based contaminant produces a blue fluorescence when excited by the ultraviolet laser radiation during the cleaning process.

Gamma spectrometry (GAMMA) was used to measure the sample activity of the substrate before and after dry laser cleaning to monitor the removal of nuclear contamination[4].

Electronic techniques have been used to analyse laser cleaning. A surface photovoltage technique (SPV) has been used to monitor the minimum diffusion length, the iron concentration in the bulk surface, and therefore the cleaning efficiency on the rear surface of a silicon wafer[81]. After laser cleaning, it was found that the minimum diffusion length increased, and the iron concentration decreased.

The emission current (ECM) of field emitter arrays can also be observed[25]. Contamination on the tips of the arrays after they are used, can cause a decrease in the current. The current can be monitored in situ, and should increase after laser cleaning and removal of the contamination.

A heterodyne interferometer (HI) method has also been developed to measure surface acceleration and displacement in the substrate caused by pulsed laser irradiation[92]. The accelerations produced in the substrate need to be able to overcome the particle-substrate adhesion force for laser-induced removal to occur, in dry laser cleaning. The heterodyne interferometer determines the maximum achievable surface acceleration (silicon was used in this example), and therefore whether dry laser cleaning is a feasible/usable method.

A fuzzy rule based system (FUZZY) was designed to predict surface damage during laser cleaning[58]. A microphone was used as an acoustic sensor and a change in acoustic intensity was detected with laser pulses and with surface damage. This information was embedded in the fuzzy rule base system, and tests showed that the prediction of damage agrees with the experimental results. The system is used during laser cleaning and allows for surface damage to the substrate to be avoided. It is also able to adjust the laser parameters to control the cleaning process efficiently.

7. A Review of Laser Cleaning of Contaminants from Surfaces

The ability of laser cleaning to remove contamination from surfaces is dependent on a number of issues, as discussed above. Firstly, the type of laser cleaning must be chosen to suit the application – for example – can liquids be applied to the substrate or is dry laser cleaning required? Can the conventional laser cleaning technique provide enough force to remove the contamination, or should new laser cleaning techniques such as shock wave cleaning or line beam cleaning be used instead? Furthermore, do these techniques result in possible damage to the surface, and how can this be monitored and avoided?

The material properties of the contaminant, the substrate, and any liquids used also need to be considered, as they all affect the energy absorption. This information can be used to either enhance the laser cleaning efficiency, or avoid substrate damage. In addition, other parameters such as the particle size and shape, and the particle and the substrate surface roughness, also affect the total adhesion force and therefore govern how easy a particle is to remove from the substrate. Smaller particles are normally harder to remove, while if surfaces are smooth and dry, the van der Waals adhesion forces dominate.

The laser parameters must also be considered – such as the angle of incidence, front or rear irradiation, the laser fluence, and the number of pulses used, to the choice of wavelength and pulse length (although shorter wavelengths and pulse lengths normally produce better results). As evidenced in Table 2 and discussed above, these parameters have a great impact on the cleaning efficiency and the avoidance of unwanted damage.

Although the use of wet and steam laser cleaning normally reduces the threshold fluence required for contaminant removal, one of the easiest types of particles removed were 800nm polystyrene spheres from a polyimide substrate, using dry laser cleaning[61]. This was done with a single pulse from a KrF excimer laser, where the threshold fluence for removal was ~5mJ/cm^2. Silica spheres (400nm, 800nm) were also easily removed from polyimide, where a pulse fluence under 20mJ/cm^2 was used to obtain cleaning efficiencies above 90%. Furthermore, silica spheres could be even more easily removed from germanium substrates. Dry laser cleaning using a frequency tripled Nd:YAG laser required a threshold fluence of 1mJ/cm^2 for cleaning to commence, while 8mJ/cm^2 was required for total removal of the 0.5μm sized spheres[28]. However, the total number of pulses used was not specified. Materials which are harder to remove using conventional laser cleaning include tungsten[77] and nuclear contamination from metal[4].

In addition, the removal of polystyrene spheres, alumina particles, and silica spheres from silicon was investigated also using dry laser cleaning[83]. This work looked at the use of different lasers to cover a range of wavelengths and also pulse lengths. This is one of the few reported studies investigating the use of femtosecond lasers where a Ti:Sapphire laser with a wavelength of 800nm, and a pulse length of 150fs was used. However, the threshold fluence required for particle removal was also the same fluence required for hole formation (ablation of the substrate). From a laser cleaning point of view, this effect is undesirable, although new work has begun investigating the use of particles to form structures or patterning of the surface.

The aim of Table 2 is to act as a guide to pulsed laser cleaning. It gives readers a collation of the different types of contamination that can be removed, and what the optimal parameters are for doing so. The table summarises the published information on particle/contaminant material, size and deposition technique, the laser cleaning technique used, the fluence used, pulse sequencing, the minimum threshold laser fluence required for cleaning, the analysis technique used to evaluate the cleaning efficiency, the reported effectiveness of the experiment, and the reference source of the data. This allows for detailed comparison between experimental results, and an understanding of how the removal efficiency can be improved for a specific requirement.

Table 2 is organised alphabetically, by the ***substrate*** material used in the laser cleaning studies in the first column, then by the ***type of laser cleaning (LC)*** technique that was used in column five, and finally by the type of ***contaminant*** being removed in column two. It is evident that each cleaning application needs to be considered carefully, as the optimal experimental parameters such as the laser wavelength, angle of incidence, and pulse length vary for different materials. It is hoped that readers will find this guide useful and beneficial to help optimise contaminant removal for their own laser cleaning requirements.

Table 2: An overview of pulsed laser cleaning of contaminants from surfaces. New references added since Kane et al. are marked with an asterisk (*).

Substrate	*Contaminant*	*Size of contam.*	*Deposition technique*	*Type of LC*	*Threshold fluence*	*Fluence used*	*% clean*	*Laser*	*No. of pulses*	*Pulse length*	*Analysis technique*	*Ref. No.*
Aluminium	-	-	-	Ablation	-	0.52J/cm^2	Eff	KrF	10	20ns	AWM	3*
Aluminium	Radionuclei-des ^{125}Sb, ^{154}Eu, ^{137}Cs	-	Nucl	Ablation	-	3J/cm^2	85	XeCl	~500	25ns	Gamma	4*, 5*
Aluminium	Finger print	N/A	-	Dry	-	375mJ/cm^2	100	KrF	1	20ns	OM,SEM, AES	6
Aluminium	Al_2O_3, SiC, BC, CeO_2	>0.1mm	EA/SCCS	Wet, WF	650mJ/cm^2	>6.2J/cm^2	Opt	$CO_2$$^{10.6}$	multiple	0.25ms	PMS	7
Aluminium (anodized)	Oxide layer	20μm	-	Dry	-	-	NE	Nd:YAG	-	10ns	SEM, OM	8
Aluminium (anodized)	Oxide layer	20μm	-	Dry	-	-	NE	Nd:YAG^2	-	10ns	SEM, OM	8
Aluminium (anodized)	Oxide layer	20μm	-	Dry	-	~1.3J/cm^2	Opt	KrF	4 -10	25ns	SEM, OM	8
Aluminium coated Si	Volcanic dust	-	Nat	Dry	-	380mJ/cm^2	90	XeCl	8	-	VB	9
Aluminium mirror coating	Dust (mainly quartz sand	several to 100's μms	Nat	Dry	50mJ/cm^2	160+/- 30mJ/cm^2 optimal	Opt	KrF (optimal)*	5 best (range 1-1000)	30ns	LS	10

* KrF was the optimal laser for cleaning, compared with ArF, XeCl, XeF and Nd:YAG[3]. The pulse lengths ranged from 50-200ns.

Table 2 (*Continued*)

Substrate	*Contaminant*	*Size of contam.*	*Deposition technique*	*Type of LC*	*Threshold fluence*	*Fluence used*	*% clean*	*Laser*	*No. of pulses*	*Pulse length*	*Analysis technique*	*Ref. No.*
Art objects	Stains, pollutants	-	-	Ablation*	-	-	-	Nd:YAG, Nd:YAG², Nd:YAG³	-	3ns	Visual	11*
CaF_2	Tungsten	1-2μm	Du	Dry	-	2.1 J/cm²	>95	Nd:YAG	100	24ns	OM, Pol	12
Carbon steel plate	Electroplated zinc coating	21μm	-	Ablation	-	$3x10^5$J	1m²	Nd:YAG	-	100ps	DEKTAK3	13*
Cast Iron & Stainless Steel	-	-	-	Dry	-	1-1.5 J/cm²	Opt	Nd:YAG	-	10-20ns	SEM	14
Chrome photomask	Fingerprints	-	-	Dry	-	3J/cm²	Eff	KrF	-	-	-	15*
Concave metal†	Medical powders	-	Tabl	Ablation	-	Up to 1.1J/cm²	Eff, NegD	Nd:YAG	2-4	20ns	DOE	16*
Concave metal†	Medical powders	-	Tabl	Ablation	-	Up to 0.54J/cm²	NE	Nd:YAG²	-	20ns	DOE	16*
Concave type storage nodes	Photoresist & metallic polymer	0.25μm thick	Proc	Ablation	-	90mJ/cm²	Ph Rem	KrF	5	20-30ns	SEM	17*
Concave type storage nodes	Photoresist & metallic polymer	0.25μm thick	Proc	Ablation	-	130mJ/cm²	100	KrF	3	20-30ns	SEM	17*
Concave type storage nodes	Photoresist & metallic polymer	0.25μm thick	Proc	Ablation	-	90mJ/cm²	100	KrF	6	20-30ns	SEM	17*
Copper	Copper oxides	0.3μm thick	Bake	Ablation	-	7.7-9.8J/cm²	Opt	Nd:YAG	-	-	OM, SEM, SIMS	18*

* Ablation cleaning could be assisted using a dye, liquid, or gas.

† Metal pharmaceutical punches.

Table 2 (*Continued*)

Substrate	*Contaminant*	*Size of contam.*	*Deposition technique*	*Type of LC*	*Threshold fluence*	*Fluence used*	*% clean*	*Laser*	*No. of pulses*	*Pulse length*	*Analysis technique*	*Ref. No.*
Copper	Copper oxides	0.3μm thick	Bake	Ablation	-	6.6-9.8J/cm^2	Opt	Nd:YAG^2	-	-	OM, SEM, SIMS	18*
Copper	Copper oxides	0.3μm thick	Bake	Ablation	-	2.7-4.2J/cm^2	Opt	Nd:YAG^4	-	-	OM, SEM, SIMS	18*
Copper	contaminants	-	Nat	Dry	-	0.8J/cm^2	Opt	KrF	10	23ns	AWM	19
Copper	Copper	-	Esus	Dry	0.15 J/cm^2	0.8 J/cm^2	80 NI	Nd:YAG^2	10	10ns	AR	20
Copper	Copper	-	Esus	Dry	0.01 J/cm^2	0.15 J/cm^2	130, BGI	Nd:YAG^2	10	10ns	AR	20
Copper	Copper particles	~10's of μm	Esus	Dry	0.15J/cm^2	1.0J/cm^2	100*	Nd:YAG^2	10	10ns	-	21*
Copper	Copper particles	~10's of μm	Esus	Dry	-	0.08J/cm^2	100†	Nd:YAG^2	10	10ns	-	21*
Copper	Copper particles	~10's of μm	Esus	Dry	-	0.15J/cm^2	130†, BGI	Nd:YAG^2	10	10ns	-	21*
Copper	Magic Marker	N/A	Dr	Dry	-	0.50J/cm^2	Opt	KrF	20	20ns	OM, SEM, AES	22
Copper	Magic Marker	N/A	Dr	Dry	-	0.42J/cm^2	Opt	KrF	20	20ns	OM, SEM, AES	6
Copper	Oil & grease layer	N/A	Lay	Dry	-	460 mJ/cm^2	100	KrF	5	20ns	OM, SEM, AES	6
Copper	Oil & grease layer	N/A	Lay	Dry	-	70 mJ/cm^2	Opt	KrF	600	20ns	OM, SEM, AES	6
Copper	Oil, grease, dust, finger-prints	-	WExp	Dry	-	625 mJ/cm^2	100	KrF	5	20ns	AWM	23*

* Damage occurs at fluences greater than 0.8J/cm^2.

† Efficiency obtained using a beam at a 'glancing' angle of 10° to the substrate surface.

Table 2 (*Continued*)

Substrate	*Contaminant*	*Size of contam.*	*Deposition technique*	*Type of LC*	*Threshold fluence*	*Fluence used*	*% clean*	*Laser*	*No. of pulses*	*Pulse length*	*Analysis technique*	*Ref. No.*
Copper	Surface contaminants / magic marker	-	-	Dry	-	-	NE	CO_2 [10.6]	-	100ns	OM, SEM, AES	6
Copper printed circuit boards	Copper oxides	-	Bake	Ablation	2.0J/cm^2	3.5J/cm^2	Eff	Nd:YAG	2	10ns	AWM	21*, 24*
Field emitter arrays	Oxide layer / surface contaminants	-	Fab	Dry	-	85mJ/cm^2	NE	Nd:YLF		15ns	ECM	25*, 26
Field emitter arrays	Oxide layer / surface contaminants	-	Fab	Dry	-	85mJ/cm^2	NE	Nd:YLF[2]		15ns	ECM	25*, 26
Field emitter arrays	Oxide layer / surface contaminants	-	Fab	Dry	-	35mJ/cm^2	Eff	Nd:YLF[3]		15ns	ECM	25*, 26
Fused silica slides	Al_2O_3	3μm	DT	Dry	815 mJ/cm^2	-	-	KrF	single	8.1ns	OM, IP	27*
GaAs	Particles & fingerprints	-	-	Dry	-	16J/cm^2	Eff	KrF	-	-	-	15*
GaAs & thin Ni-Cr pattern	Au flakes	-	-	Dry	-	8J/cm^2	Eff	KrF	-	-	-	15*
Germanium	silica spheres	0.5μm	-	Dry	1 mJ/cm^2	8 mJ/cm^2	~100	Nd:YAG[3]	-	-	-	28*, 29
Germanium	silica spheres	0.5μm	-	Dry	2 mJ/cm^2	10mJ/cm^2	~100	Nd:YAG[2]	-	-	-	28*, 29
Germanium	silica spheres	0.5μm	-	Dry	15mJ/cm^2	39mJ/cm^2	~100	Nd:YAG	-	-	-	28*, 29

Table 2 (*Continued*)

Substrate	*Contaminant*	*Size of contam.*	*Deposition technique*	*Type of LC*	*Threshold fluence*	*Fluence used*	*% clean*	*Laser*	*No. of pulses*	*Pulse length*	*Analysis technique*	*Ref. No.*
Germanium	silica spheres	1μm	WS/AJ	Dry	30 mJ/cm^2	90mJ/cm^2	~100	KrF	200	23ns	SEM	28*, 29
Glass	Al_2O_3	0.3-3μm	DT	Dry	62-93mJ/cm^2	-	-	UV CVL	single	35ns	CCD, IP	30
Glass	Aluminium silicate hydroxide	1-10μm	-	Dry	-	-	-	Nd:YAG	-	7.5ns	Visual /PDT	31*
Glass	Diamond	5-7μm	Esus	Wet, Dry	-	0.8J/cm^2	NE	KrF	50	20ns	OM	32*
Glass microscope slides	Al_2O_3	1μm	DT	Dry	90-115mJ/cm^2	-	100	UV CVL	single	35ns	OM, IP	33
Glass microscope slides	Al_2O_3	1μm	DT	Dry	330-380mJ/cm^2	-	95	XeCl	single	8ns	OM, IP	33
Glass microscope slides	Al_2O_3	1μm	DT	Dry	200-400mJ/cm^2	-	97	KrF	single	8ns	OM, IP	33
Glass microscope slides	Al_2O_3	1μm	LAD	Dry	320-350mJ/cm^2	-	100	XeCl	single	8ns	OM, IP	33
Glass microscope slides	Al_2O_3	3μm	DT	Dry	100mJ/cm^2	-	-	KrF	single	8.1ns	OM, IP	27*
Glass microscope slides	Paraffin wax (particles)	1-20μm	LPar	Dry	~4-5 mJ/cm^2	218mJ/cm^2	Most*	KrF	single	25ns	OM	34*
Gold/SiO_2	Al_2O_3, SiC, BC, CeO_2	>0.1μm	EA/SCCS	Wet, WF	650mJ/cm^2	3.2-6.2J/cm^2	Opt	$CO_2^{10.6}$	multiple	0.25ms	PMS	7

* All but 5 of 406 particles removed.

Table 2 (*Continued*)

Substrate	*Contaminant*	*Size of contam.*	*Deposition technique*	*Type of LC*	*Threshold fluence*	*Fluence used*	*% clean*	*Laser*	*No. of pulses*	*Pulse length*	*Analysis technique*	*Ref. No.*
IC mould	Organic (grease / wax / resin)	-	Pack	Dry	$100mJ/cm^2$	$400mJ/cm^2$	Opt	KrF	10	23ns	OM	35, 36*
IC packages	Flash	-	-	Ablation	-	$300mJ/cm^2$	Eff	$Nd:YAG^2$	4	7ns	OMG	36*
Inconel half tubes	Radionucleides ^{60}Fe	-	Nucl	Ablation	-	$3J/cm^2$	47	XeCl	700	25ns	Gamma	4*, 5*
Iron	Iron oxide	1020nm layer	-	Dry	-	$0.5J/cm^2$	Eff	$Nd:YAG^2$	1	14.5ns	SEM, OM	37
Iron	Iron oxide	1020nm layer	-	Dry	-	$0.5J/cm^2$	NAG SW	Nd:YAG	1	14.5ns	SEM, OM	37
Iron	Iron oxide	1020nm layer	-	Wet, Lyte	-	$0.5J/cm^2$	IE cf Dry	Nd:YAG	1	14.5ns	SEM, OM	37
Iron plates - rough	Resin containing $PbCrO_4$	25μm	Spra	Ablation (liquid assisted)	-	-	Eff	$CO_2{}^{10.6}$	15	-	EDXA	38*
Iron plates - smooth	Resin containing $PbCrO_4$	25μm	Spra	Ablation (liquid assisted)	-	-	Eff	$CO_2{}^{10.6}$	5	-	EDXA	38*
Ivory	Ink stains	-	Dr	Ablation (liquid assisted)	-	-	Comb	Multiple*	-	3ns, 25ns, 25ns	Visual, PLM	39*
KAPTON	Carbon	-	-	Ablation	-	$100mJ/cm^2$	Eff	$Nd:YAG^2$	10	7ns	OMG	36*
Limestone	Water repellent treatments†	-	-	Ablation	-	$2.3J/cm^2$	-	Nd:YAG	1800 (10Hz)	5ns	Spect, Surf, SEM	40*

* Multiple lasers used: Nd:YAG, $Nd:YAG^2$, $Nd:YAG^3$, KrF, ArF, CWD, and WTO.

† Water repellant treatments used: PB-72 and HL-100.

Table 2 (*Continued*)

Substrate	*Contaminant*	*Size of contam.*	*Deposition technique*	*Type of LC*	*Threshold fluence*	*Fluence used*	*% clean*	*Laser*	*No. of pulses*	*Pulse length*	*Analysis technique*	*Ref. No.*
Limestone	Water repellent treatments†	-	-	Ablation	-	$0.9J/cm^2$	-	Nd:YAG[2]	1800 (10Hz)	5ns	Spect, Surf, SEM	40*
Limestone	Water repellent treatments†	-	-	Ablation	-	$0.4J/cm^2$	Part	Nd:YAG[3]	1800 (10Hz)	5ns	Spect, Surf, SEM	40*
Limestone	Water repellent treatments†	-	-	Ablation	-	$0.2J/cm^2$	Part	Nd:YAG[4]	1800 (10Hz)	5ns	Spect, Surf, SEM	40*
Limestone sculpture	Black crusts	0.05mm	Nat	Dry	-	$400mJ/cm^2$	Opt	Nd:YAG	30	6ns	LS, AWM	41
Limestone sculpture	Black crusts	0.05mm	Nat	Wet, WFt	-	$400mJ/cm^2$	Opt	Nd:YAG	10	6ns	LS, AWM	41
$LiNbO_3$	Epoxy particles	1-2μm	Du/LiqC	Dry	-	650 mJ/cm^2	<5	Nd:YAG	600	20ns	OM, Pol	12
$LiNbO_3$	Tungsten	down to 1-2 mm	Du/LiqC	Dry	-	650 mJ/cm^2	>95	Nd:YAG	100	20ns	OM, Pol	12
$LiNbO_3$	Tungsten	down to 1-2 μm	Du/LiqC	Dry	-	$0.16J/cm^2$	<50	Nd:YAG	100	20ns	OM, Pol	12
Magnetic disk	SiC	-	Sol	Dry	-	$200mJ/cm^2$	100	KrF	50*	23ns	-	42
Magnetic head	Sn	-	Asus/AJ	Dry	$25mJ/cm^2$	100 mJ/cm^2	100	KrF	100	23ns	SEM	43
Magnetic head	Al	-	Asus/AJ	Dry	$25mJ/cm^2$	100 mJ/cm^2	90	KrF	100	23ns	SEM	43
Magnetic head	Epoxy resin	several μms thick	Glue	Dry	-	60 mJ/cm^2	100	KrF	2500	20ns	OM, SEM, AES, EPM	44

* 100% cleaning obtained using 50 pulses at $200mJ/cm^2$, and also with 5 pulses at $250mJ/cm^2$.

Table 2 (*Continued*)

Substrate	*Contaminant*	*Size of contam.*	*Deposition technique*	*Type of LC*	*Threshold fluence*	*Fluence used*	*% clean*	*Laser*	*No. of pulses*	*Pulse length*	*Analysis technique*	*Ref. No.*
Magnetic head slider	Epoxy resin	N/A	Glue	Dry	-	450mJ/cm^2	Opt	KrF	25	20ns	OM, SEM, AES	22
Magnetic head slider	Metal particles	micron sized	-	Dry	-	330 mJ/cm^2	Opt	KrF	30	20ns	OM, SEM, AES	22
Magnetic head slider	Metal particles	micron sized	-	Dry	-	187.5 mJ/cm^2	Opt	KrF	150	20ns	OM, SEM, AES	22
Marble	Black encrustation	-	-	Ablation	-	0.8J/cm^2	Opt	Nd:YAG	7	10ns	OCM	45*
Marble	Black encrustation*	-	-	Dry	-	~1J/cm^2 (normal incidence)	100	Nd:YAG	5	-	-	46*
Marble	Black encrustation*	-	-	Dry	-	150mJ/cm^2 (10° to surface†)	100	Nd:YAG	5	-	-	46*
Marble	Black encrustation*	-	-	Dry, LAS	-	2J/cm^2 (shock cleaning)	100‡	Nd:YAG	3	-	-	46*
Microcavi-ties in steel	Oil-based organics	-	-	Ablation	-	100mJ/cm^2	Eff	Nd:YAG^2	single	4-6ns	OM, Fluor	47*
Microscope Cover Slide	Finger prints	N/A	-	Dry	-	450 mJ/cm^2	100	KrF	2	20ns	EPM	48
Microscope Cover Slide	finger prints	N/A	-	Dry	-	70 mJ/cm^2	IR	KrF	18000	20ns	EPM	48
Microscope Cover Slide	finger prints	N/A	-	Dry	-	420 mJ/cm^2	IR	$CO_2^{10.6}$	50	100ns	EPM	48

* Model assumes crust is made up of particles.

† Laser beam at 10° to substrate surface, optimal cleaning occurs at 'glancing' angles (up to 30°) to substrate surface.

‡ The optimal gap for shock cleaning is 1-2mm between the laser focus and the target material. This is not as effective a method as using the laser beam at a glancing angle of 10°, but is better than using the laser beam at normal incidence.

Table 2 (*Continued*)

Substrate	*Contaminant*	*Size of contam.*	*Deposition technique*	*Type of LC*	*Threshold fluence*	*Fluence used*	*% clean*	*Laser*	*No. of pulses*	*Pulse length*	*Analysis technique*	*Ref. No.*
NiP	Quartz	5μm	-	Dry	16mJ/cm^2	140mJ/cm^2	~65	KrF	100	23ns	OM	49, 50, 51, 52, 53
NiP	Silica spheres	1μm	WS/AJ	Dry	8 mJ/cm^2	20mJ/cm^2	~100	KrF	200	23ns	SEM	28*, 29
NiP	Alumina	0.3μm avg size	-	Steam	86mJ/cm^2	-	-	Nd:YAG	10	6ns	OM	54*
NiP	Alumina	0.3μm avg size	-	Steam	63mJ/cm^2	-	-	Nd:YAG[3]	10	6ns	OM	54*
NiP	Al_2O_3	0.3μm	-	Steam, WI	Complete removal*	86mJ/cm^2	100	Nd:YAG	10	6ns	OR	55*
NiP	Al_2O_3	0.3μm	-	Steam, WI	Complete removal§	63mJ/cm^2	100	Nd:YAG[3]	10	6ns	OR	55*
NiP	Al_2O_3	1μm	-	Wet, IF	30mJ/cm^2	70mJ/cm^2	Opt	KrF	-	23ns	OM	50, 51, 52, 53, 56, 57
Paper	Photocopier toner	-	-	Ablation	-	1.5J/cm^2	Opt	Nd:YAG	7	10ns	OCM	45*
Paper	Photocopier toner	-	-	Ablation	-	1.5J/cm^2	Opt	Nd:YAG	7	10ns	Fuzzy	58*

* The threshold fluence is defined here to be when complete removal is achieved.

Table 2 (*Continued*)

Substrate	*Contaminant*	*Size of contam.*	*Deposition technique*	*Type of LC*	*Threshold fluence*	*Fluence used*	*% clean*	*Laser*	*No. of pulses*	*Pulse length*	*Analysis technique*	*Ref. No.*
Paper with ink lines	Carbon particles	~0.6μm	Du	Dry, Rear	-	front: 0.17J/cm^2, rear: 0.23J/cm^2	Same cleaning*	Nd:YAG^2	100	7ns	Visual	59*
Parchment	Glue spots	-	-	Ablation	-	0.8-1.4J/cm^2	Eff	XeCl	300	17ns	LIPS, Fluor	60*
PMMA	PS	1700nm	SCCS	Dry	35mJ/cm^2	160mJ/cm^2	>80	KrF	single	31ns	CCD, IP	61
PMMA	PS	800nm	SCCS	Dry	60mJ/cm^2	160mJ/cm^2	>80	KrF	single	31ns	CCD, IP	61
PMMA	PS	320nm	SCCS	Dry	110mJ/cm^2	160mJ/cm^2	>80	KrF	single	31ns	CCD, IP	61
PMMA	SiO_2	800nm	SCCS	Dry	-	160mJ/cm^2	0	KrF	single	31ns	CCD, IP	61
PMMA	SiO_2	400nm	SCCS	Dry	-	160mJ/cm^2	0	KrF	single	31ns	CCD, IP	61
PMMA	SiO_2	400nm	SCCS	Dry	18.5 mJ/cm^2	25.5 mJ/cm^2	~35	ArF	single	>31ns	CCD, IP	61
PMMA	SiO_2	400nm	SCCS	Dry	-	25.5 mJ/cm^2	~80	ArF	20	>31ns	CCD, IP	61
PMMA	PS	320nm	SCCS	Dry, RH	~110 mJ/cm^2 †	-	-	KrF	single	31ns	CCD, IP	61
PMMA	SiO_2	400nm	SCCS	Dry, RH	no cleaning	n/a	n/a	KrF	single	31ns	CCD, IP	61
Polyimide	Polystyrene spheres	320nm	SCCS	Dry	6mJ/cm^2	-	-	Dye^2	-	30ps	VR‡	62*

* The same cleaning efficiency was obtained using front and rear irradiation at the stated fluences. However, damage occurred when using front irradiation, and no damage was observed when using rear irradiation.

† There was no change in the threshold fluence for removing 320nm PS spheres from PMMA in a raised humidity environment, compared to normal conditions.

‡ The voltage readings were obtained using reference and transmission detectors.

Table 2 (*Continued*)

Substrate	*Contaminant*	*Size of contam.*	*Deposition technique*	*Type of LC*	*Threshold fluence*	*Fluence used*	*% clean*	*Laser*	*No. of pulses*	*Pulse length*	*Analysis technique*	*Ref. No.*
Polyimide	Polystyrene spheres	800nm	SCCS	Dry	-	10mJ/cm^2	100	Dye^2	-	30ps	VR ‡‡	62*
Polyimide	PS	800nm	SCCS	Dry	~5mJ/cm^2	17.5 mJ/cm^2	69+/- 5	KrF	single	31ns	CCD, IP	61
Polyimide	PS	320nm	SCCS	Dry	~12 mJ/cm^2	17.5 mJ/cm^2	31+/- 7	KrF	single	31ns	CCD, IP	61
Polyimide	PS	110nm	SCCS	Dry	~14 mJ/cm^2	17.5 mJ/cm^2	10+/- 10	KrF	single	31ns	CCD, IP	61
Polyimide	SiO_2	800nm	SCCS	Dry	~7 mJ/cm^2	17.5 mJ/cm^2	95+/- 5	KrF	single	31ns	CCD, IP	61
Polyimide	SiO_2	400nm	SCCS	Dry	~14 mJ/cm^2	17.5 mJ/cm^2	95+/- 5	KrF	single	31ns	CCD, IP	61
Polyimide	SiO_2	400nm	SCCS	Dry	~11 mJ/cm^2	17.5 mJ/cm^2	92+/- 5	ArF	single	>31ns	CCD, IP	61
Polyimide	SiO_2	400nm	SCCS	Dry	-	17.5 mJ/cm^2	95+/- 5	ArF	20	>31ns	CCD, IP	61
Polyimide	SiO_2	400nm	SCCS	Dry	-	12.5 mJ/cm^2	80+/- 5	ArF	20	>31ns	CCD, IP	61
Polyimide	SiO_2	400nm	SCCS	Dry	-	10 mJ/cm^2	40+/- 7	ArF	20	>31ns	CCD, IP	61
Polyimide	SiO_2	400nm	SCCS	Dry, RH	10mJ/cm^{2*}	-	-	KrF	single	31ns	CCD, IP	61
Polyimide	SiO_2	800nm	SCCS	Dry, RH	~5mJ/cm^{2§} §	-	-	KrF	single	31ns	CCD, IP	61
Printed circuit boards	Resin	-	Proc	Dry	75mJ/cm^2	400mJ/cm^2	Eff	Nd:YAG	-	7ns	OM, XPS, AWM	63*

* The threshold fluence for cleaning to occur was 30% lower in a raised humidity environment, compared to normal conditions.

Table 2 (*Continued*)

Substrate	*Contaminant*	*Size of contam.*	*Deposition technique*	*Type of LC*	*Threshold fluence*	*Fluence used*	*% clean*	*Laser*	*No. of pulses*	*Pulse length*	*Analysis technique*	*Ref. No.*
Printed circuit boards	Resin	-	Proc	Dry	-	150mJ/cm^2	Eff	Nd:YAG	5	7ns	OM, XPS, AWM	36*
Pyrex	Fakes & fingerprints	-	-	Dry	-	80J/cm^2	Eff	KrF	-	-	-	15*
Quartz	Aluminium	<20μm	Esus	Dry	40mJ/cm^2	150mJ/cm^2	Opt	Nd:YAG	500	7ns	OM	51, 64
Quartz	Aluminium	<20μm	Esus	Dry	30mJ/cm^2	120mJ/cm^2	Opt	Nd:YAG[2]	500	7ns	OM	51, 64
Quartz	Aluminium	<20μm	Esus	Dry	10mJ/cm^2	110mJ/cm^2	Opt	Nd:YAG[4]	500	7ns	OM	51, 64
Quartz	Aluminium	<3μm	Msus/AJ	Dry	40mJ/cm^2	150mJ/cm^2	100	Nd:YAG	500	7ns	OM	65
Quartz	Aluminium	<3μm	Msus/AJ	Dry	30mJ/cm^2	120mJ/cm^2	100	Nd:YAG[2]	500	7ns	OM	65
Quartz	Aluminium	<3μm	Msus/AJ	Dry	10mJ/cm^2	110mJ/cm^2	100	Nd:YAG[4]	500	7ns	OM	65
Quartz	Aluminium	<3μm	Msus/AJ	Dry	50mJ/cm^2	100mJ/cm^2	24	KrF front irradiation	100	23ns	OM	65, 66, 67
Quartz	Aluminium	<3μm	Msus/AJ	Dry	50mJ/cm^2	100mJ/cm^2	100	KrF back irradiation	100	23ns	OM	65, 66, 67
Quartz	Copper	-	Sus	Dry	95mJ/cm^2	-	-	Nd:YAG	5	7ns	PC	68*
Quartz	Copper	-	Sus	Dry	80J/cm^2	-	-	Nd:YAG	5	7ns	OMG	68*
Quartz	Copper	-	Sus	Dry	40mJ/cm^2	-	-	Nd:YAG[2]	5	7ns	PC, OMG	68*
Quartz	Copper	-	Sus	Dry	17mJ/cm^2	-	-	Nd:YAG[3]	5	7ns	PC	68*
Quartz	Copper	-	Sus	Dry	20mJ/cm^2	-	-	Nd:YAG[3]	5	7ns	OMG	68*
Quartz	Copper	<20μm	Esus	Dry	80mJ/cm^2	400mJ/cm^2	~65	Nd:YAG	500	7ns	OM	51, 52, 56, 64

Table 2 (*Continued*)

Substrate	*Contaminant*	*Size of contam.*	*Deposition technique*	*Type of LC*	*Threshold fluence*	*Fluence used*	*% clean*	*Laser*	*No. of pulses*	*Pulse length*	*Analysis technique*	*Ref. No.*
Quartz	Copper	<20μm	Esus	Dry	$40mJ/cm^2$	$200mJ/cm^2$	Opt	Nd:YAG^2	500	7ns	OM	51, 52, 56, 64
Quartz	Copper	<20μm	Esus	Dry	$20mJ/cm^2$	$180mJ/cm^2$	Opt	Nd:YAG^4	500	7ns	OM	51, 52, 56, 64
Quartz	Finger prints	N/A	-	Dry	-	400 mJ/cm^2	100	KrF back irradiation	600	20ns	EPM	48
Quartz	Finger prints	N/A	-	Dry	-	400 mJ/cm^2	IR	KrF front irradiation	1200	20ns	EPM	48
Quartz	Glass	1μm	SCCS	Dry	-	$95mJ/cm^2$	FIEff	Nd:YAG^2, KrF	10	7ns & 23ns	OMG	69*
Quartz	Plate-like Al particles	-	SCCS	Dry	$30mJ/cm^2$	-	-	Nd:YAG^2	-	7ns	OMG	69*
Quartz	Plate-like Si	-	SCCS	Dry, rear	-	$275mJ/cm^2$	RIEff	Nd:YAG^2	10	7ns	OMG	69*
Quartz	Paint, fingerprints, particles & haze	-	-	Dry	-	2-20J/cm^2 average flux	Opt	KrF	multiple	34ns - duty cycle	-	70
Silicon	Gold	-	-	-	-	-	-	Nd:YAG^O	-	6ns	SEM	71
Silicon	Native oxide	-	-	Ablation	-	$0.8J/cm^2$	Cle-Dam	KrF	-	20ns	OM, AES, LEED	72*
Silicon	Photoresist	0.82μm thick	SCCS	Ablation	-	$300mJ/cm^2$	100	KrF	6	20ns	SEM, AFM	73*
Silicon	Al_2O_3	0.3 - 1.5μm	Esus	Dry	-	$300mJ/cm^2$	Opt	KrF	4	15ns	SEM	74, 75

Table 2 (*Continued*)

Substrate	*Contaminant*	*Size of contam.*	*Deposition technique*	*Type of LC*	*Threshold fluence*	*Fluence used*	*% clean*	*Laser*	*No. of pulses*	*Pulse length*	*Analysis technique*	*Ref. No.*
Silicon	Al_2O_3	0.1-0.3μm	EWsus	Dry	-	200-600mJ/cm^2	100	KrF	50	-	SEM	76
Silicon	Al_2O_3	0.1-0.3μm	EWsus	Dry, LB	-	50-100 mJ/cm^2	100	KrF	Scanned line*	-	SEM	76
Silicon	Copper	1μm	SCCS	Dry	-	180 mJ/cm^2	100	Nd:YAG^4	10	10ns	OMG	77
Silicon	Copper	1μm	SCCS	Dry	-	0.20J/cm^2	100	Nd:YAG^4	10	10ns	OMG	78*
Silicon	Copper	0.5-5.5μm	Esus	Dry	-	0.3J/cm^2	71	KrF	100	~ 30ns	OM, SEM	79*, 80*
Silicon	Fe_2O_3 - front surface	0.1-2μm	PMS	Dry	-	350mJ/cm^2	LT	KrF	-	22ns	PMS	81
Silicon	Gold	0.6-1.8μm	Esus	Dry	-	0.3J/cm^2	91	KrF	100	~ 30ns	OM, SEM	79*, 80*
Silicon	Gold	1μm	SCCS	Dry	-	0.08J/cm^2	100	Nd:YAG^4	10	10ns	OMG	78*
Silicon	Particles & fingerprints	-	EWsus	Dry	-	40J/cm^2	Eff	KrF	-	-	-	15*
Silicon	Polystyrene	0.2μm	EWsus	Dry	0.16J/cm^2	-	-	KrF	single	20ns	DFM, SEM	82*
Silicon	Polystyrene	0.5μm	EWsus	Dry	0.1J/cm^2	-	-	KrF	single	20ns	DFM, SEM	82*
Silicon	Polystyrene	1.1μm	EWsus	Dry	0.049J/cm^2	-	-	KrF	single	20ns	DFM, SEM	82*
Silicon	Polystyrene spheres (PS), SiO_2, Al_2O_3	320-1700nm	Isus/SCCS	Dry	11-80 mJ/cm^2†	-	Holes formed‡	Ti: Sapph^{800} §	-	150fs	SEM, AFM	83*

* The surface was scanned with a line beam. As the threshold fluence for removing the contaminants was reduced, the efficiency was improved.

† The threshold cleaning fluence was 11mJ/cm^2 for 1700nm PS spheres, 25mJ/cm^2 for 800nm PS spheres, and 80mJ/cm^2 for 320nm PS spheres.

‡ The threshold fluence for cleaning is equal to the threshold fluence for hole formation (ablation).

§ Results are for Ti:Sapph^{800}. Other lasers used include Ti:Sapph^{400} (150fs), Dye (30ps and 2.5ps), Nd:YAG^2 (6.5ns) and Nd:YAG^{OPO} (800nm, ns).

Table 2 (*Continued*)

Substrate	*Contaminant*	*Size of contam.*	*Deposition technique*	*Type of LC*	*Threshold fluence*	*Fluence used*	*% clean*	*Laser*	*No. of pulses*	*Pulse length*	*Analysis technique*	*Ref. No.*
Silicon	Quartz	mixed	Msus/AJ	Dry	135mJ/cm^2	300mJ/cm^2	~60	KrF	100	23ns	OM	51, 56, 84
Silicon	Si_3N_4	0.15-0.30μm	WS/Spin	Dry	-	-	LA	KrF	-	30ns	CANS	85
Silicon	Silica	0.3μm	WS/Spin	Dry	-	-	NC	KrF	-	30ns	CANS	85
Silicon	Silica spheres	0.5μm	WS/AJ	Dry	225mJ/cm^2	325mJ/cm^2	~75	KrF	200	23ns	SEM	28*, 29, 86
Silicon	Silica spheres	1μm	WS/AJ	Dry	100mJ/cm^2	325mJ/cm^2	~95	KrF	200	23ns	SEM	28*, 29, 86
Silicon	Silica spheres	2.5μm	WS/AJ	Dry	<5mJ/cm^2	25mJ/cm^2	100	KrF	200	23ns	SEM	28*, 29, 86, 87
Silicon	Silica spheres	5μm	WS/AJ	Dry	<5mJ/cm^2	25mJ/cm^2	100	KrF	200	23ns	SEM	28*, 29, 86, 87
Silicon	Silica spheres	0.5μm	-	Dry	<5 mJ/cm^2	~10 mJ/cm^2	~100	Nd:YAG^3	-	-	-	28*, 29
Silicon	Silica spheres	0.5μm	-	Dry	<5 mJ/cm^2	~15 mJ/cm^2	~100	Nd:YAG^2	-	-	-	28*, 29
Silicon	Silica spheres	1500nm	Isus/SCCS	Dry	65±5 mJ/cm^2 after 4hrs	180mJ/cm^2	~100	KrF	-	28ns	IP,PC	88*
Silicon	Silica spheres	1500nm	Isus/SCCS	Dry	125±8 mJ/cm^{2*}	180mJ/cm^2	~100	KrF	-	28ns	IP,PC	88*

* The threshold fluence increased after 400 hours sample storage time. There was no further increase in the threshold fluence for up to 800 hours storage.

Table 2 (*Continued*)

Substrate	*Contaminant*	*Size of contam.*	*Deposition technique*	*Type of LC*	*Threshold fluence*	*Fluence used*	*% clean*	*Laser*	*No. of pulses*	*Pulse length*	*Analysis technique*	*Ref. No.*
Silicon	Silica spheres	2.5μm	WS/AJ	Dry	5 mJ/cm^2	43 mJ/cm^2	100*	KrF	200	23ns	SEM	28*, 29, 89
Silicon	Silica spheres	500nm	Isus/SCCS	Dry	200±10 mJ/cm^2 after 4hrs	350mJ/cm^2	~100	KrF	-	28ns	IP,PC	88*
Silicon	Silica spheres	500nm	Isus/SCCS	Dry	228±10 mJ/cm^2 (364hrs storage)	400mJ/cm^2	~100	KrF	-	28ns	IP,PC	88*
Silicon	Silicone	0.3-5μm	-	Dry	20mJ/cm^2	-	-	KrF	1	23ns	PC	68*
Silicon	Silicone	0.3-5μm	-	Dry	16mJ/cm^2	-	-	KrF	1	23ns	OMG	68*
Silicon	Silicone	0.3-5μm	-	Dry	540mJ/cm^2	-	-	Nd:YAG	1	7ns	PC	68*
Silicon	Silicone	0.3-5μm	-	Dry	122J/cm^2	-	-	Nd:YAG[2]	1	7ns	PC	68*
Silicon	Silicone	0.3-5μm	-	Dry	92J/cm^2	-	-	Nd:YAG[3]	1	7ns	PC	68*
Silicon	Tungsten	1-2μm	Du/LiqC	Dry	-	2.1 J/cm^2	>95	Nd:YAG	100	24ns	OM, Pol	12
Silicon	Tungsten	1μm	SCCS	Dry	-	280 mJ/cm^2	NE	Nd:YAG[4]	10	10ns	OMG	77
Silicon	Tungsten	1μm	SCCS	Dry	-	0.65J/cm^2	NE	Nd:YAG[4]	10	10ns	OMG	78*
Silicon	Tungsten	0.7±0.3 μm	Esus	Dry	-	0.3J/cm^2	59	KrF	100	~ 30ns	OM, SEM	79*, 80*
Silicon	Tungsten	1μm	SCCS	Dry, LAS	-	$10^{12}W/cm^2$ at beam focal point	100	Nd:YAG[4]	3 pulses (3 shock waves)	10ns	OMG	77
Silicon	SiO_2	0.05-3μm	Scat	Dry, LB	-	100-300mJ/cm^2	Eff	KrF	-	20ns	SEM, AFM	73*

* 100% cleaning is obtained at normal incidence. This drops to zero when angles greater than 15° from the normal incidence are used.

Table 2 (*Continued*)

Substrate	*Contaminant*	*Size of contam.*	*Deposition technique*	*Type of LC*	*Threshold fluence*	*Fluence used*	*% clean*	*Laser*	*No. of pulses*	*Pulse length*	*Analysis technique*	*Ref. No.*
Silicon	Si_3N_4	0.3μm	WS/Spin	Dry, RH100	-	300mJ/cm^2	78	KrF	12	30ns	CANS	90
Silicon	SiO_2	1μm	WS/Spin	Dry, RH100	-	300mJ/cm^2	72+/-10	KrF	1	30ns	CANS	90
Silicon	SiO_2	0.3μm	WS/Spin	Dry, RH100	-	300mJ/cm^2	88+/-6	KrF	1	30ns	CANS	90
Silicon	Si_3N_4	0.3μm	WS/Spin	Dry, RH40	-	300mJ/cm^2	33+/-4	KrF	12	30ns	CANS	90
Silicon	SiO_2	1μm	WS/Spin	Dry, RH40	-	300mJ/cm^2	84+/-8	KrF	1	30ns	CANS	90
Silicon	SiO_2	0.3μm	WS/Spin	Dry, RH40	-	300mJ/cm^2	12+/-7	KrF	1	30ns	CANS	90
Silicon	Alumina	1-10μm mixture	Du	Dry, SAW	-	10^{-3}J/cm^2	Acc	Nitrogen	50	10ns	OSA	91
Silicon	Alumina	1-10μm mixture	Du	Dry, SAW	-	10^{-3}J/cm^2	Acc	Nd:YAG	50	10ns	OSA	91
Silicon	-	-	-	Dry, SuD	-	-	-	Nd:YAG^2	-	-	HI	92
Silicon	Native oxide	-	-	EtAb	-	0.8J/cm^2	100, no damage	KrF	15	20ns	OM, AES, LEED	72*
Silicon	Alumina	0.2-2μm	Sus	LFE	280-350 mJ/cm^2	350 mJ/cm^2	99	XeCl	44	-	DFM	93
Silicon	Alumina	0.2-2μm	Sus	LFE	260-320 mJ/cm^2	-	-	KrF	-	-	DFM	93
Silicon	Silica spheres	1200nm	Sus	LFE	260 mJ/cm^2	-	-	XeCl	-	-	DFM	93
Silicon	Silica spheres	500nm	Sus	LFE	260 mJ/cm^2	260 mJ/cm^2	99.7	XeCl	1	-	DFM	93

Table 2 (*Continued*)

Substrate	*Contaminant*	*Size of contam.*	*Deposition technique*	*Type of LC*	*Threshold fluence*	*Fluence used*	*% clean*	*Laser*	*No. of pulses*	*Pulse length*	*Analysis technique*	*Ref. No.*
Silicon	Silica spheres	250nm	Sus	LFE	<400 mJ/cm^2	-	-	XeCl	-	-	DFM	93
Silicon	Au, Cu, W	0.5-5.5μm	Esus	Steam	-	0.3J/cm^2	100	KrF	5	~ 30ns	OM, SEM	79*, 80*
Silicon	Al_2O_3 & epoxy film	1μm particles, 300nm film	-	Steam(WA), Dry	-	110mJ/cm^2	Opt	KrF	Multi-ple*	16ns	OM	94
Silicon	Alumina	1μm	-	Steam, AF	80 mJ/cm^2	-	-	Excimer	-	-	-	52
Silicon	PS	800nm	Isus/SCCS	Steam, AlF	50mJ/cm^2 universal threshold	100mJ/cm^2	>90	Nd:YAG[2]	20	8ns	LS, SR	95, 96
Silicon	PS	800nm	Isus/SCCS	Steam, AlF	50mJ/cm^2 universal threshold	75mJ/cm^2	>90	Nd:YAG[2]	20	2.5ns	LS, SR	95, 96
Silicon	PS	800nm	Isus/SCCS	Steam, AlF	50mJ/cm^2 universal threshold	75mJ/cm^2	>90	Dye	20	2.5ns	LS, SR	95, 96
Silicon	PS	800nm	Isus/SCCS	Steam, AlF	20mJ/cm^2 universal threshold	75mJ/cm^2	>90	Dye	20	30ps	LS, SR	95, 96
Silicon	Al_2O_3 particles, epoxy film	1μm particles, 1μm film	-	Steam, Dry, U	-	110mJ/cm^2	Opt	KrF	multiple	-	OM	97
Silicon	Alumina	1μm	-	Steam, EF	90 mJ/cm^2	-	-	Excimer	-	-	-	52

* Steam laser cleaning was more effective than dry laser cleaning.

Table 2 (*Continued*)

Substrate	*Contaminant*	*Size of contam.*	*Deposition technique*	*Type of LC*	*Threshold fluence*	*Fluence used*	*% clean*	*Laser*	*No. of pulses*	*Pulse length*	*Analysis technique*	*Ref. No.*
Silicon	Polystyrene	140-1300nm	SCCS	Steam, IF	~100 mJ/cm^2 *	-	-	Nd:YAG[2]	-	8ns	PM, OM, AFM	98*
Silicon	PS, SiO_2, Al_2O_3	60-800 nm†	Isus/SCCS	Steam, WA	110mJ/cm^2 universal threshold	170mJ/cm^2	>90	Nd:YAG[2]	20	7ns	OM, SEM, LS	99
Silicon	Al_2O_3	0.1μm	Esus	Steam, WAV	-	120mJ/cm^2	Opt	KrF	20	15ns	SEM	74, 75
Silicon	Al_2O_3	0.1μm	Esus	Steam, WAV	-	120 mJ/cm^2	Opt	KrF	20	16ns	SEM	74, 75, 100
Silicon	PS	800nm	WS	Steam, WAV	110mJ/cm^2	-	-	Nd:YAG[2]	Single	7ns	LS, PM	101
Silicon	silica spheres	800nm	WS	Steam, WAV	110mJ/cm^2	180mJ/cm^2	>90	Nd:YAG[2]	Single	7ns	LS, PM	101
Silicon	Fe_2O_3 - back surface	0.3-2μm	PMS	Steam, WV	-	200mJ/cm^2	Opt	KrF front irradiation	2 LCSC‡	22ns	PMS	81
Silicon	Fe_2O_3 - back surface	0.3-2μm	PMS	Steam, WV, rear	-	200mJ/cm^2	Opt	KrF back irradiation	2 LCSC†† †††	22ns	SPV	81
Silicon	Polystyrene	140-1300nm	SCCS	Steam, WV	~150 mJ/cm^2 §	-	-	Nd:YAG[2]	-	8ns	PM, OM, AFM	98*
Silicon	Al_2O_3	mixed	Alsus	Wet	-	30J/cm^2	Opt	CO_2[10.6]	5**	1ms	OM	102

* An universal threshold fluence was determined for all diameters used. All particles were removed without damage when using IF instead of WV, due to less pronounced near-field focusing effects.

† Particles used included 60nm, 235nm, 300nm, 500nm and 800nm Polystyrene (PS) spheres, 500nm and 800nm SiO_2 spheres, and 300nm Al_2O_3 particles.

‡ Laser cleaning scanning cycles.

§ A universal threshold cleaning fluence was found for all diameters used. Damage free removal only occurred for particles with diameters ≤510nm.

** Multiple pulses were required due to particle re-deposition.

Table 2 (*Continued*)

Substrate	*Contaminant*	*Size of contam.*	*Deposition technique*	*Type of LC*	*Threshold fluence*	*Fluence used*	*% clean*	*Laser*	*No. of pulses*	*Pulse length*	*Analysis technique*	*Ref. No.*
Silicon	Al_2O_3	9.5μm	Alsus	Wet	2.14J/cm^2	-	-	$CO_2^{9.6}$	single	-	OM	103
Silicon	Al_2O_3	5μm	Alsus	Wet	2.23J/cm^2	-	-	$CO_2^{9.6}$	single	-	OM	103
Silicon	Al_2O_3	9.5μm	Alsus	Wet	1.45J/cm^2	-	-	$CO_2^{10.6}$	single	-	OM	103
Silicon	Al_2O_3	5μm	Alsus	Wet	1.49J/cm^2	-	-	$CO_2^{10.6}$	single	-	OM	103
Silicon	PS	1μm	Alsus	Wet	2.22J/cm^2	2.43J/cm^2	92	$CO_2^{9.6}$	single	-	OM	103
Silicon	PS	1μm	Alsus	Wet	1.41J/cm^2	-	-	$CO_2^{10.6}$	single	-	OM	103
Silicon	Diamond	1-7μm	Esus	Wet, Dry	-	0.75J/cm^2	Wet*	KrF	10	20ns	OM	32*
Silicon	Al_2O_3, SiC, BC, CeO_2	>0.1μm	EA/SCCS	Wet, WF	650mJ/cm^2	2.9-3.2 J/cm^2	Opt	$CO_2^{10.6}$	multiple	0.25ms	PMS	7
Silicon	Particles	1 – 9μm	-	Wet, WVC	-	2J/cm^2	>90	$CO_2^{10.6}$	single	200ns	SR	104
Silicon (hydro-philic)	Al_2O_3	0.2μm	PMS	Dry	-	314mJ/cm^2	~40	KrF	4 LCSC†	22ns	PD	105, 106, 107
Silicon (hydro-philic)	CML	0.1μm	PMS	Dry	-	353mJ/cm^2	~99	KrF	2 LCSC†† ††††	22ns	PD	105, 108
Silicon (hydro-philic)	PSL	0.1μm	PMS	Dry	76mJ/cm^2	320mJ/cm^2	~100	KrF	2 LCSC†† ††††	22ns	PD	105, 106, 107, 108
Silicon (hydro-philic)	SiO_2	0.1-0.2μm	PMS	Dry	-	326mJ/cm^2	~15	KrF	4 LCSC†† ††††	22ns	PD	105, 107

* Wet laser cleaning was more efficient than dry laser cleaning.
† Laser cleaning scanning cycles.

Table 2 (*Continued*)

Substrate	*Contaminant*	*Size of contam.*	*Deposition technique*	*Type of LC*	*Threshold fluence*	*Fluence used*	*% clean*	*Laser*	*No. of pulses*	*Pulse length*	*Analysis technique*	*Ref. No.*
Silicon (hydro-philic)	Alumina	0.1μm	PMS	Steam	143mJ/cm^2	154mJ/cm^2	~90	KrF	4 LCSC†† ††††	22ns	PD	105, 108
Silicon (hydro-philic)	SiO_2	0.1μm	PMS	Steam	143mJ/cm^2	180mJ/cm^2	~95	KrF	5 LCSC†† ††††	22ns	PD	105, 106, 108
Silicon membrane	Al_2O_3	down to 0.35 μm	Esus	Dry	-	350 mJ/cm^2	Opt	KrF	4	16ns	SEM	109
Silicon membrane stencil masks	Alumina	0.2-2μm	Sus	LFE	-	690 mJ/cm^2	Eff	KrF	1	-	DFM	93
Silicon wafer	Acrylate residues	-	-	Ablation	200±20 mJ/cm^2	400±20 mJ/cm^2	100	KrF	40-60	<20ns	VM	110*
Silicon wafer	Photoresist (organic compounds)	1.2μm thick	SCCS	Ablation	0.037J/cm^2	0.3J/cm^2	100	KrF	20	30ns	profile, SEM, AES, FTIRS	111*
Silicon wafer	Copper	1μm	-	Dry	-	0.60J/cm^2	NAG SW	Nd:YAG	10	10ns	PC	112*
Silicon wafer	Copper	1μm	-	Dry	-	0.46J/cm^2	100	Nd:YAG^2	10	10ns	PC	112*
Silicon wafer	Copper	1μm	-	Dry	-	0.18J/cm^2	100	Nd:YAG^4	10	10ns	PC	112*
Silicon wafer	Particles	0.1-10μm	-	Dry	-	3.34J/cm^2 *	-	Excimer	-	-	-	113
Slider	Al	-	-	Dry	-	100mJ/cm^2	IN	Excimer	Varies	-	-	52
Slider	Sn	-	-	Dry	-	50 mJ/cm^2	IN	Excimer	Varies	-	-	52
Slider	-	-	-	Steam	-	130 mJ/cm^2	IN	Excimer	Varies	-	OMG	52

* Integrated fluence (average energy flux).

Table 2 (*Continued*)

Substrate	*Contaminant*	*Size of contam.*	*Deposition technique*	*Type of LC*	*Threshold fluence*	*Fluence used*	*% clean*	*Laser*	*No. of pulses*	*Pulse length*	*Analysis technique*	*Ref. No.*
SS 316L	fingerprints	-	-	Dry	-	31J/cm^2	Eff	KrF	-	-	-	15*
Stainless steel	Al oxide	-	-	Dry	1.0J/cm^2	-	-	Nd:YAG, Nd:YAG^2	Single	14.5ns	-	114*
Stainless steel	aluminium silicate hydroxide	1-10μm	-	Dry	-	0.55J/cm^2	100	XeCl	35	28ns	Visual/PDT	31*
Stainless steel	Cr oxide	-	-	Dry	0.8J/cm^2	-	-	Nd:YAG, Nd:YAG^2	Single	14.5ns	-	115
Stainless steel	Iron oxide	-	-	Dry	0.5J/cm^2	-	-	Nd:YAG, Nd:YAG^2	Single	14.5ns	-	115
Stainless steel	Oxide layer	-	Temp	Dry	-	1-2J/cm^2	Eff	Nd:YAG	-	10ns	SEM, EDS	116*
Stainless steel (SUS)	Magic marker	N/A	Dr	Dry	-	500 mJ/cm^2	Opt	KrF	20	20ns	OM, SEM, AES	22
Stainless steel (SUS)	Magic marker	N/A	Dr	Dry	-	830mJ/cm^2	Opt	KrF	20	20ns	OM, SEM, AES	6
Stainless steel flat samples	Radionucleides ^{60}Cu, ^{60}Ni	-	Nucl	Ablation	-	2.5J/cm^2	97	XeCl	1000	25ns	Gamma	4*, 5*
Steel (A3)	Rust	-	-	Ablation	13mJ/cm^2	32mJ/cm^2	92	Nd:YAG	2.7kHz*	200ns	IP	117*
Steel (A3)	Rust	-	-	Ablation	11mJ/cm^2	26mJ/cm^2	100	Nd:YAG	2.9kHz§ §§§§§	200ns	IP	117*
Steel (A3)	Rust	-	-	Ablation	10mJ/cm^2	25.3 mJ/cm^2	100	Nd:YAG	3.3kHz§ §§§§§	200ns	IP	117*
Stone, ceramic, glass	Epoxy resins and systems	-	Film	Ablation	-	-	Opt using Er: YAG	Nd:YAG, Nd:YAG^2, Nd:YAG^3, Er:YAG	-	3ns, 200μs	-	118*

* Laser repetition rate.

Table 2 (*Continued*)

Substrate	*Contaminant*	*Size of contam.*	*Deposition technique*	*Type of LC*	*Threshold fluence*	*Fluence used*	*% clean*	*Laser*	*No. of pulses*	*Pulse length*	*Analysis technique*	*Ref. No.*
SUS plates - rough	Resin containing $PbCrO_4$	25μm	Spra	Ablation (liquid assisted)	-	-	Eff	$CO_2^{10.6}$	8	-	EDXA	38*
SUS plates - smooth	Resin containing $PbCrO_4$	25μm	Spra	Ablation (liquid assisted)	-	-	Eff	$CO_2^{10.6}$	2	-	EDXA	38*
TiN coating	Polymer*	-	Res	Dry	-	250mJ/cm^2	Opt	KrF at 45°	60	23ns	SEM, AFM	119
TiN coating	Poly-mer*******	-	Res	Dry	-	100mJ/cm^2	Opt	KrF at 45°	1200	23ns	SEM, AFM	119
Vias in polyimide	Soot	<500nm	-	Dry	-	0.48J/cm^2	100	$CO_2^{9.2}$	75 (5Hz)	-	OM	120*
Vias in polyimide	Soot	<500nm	-	Dry	-	0.6J/cm^2	100, Opt	$CO_2^{9.3}$	50 (20Hz)	-	OM	120*

* Polymers used were M-CxFyOz and M-CFx where M=Ti, Al, and Cu.

Acknowledgments

This research has been supported by the Australian Research Council and Macquarie University. A. Fernandes would like to acknowledge the assistance of a Macquarie University Research Award for Areas and Centres of Excellence scholarship, and a Macquarie University Postgraduate Research Fund Grant.

References

1. B. Luk'yanchuk, ed., *Laser Cleaning,* World Scientific, Singapore, (2002).
2. D.M. Kane, A.J. Fernandes and DR Halfpenny, *'Pulsed Laser Cleaning of Particles from Surfaces and Optical Materials'*, (Chapter 4, *Laser Cleaning,* Ed. B. Luk'yanchuk), World Scientific, Singapore, (2002).
3. Y.F. Lu, M.H. Hong, S.J. Chua, B.S. Teo and T.S. Low, *J. Appl. Phys.* **79**, 2186 (1996).
4. Ph. Delaporte, M. Gastaud, W. Marine, M. Sentis, O. Uteza, P. Thouvenot, J.L. Alcaraz, J.M. Le Samedy and D. Blin, *Appl. Surf. Sci.* **208-209**, 298 (2003).
5. Ph. Delaporte, M. Gastaud, W. Marine, M. Sentis, O. Uteza, P. Thouvenot , J.L. Alcaraz, J.M. Le Samedy and D. Blin, *Appl. Surf. Sci.* **197-198**, 826 (2002).
6. Y.F. Lu, M. Takai, S. Komuro, T. Shiokawa and Y. Aoyagi, *Appl. Phys. A* **59**, 281 (1994).
7. J.B. Heroux, S. Boughaba, I. Ressejac, E. Sacher and M. Meunier, *J. Appl. Phys.* **79**, 2857 (1996).
8. P. Meja, M. Autric, P. Delaporte and P. Alloncle, *Appl Phys A* **69**, S343 (1999).
9. W.D. Kimura, G.H. Kim and B. Balick, [Conference Paper] CLEO '94. Summaries of Papers Presented at the Conference on Lasers and Electro-Optics. Vol.8. 1994 Technical Digest Series. Conference Edition (Cat. No.94CH3463-7), Opt. Soc. America, USA, 424 (1994).
10. K. Mann, B. Wolff-Rottke and F. Muller, *Appl. Surf. Sci.* **96-98**, 463 (1996).
11. M. Abraham, O. Madden and S. Scheerer, *J. Cult. Heritage* **4**, 92s (2003).
12. J.D. Kelley and F.E. Hovis, *Microelectronic Engineering* **20**, 159 (1993).
13. X. Zhou, K. Imasaki, H. Furukawa, S. Nakai and C. Yamanaka, *Int. J. Adv. Manuf. Technol.* **19**, 642 (2002).
14. J.P. Boquillon and R. Oltra, [Conference Paper] *Proceedings CLEO Europe '94,* Opt. Soc. America, 6 (1994).

15. A.C. Engelsberg, '*A dry laser-assisted process for the removal of surface contamination*', 40th Annual Institute of Environmental Sciences meeting, IEST, Illinois, (1994).
16. V. Hyvärinen, M. Sorjonen, K.E. Peiponen, R. Silvennoinen, T. Niskanen and J. Kalliokoski, *Optics &Lasers in Engineering* **33**, 311 (2000).
17. H. J. Kim, D.J. Kim, J.K. Ryu and S.S. Pak, *Appl. Surf. Sci.* **228**, 100 (2004).
18. A. Kearns, C. Fischer, K.G. Watkins, M. Glasmacher, H. Kheyrandish, A. Brown, W.M. Steen and P. Beahan, *Appl. Surf. Sci.* **127-129**, 773 (1998).
19. Y.F. Lu, Y.P. Lee, M.H. Hong and T.S. Low, *Appl. Surf. Sci.* **119,** 137 (1997).
20. J.M. Lee, K.G. Watkins and W.M. Steen, *Appl. Phys. A* **71**, 671 (2000).
21. J.M. Lee and K.G. Watkins, *Optics Express* **7**, 68 (2000).
22. Y.F. Lu, Y. Aoyogi, M. Takai and S. Namba, *Jpn. J. Appl. Physics* **33,** 7138 (1994).
23. Y.F. Lu and Y. Aoyagi, *Jpn. J. Appl. Phys. 2 (Lett.)* **34**, L1557 (1995).
24. J.M. Lee and K.G. Watkins, *Optics and Lasers in Engineering* **34,** 429 (2000).
25. M. Takai, N. Suzuki, O. Yavas, A. Hosono and S. Okuda, *Proc. SPIE* **4426,** 302 (2002).
26. O. Yavas, N. Suzuki and M. Takai, *Appl. Phys. Lett.* **72**, 2797 (1998).
27. S. Pleasants and D.M. Kane, *J. Appl. Phys*. **93**, 8862 (2003).
28. Y.F. Lu, Y.W. Zheng, L. Zhang, B.S. Luk'yanchuk, W.D. Song, W.J. Wang, M.H. Hong and T.C. Chong, *Mat. Res. Soc. Symp. Proc.* **704**, 23 (2002).
29. Y.W. Zheng, B.S. Luk'yanchuk, Y.F. Lu, W.D. Song and Z.H. Mai, *J. Appl. Phys.* **90**, 2135 (2001).
30. D.R. Halfpenny and D.M. Kane, *J. Appl. Phys.* **86**, 6641 (1999).
31. V.B. Bregar and J. Mozina, *Appl. Phys. A* **A77**, 633 (2003).
32. V.I. Konov, V.V. Kononenko, A.M. Lomonosov, V.G. Mihalevich, S.M. Pimenov and K.V. Zhuk, *Proc. SPIE* **4426,** 347 (2002).
33. D.M. Kane and A.J. Fernandes, *Proc. SPIE* **4426**, 334 (2002).
34. D. Hirschausen and D.M. Kane, *J. Appl. Phys*. **92**, 4201 (2002).
35. Y.F. Lu, W.D. Song, M.H. Hong, Z.M. Ren, Q. Chen and T.C. Chong, *Jpn. J. Appl. Phys.* **39** 4811 (2000).
36. W.D. Song, M.H. Hong, L. Zhang, Y.F. Lu and T.C. Chong, *Proc. SPIE* **4426,** 280 (2002).
37. P. Meja, M. Autric, P. Alloncle, P. Pasquet, R. Oltra and J.P. Boquillon, *Appl. Phys. A* **69 (Suppl),** S687 (1999).

38. A. Tsunemi, K. Hagiwara, N. Saito, K. Nagasaka, Y. Miyamoto, O. Suto and H. Tashiro, *Appl. Phys. A* **63**, 435 (1996).
39. O. Madden, P. Pouli, M. Abraham and C. Fotakis, *J. Cult. Heritage* **4**, 98s (2003).
40. M. Gómez-Heras, M. Alvarez de Buergo, E. Rebollar, M. Oujja, M. Castillejo and R. Fort, *Appl. Surf. Sci.* **219**, 290 (2003).
41. M.I. Cooper, D.C. Emmony and J. Larson, *Optics & Laser Technology* **27**, 69 (1995).
42. W.D. Song, Y.F. Lu, M.H. Hong and T.S. Low, *Proc. SPIE* **3550**, 19 (1998).
43. Y.F. Lu, W.D. Song, M.H. Hong, B.S. Teo, T.C. Chong and T.S. Low, *J. Appl. Phys.* **80**, 499 (1996).
44. Y.F. Lu and Y. Aoyogi, *Jpn. J. Appl. Phys.* **33** L430 (1994).
45. J.M. Lee, W.M. Steen, *Int. J. Adv. Manuf. Technol.* **17**, 281 (2001).
46. K.G. Watkins, C. Curran and J.M. Lee, *J. Cult. Heritage* **4**, 59s (2003).
47. S.M. Durbin, A.V. Deshmukh, T.D. Brooks and L.J. van de Burgt, *Appl. Surf. Sci.* **127-129**, 810 (1998).
48. Y.F. Lu, S. Komuro and Y. Aoyogi, *Jpn. J. Appl. Phys.* **33,** 4691 (1994).
49. Y.F. Lu, Y.W. Zheng and W.D. Song, *Appl. Phys. A* **A68**, 569 (1999).
50. Y.F. Lu, W.D. Song, M.H. Hong, Z.M. Ren and Y.W. Zheng, *Proc. SPIE* **4088**, 371 (2000).
51. Y.F. Lu, W.D. Song, Y. Zhang and T.S. Low, *Proc. SPIE* **3550,** 7 (1998).
52. Y.F. Lu, W.D. Song, Y. Zhang, M.H. Hong and T.C. Chong, *Riken Review* **32**, 64 (2001).
53. Y.F. Lu, W.D. Song, M.H. Hong and Y.W. Zheng, *Proc. SPIE* **4070,** 331 (2000).
54. C.P. Grigoropoulos and D. Kim, *Proc. SPIE* **4426,** 315 (2002).
55. M. She, D. Kim and C.P. Grigoropoulos, *J. Appl. Phys.* **86**, 6519 (1999).
56. Y.F. Lu, W.D. Song, Y. Zhang and Y.W. Zheng, *Proc. SPIE* **3618** 278 (1999).
57. Y.F. Lu, L. Zhang, W.D. Song and D.S.H. Chan, *Jpn. J. Appl. Phys.* **37,** L1330 (1998).
58. J.M. Lee, K.G. Watkins, W.M. Steen, *Int. J. Adv. Manuf. Technol.* **16,** 649 (2000).
59. A. Barone, F. Bloisi and L. Vicari, *Appl. Surf. Sci.* **208-209**, 468 (2003).
60. W. Kautek, S. Pentzien, P. Rudolph, J. Krüger and E. König, *Appl. Surf. Sci.* **127-129**, 746 (1998).
61. T. Fourrier, G. Schrems, T. Muhlberger, J. Heitz, N. Arnold, D. Bäuerle, M. Mosbacher, J. Boneberg and P. Leiderer, *Appl Phys A* **72**, 1 (2001).

62. N. Chaoui, J. Solis, C.N. Afonso, T. Fourrier, T. Muehlberger, G. Schrems, M. Mosbacher, D. Bäuerle, M. Bertsch and P. Leiderer, *Appl Phys A* **76**, 767 (2003).
63. W.D. Song, M.H. Hong, Y.F. Lu and T.C. Chong, *Appl. Surf. Sci.* **208-209**, 463 (2003).
64. Y.F. Lu, W.D. Song, C.K. Tee, D.S.H. Chan and T.S. Low, *Jpn. J. Appl. Phys.* **37**, 840 (1998).
65. Y.F. Lu, W.D. Song, M.H. Hong, D.S.H. Chan and T.S. Low, *Proc. SPIE* **3097**, 352 (1997).
66. Y.F. Lu, W.D. Song, B.W. Ang, M.H. Hong, D.S.H. Chan and T.S. Low, *Appl. Phys. A* **65**, 9 (1997).
67. Y.F. Lu, W.D. Song and T.S. Low, *Materials Chemistry & Physics* **54**, 181 (1998).
68. W.D. Song, M.H. Hong, S.H. Lee, Y.F. Lu and T.C. Chong, *Appl. Surf. Sci.* **208-209**, 306 (2003).
69. W.D. Song, M.H. Hong, H.L. Koh, W.J. Wang, Y.W. Zhen, YF Lu and T.C. Chong, *Appl. Surf. Sci.* **186**, 69 (2002).
70. A.C. Engelsberg, *Precision Cleaning Magazine* **3** (May), 35 (1995).
71. C. Curran, J.M. Lee and K.G. Watkins, [Conference Paper] *Conference Digest, 2000 Conference on Lasers and Electro-Optics Europe (Cat. No.00TH8505),* IEEE. Piscataway, NJ, USA, 1 (2000).
72. R. Larciprete, E. Borsella and P. Cinti, *Appl. Phys. A* **A62**, 103 (1996).
73. D.J. Kim, Y.K. Kim, J.K. Ryu and H.J. Kim, *Jpn. J. Appl. Phys.* **41**, 4563 (2002).
74. W. Zapka, W. Ziemlich, W.P. Leung and A.C. Tam, *Microelectronic Engineering* **20**, 171 (1993).
75. W. Zapka, W. Ziemlich, W.P. Leung and A.C. Tam, *Advanced Materials for Optics and Electronics* **2**, 63 (1993).
76. D. Kumar, A. Ata, U. Mahajan and R.K. Singh, *J. Electronic Materials* **27**, 1104 (1998).
77. J.M. Lee and K.G. Watkins, *J. Appl. Phys.* **89,** 6496 (2001).
78. C. Curran, J.M. Lee and K.G. Watkins, *Optics and Lasers in Engineering* **38,** 405 (2002).
79. M. Arronte, P. Neves and R. Vilar, *J. Appl. Phys.* **92**, 6973 (2002).
80. P. Neves, M. Arronte, R. Vilar and A.M. Botelho do Rego, *Appl. Phys. A* **74,** 191 (2002).
81. F. Beaudoin, M. Meunier, M. Simard-Normandin and D. Landheer, *J. Vac. Sci. Technol. A* **16,** 1976 (1998).
82. S.I. Kudryashov and S.D. Allen, *J. Appl. Phys.* **92**, 5159 (2002).

83. M. Mosbacher, H.J. Munzer, J. Zimmermann, J. Solis, J. Boneberg and P. Leiderer, *Appl. Phys. A* **72**, 41 (2001).
84. Y.F. Lu, W.D. Song, K. Ye, Y. Lee, D. Chan and T. Low, *Jpn. J. Appl. Phys.* **36,** L1304 (1997).
85. G. Vereecke, E. Rohr and M.M. Heyns, *Appl. Surf. Sci.* **157**, 67 (2000).
86. Y.F. Lu, Y.W. Zheng and W.D. Song, *J. Appl. Phys.* **87,** 1534 (2000).
87. Y.F. Lu, Y.W. Zheng and W.D. Song, *J. Appl. Phys.* **87**, 549 (2000).
88. G. Schrems, M.P. Delamare, N. Arnold, P. Leiderer and D. Bäuerle, *Appl. Phys. A* **76**, 847 (2003).
89. Y.W. Zheng, Y.F. Lu and W.D. Song, *J. Appl. Phys.* **90**, 59 (2001).
90. G. Vereecke, E. Rohr and M.M. Heyns, *J. Appl. Phys.* **85**, 3837 (1999).
91. A.A. Kolomenskii, H.A. Schuessler, V.G. Mikhalevich and A.A. Maznev, *J. Appl. Phys.* **84**, 2404 (1998).
92. V. Dobler, R. Oltra, J.P. Boquillon, M. Mosbacher, J. Boneberg and P. Leiderer, *Appl. Phys. A* **69 (Suppl)**, S335 (1999).
93. W. Zapka, R. Lilischkis and K.F. Zapka, *Proc. SPIE* **3996**, 92 (2000).
94. H.K. Park, C.P. Grigoropoulos, W.P. Leung and A.C. Tam, *IEEE Transactions on Components, Packaging & Manufact. Technology A* **17**, 631 (1994).
95. M. Mosbacher, N. Chaoui, J. Siegel, V. Dobler, J. Solis, J. Boneberg, C.N. Afonso and P. Leiderer, *Appl. Phys. A* **69**, S331 (1999).
96. R. Oltra, E. Arenholz, P. Leiderer, W. Kautek, C. Fotakis, M. Autric, C. Afonso and P. Wazen, *Proc. SPIE* **3885,** 499 (2000).
97. A.C. Tam, H.K. Park and C.P. Grigoropoulos, *Appl. Surf. Sci.* **127-129**, 721 (1998).
98. F. Lang, M. Mosbacher and P. Leiderer, *Appl. Phys. A*, **A77**, 117 (2003).
99. M. Mosbacher, V. Dobler, J. Boneberg and P. Leiderer, *Appl. Phys. A* **A70,** 669 (2000).
100. A.C. Tam, W.P. Leung, W. Zapka and W. Ziemlich, *J. Appl. Phys.* **71**, 3515 (1992).
101. P. Leiderer, J. Boneberg, M. Mosbacher, A. Schilling and O. Yavas, *Proc. SPIE* **3274,** 68 (1998).
102. K. Imen, S.J. Lee and S.D. Allen, *Appl. Phys. Lett.* **58**, 203 (1991).
103. S.J. Lee, K. Imen and S.D. Allen, *Microelectronic Engineering* **20**, 145 (1993).
104. S.D. Allen, A.S. Miller and S.J. Lee, *Materials Science & Engineering* **B49**, 85 (1997).
105. M. Meunier, X. Wu, F. Beaudoin, E. Sacher and M. Simard-Normandin, *Proc. SPIE* **3618,** 290 (1999).

106. X. Wu, E. Sacher and M. Meunier, *J. Adhesion* **70**, 167 (1999).
107. X. Wu, E. Sacher and M. Meunier, *J. Appl. Phys.* **86**, 1744 (1999).
108. X. Wu, E. Sacher and M. Meunier, *J. Appl. Phys.* **8**, 3618 (2000).
109. W. Zapka, W. Ziemlich and A.C. Tam, *Appl. Phys. Lett.* **58**, 2217 (1991).
110. A. Braun, K. Otte, K. Zimmer, St Rudschuck, D. Hirsch and F. Bigl, *Appl. Phys. A* **69**, S339 (1999).
111. Y. Feng, Z. Liu, R. Vilar and X.S. Yi, *Appl. Surf. Sci.* **150**, 131 (1999).
112. J.M. Lee, C. Curran and K.G. Watkins, *Appl. Phys. A* **73**, 219 (2001).
113. T.E. Lizotte and T.R. O'Keefe, *Proc. SPIE* **2703**, 279 (1996).
114. A. Sansonetti and M. Realini, *J. Cult. Heritage* **1**, S189 (2000).
115. P. Psyllaki, P. Pasquet, P. Meja, R. Oltra and M. Autric, [Conference Paper] *Conference Digest. 2000 Conference on Lasers and Electro-Optics Europe (Cat. No.00TH8505),* IEEE, Piscataway, NJ, USA, 1 (2000).
116. P. Psyllaki and R. Oltra, *Mat. Sci. & Eng. A* **282**, 145 (2000).
117. Z. Wang, X. Zeng and W. Huang, *Surface and Coatings Technology* **166**, 10 (2003).
118. S. Scheerer, M. Abraham and O. Madden, *J. Cult. Heritage* **4**, S223 (2003).
119. Y.F. Lu, Y.P. Lee and M.S. Zhou, *J. Appl. Phys.* **83**, 1677 (1998).
120. K. Coupland, P.R. Herman and B. Gu, *Appl. Surf. Sci.* **127-129**, 731 (1998).

Chapter 3

PARTICLE ON A SURFACE: ABOUT POSSIBLE ACOUSTIC AND PLASMONICS EFFECTS IN DRY LASER CLEANING

B S LUK'YANCHUK, Z B WANG, Y ZHOU, M H HONG,
W D SONG, T C CHONG

Data Storage Institute, Agency for Science, Technology and Research, Singapore

A laser irradiated particle on a surface produces a complicated intensity distribution on the substrate. In the case of a transparent particle, and normal incidence of radiation, the intensity has near-field focusing near the center and additional ring(s) around the basic maximum. This ring of intensity produces a cylindrical convergent surface acoustic wave, which can enhance the particle removal. Another effect is discussed for metallic particles. The intensity on the surface under a metallic particle is typically diminished due to the screening effect, in contrast to transparent particles, which act as a near-field lens. Nevertheless, if one uses radiation frequencies near the surface plasmon resonance, the conditions for efficient coupling of radiation with a metallic surface can be provided. This plasmonic effect can permit laser cleaning of metallic nanoparticles from a silicon surface.

1. Introduction

The initial theoretical examinations of dry laser cleaning were done in the frame of a so-called one-dimensional model, which ignored the influence of the particle on the intensity distribution on the substrate. The resulting 1-D model provides correct qualitative understanding of the main phenomena of the dry laser cleaning problem (Kelley, 1993; Lu, 1997; Bäuerle, 2000; Arnold, 2002 a, b). Nevertheless this simple 1-D model is often insufficient to explain the experimental values of the laser cleaning threshold fluence, which can be up to three orders of magnitude lower than that predicted by the 1-D model. The reason for this discrepancy is related to the scattering of radiation by the contaminant particle, which strongly changes the local distribution of absorbed laser intensity. For example, a small transparent particle can work as a near-field lens, which leads to a strong field enhancement (Luk'yanchuk, 2000, 2001, 2002 a, b, c, d). This produces a nonstationary 3-D distribution of temperature and a nonstationary 3D thermal deformation of the surface. It requires a more detailed theoretical analysis of the 3-D effects in dry laser cleaning. This analysis was done on the basis of the Mie theory

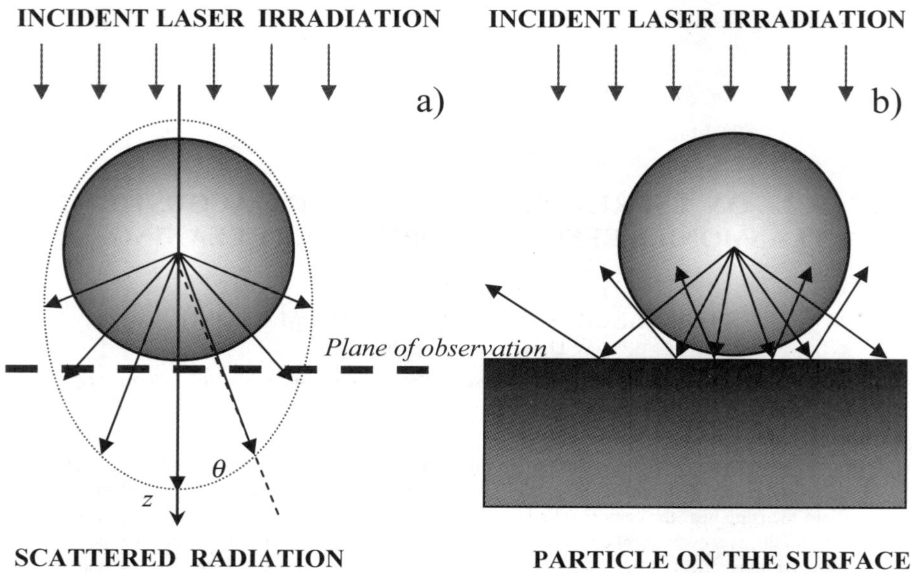

Fig.1. Schematic for the particle scattering within the Mie theory, where the distribution of field is studied in arbitrary points, for example, in some observation plane (a). In a typical consideration of the Mie theory (Born & Wolf, 1999) the incident plane wave propagates along the z-coordinate, and electric vector is directed along the x-coordinate. Particle on surface (b) – the scattered radiation reflects from the surface and participates in the secondary scattering (Bobbert & Vlieger, 1986 a, b).

(Luk'yanchuk, 2002c, 2003) and on the basis of a more complicated particle on surface theory (Luk'yanchuk, 2004). Fig. 1 illustrates the difference in the latest approximations: in the case of the Mie theory radiation scattering by the particle is taken into account, but the influence of the substrate is neglected. In the case of the particle-on-surface (POS) theory the secondary scattering of reflected radiation is taken into account.

There are two important issues which follow from the 3-D theory. The first one refers to the strong enhancement (about few tens) of laser intensity, which can be obtained within the near-field region with size < 100 nm on the substrate under the particle. This strong, near-field focusing effect with submicrometer particles follows from the theoretical analysis (Luk'yanchuk et al, 2000). High localization of the near-field was confirmed experimentally (Lu, 2000 a; Mosbacher, 2001; Münzer, 2001; Luk`yanchuk, 2002 d, Huang, 2002 a, b, 2003) and it is used at present for many applications (Denk, 2003, Hong, 2003; Leiderer, 2004).

The second issue refers to the high temperature which it is necessary to reach for a small particle near the cleaning threshold. This temperature may exceed the melting and vaporization temperature, which in fact leads to a change in the cleaning mechanism. It can be splashing of liquid, see in Fig. 2 (Luk`yanchuk, 2003, Wang, 2005) or even developed ablation (Münzer, 2002; Luk`yanchuk, 2002 d, Arnold, 2004).

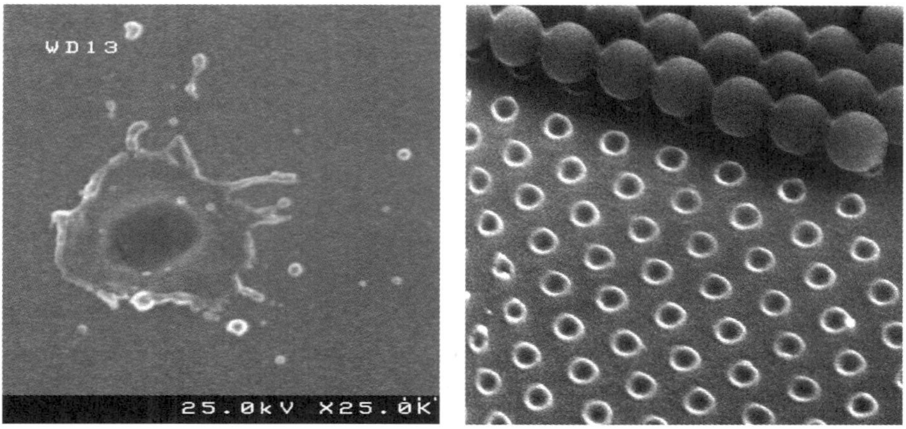

Fig. 2. Effects on the substrate surface during laser cleaning of 1 μm PS particles with different fluences of KrF-laser. (a) splashing effect on the surface of Si, (b) localized melting on the surface of GeSbTe -film.

The 3D model predicts results close to the experimental ones, while 1D model disagrees with experiments by one–two orders of magnitude. At the same time laser cleaning is a quite complicated phenomenon which may include many different mechanisms. In this paper we are discussing two new (acoustic and plasmonic) effects which can enhance laser cleaning.

The paper is organized as follows: the results of theoretical and experimental investigations of particle-on-surface effects are presented in Sect. 2. The effect related to the formation of a cylindrical convergent surface acoustic wave is discussed in Sect. 3. In Sect. 4 the plasmonics effect in laser cleaning of metallic particles is studied. The most important results are summarized in Sect. 5.

2. Particle on the Surface

In free space the distribution of laser intensity around the spherical particle can be found from the Mie theory. This theory is discussed in detail in many books (Born & Wolf, 1999; Barber & Hill, 1990; Stratton, 1941; Kerker, 1969, 1989; Van de Hulst, 1981; Bohren & Huffman, 1983). This theory can be used as a first approach for the intensity distribution on the surface of a substrate under the particle. From this analysis one can clearly see the near-field focusing effect, mentioned above.

Application of the Mie theory to dry laser cleaning (Luk`yanchuk, 2002 d, 2003) shows that for some particle sizes the theory yields results for the predicted laser cleaning threshold fluence which are much closer to the experimental results than the 1-D theory.

A better estimation of the intensity distribution follows from the POS theory, which takes into account the secondary scattering of radiation reflected from the substrate. This problem has an exact solution, found by Bobbert & Vliger, 1986 a. Although this solution is rigorous, the physical effect is quite clear. The substrate plays the role of a mirror coupled with a spherical resonator (particle). It should lead to the increase in the maximal intensity and to the sharpening in the intensity distribution. These evident effects were confirmed during the first calculations for a polystyrene particle on the silicon substrate (Luk`yanchuk, 2000). However, the calculations take a long time. The problem is that in order to find a scattered field one should calculate some inverse matrix $\left(\hat{\mathbf{1}}-\hat{\mathbf{B}}\hat{\mathbf{A}}\right)^{-1}$, where matrix $\hat{\mathbf{B}}$ describes scattering of radiation by a particle (elements of this matrix follow the Mie theory). Another matrix $\hat{\mathbf{A}}$ describes reflection of the spherical wave from the substrate. This matrix is quite complicated. Each element of this matrix is expressed through the so-called Weyl type integrals (Stratton, 1941; Morse & Feshbach, 1953; Bobbert & Vlieger, 1986 a) defined by

$$\Pi_\ell^m(r)=\frac{i^{-\ell}}{2\pi}\int_0^{2\pi} d\beta \int_0^{\pi/2-i\infty} \sin\alpha\, e^{ikr\cos\gamma}\, Y_\ell^m(\alpha,\beta)\, d\alpha\,, \tag{1}$$

where $Y_\ell^m(\alpha,\beta)$ are the spherical harmonics, related to associate Legendre polynomials (Wolfram, 1999). Integration in (1) is performed along some contour of integration on the plane of complex variable α. In our calculations we construct the integration contour as

$$\alpha=\frac{\pi}{2}\frac{cx}{1+cx}-i\,cx\,, \tag{2}$$

where c is a positive real constant, and x varies from zero to infinity. Along this integration contour we found that, at the cost of longer calculation time, a stable value can be achieved. Formula (1) presents an integration of a highly oscillatory function, thus it needs long computation times. The Laguerre polynomial approximation method, which was recommended by Bobbert, 1986 b, for the typical calculations in the laser cleaning problem, yields insufficient accuracy. With Alpha Power Station, which we used in our first calculations, a 20×20 $\hat{\mathbf{A}}$-matrix has been calculated for 8 hours. However, this permits calculations of the intensity near the central part, only. To calculate the intensity distribution off the central axis one needs a higher dimension matrix.

Recently a more efficient technique of calculation of Debye potentials, which does not involve operation of matrix inversion was suggested (Wang, 2004, 2005). Details of this technique can be found in original papers. With this technique significantly enhanced computation speed has been achieved (typical calculations take of the order of 10 minutes with a Pentium IV computer). We were also able to calculate effects in the intensity distribution away from the center (on the periphery outside of the shadow of the particle).

In Fig. 3 the intensity distribution within the xz plane is shown for a polystyrene particle (refraction index $n = 1.6$, particle diameter $2a = 1.0$ μm), on a silicon surface, under 248 nm laser irradiation. Gradations of the intensity are given from negative (dark) to positive (light) values. The dark area on the top of the particle corresponds to energy flux directed up, while the white area under the particle corresponds to energy flux directed to the substrate. From the figure one can see enhanced radiation intensity, which came through the transparent particle after its reflection by the surface.

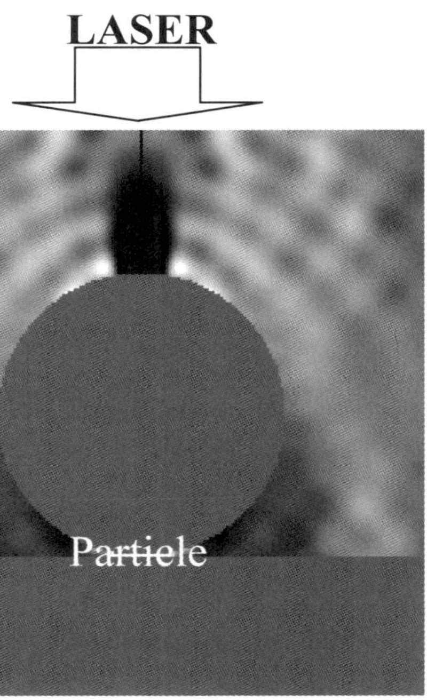

Fig. 3. The intensity $I = S_z$ (z-component of the Poynting vector) distribution within the *x-z* plane for radiation with $\lambda = 248$ nm, scattered by polystyrene particle ($n = 1.6$, $a = 0.5$ μm) on Si surface. Gradations of the intensity are given from negative (dark) to positive (light) values. The dark area on the top of the particle corresponds to energy flux directed up, while the white area under the particle corresponds to energy flux directed to the substrate.

In Fig. 4 the distribution of laser intensity on the substrate is shown. For comparison we present in the figure the distribution which follows from the Mie theory.

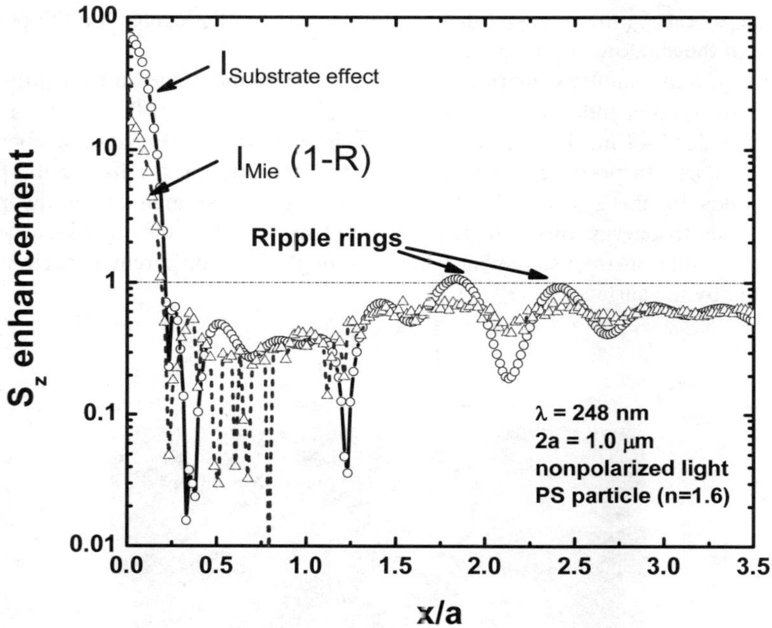

Fig. 4. Calculated S_z intensity fields on substrate surface for a 1.0 μm PS particle on GeSbTe (GST) substrate. The laser light is nonpolarized and radiation incident normally on the substrate.

One can see from Fig. 4 that the solution of the particle-on-surface (POS-solution) demonstrates 1.5 times higher S_z intensity in the center than the Mie solution. This result was expected, because the multiple reflection of the Poynting vector between the particle and the substrate results in more energy flowing into the substrate. On the particle periphery the scattered field becomes unimportant and the enhancement factor tends to unity (nondisturbed incident radiation). However, at distances smaller than 5 diameters of the particle the enhancement is still smaller than 1, i.e. the particle creates a big "shadow", similar to semitransparent material. In fact the energy from the "shadow region" is redistributed near the center (field enhancement region) to provide the overal energy conservation (Luk'yanchuk, 2002 d). The intensity distribution near the center can be approximated by Gaussian distribution, in many cases and the overall behaviour of the intensity in the shadow region also can be assumed to be in the Gaussian form. Thus, the total intensity distribution arround the particle can be written as

$$I(r,t) = I_0(t)\left[S_0\, e^{-r^2/r_0^2} + 1 - e^{-r^2/r_s^2}\right], \tag{3}$$

where $I_0(t)$ describes the input laser intensity, S_0 is the field enhancement at the center, r_0 is characteristic radius of the field enhancement region and r_s is characteristic radius of the "shadow" region. Due to the overall energy conservation, the radius of the shadow region is given by $r_s = r_0\sqrt{S_0}$.

One can consider a slightly different distribution with a fixed radius of shadow, e.g. $r_s = a + \lambda$, then one should introduce the shadow enhancement factor S_1

$$I(r,t) = I_0(t)\left[1 + S_0\, e^{-r^2/r_0^2} - S_1\, e^{-r^2/r_s^2}\right]. \tag{3a}$$

The value $S_1 = \dfrac{r_0^2}{r_s^2} S_0$ follows from the overall energy conservation.

We used the simplified distribution (3) in calculations with the 3-D model of laser cleaning. Parameters S_0 and r_0 were calculated either from the Mie theory (Luk′yanchuk, 2003) or from the POS-theory (Luk′yanchuk, 2004). Calculations were performed by the following steps. Firstly we search for the values of S_0 and r_0 from the solution of the electrodynamic problem. The next step includes calculation of the temperature distribution $T(r,z,t)$ using distibution (3) and solving the heat equation.

In the case of a single Gaussian profile, the solution of the linear heat equation can be expressed by the formula:

$$T(r,z,t) = S_0 \frac{A_0\alpha}{c_s\rho_s}\int_0^t dt_1\; I_0(t-t_1)\frac{e^{-\frac{r^2}{r_0^2+4\chi t_1}}}{1+4\chi t_1/r_0^2} F(z,t_1), \tag{4}$$

where $\chi = \kappa_s / c_s\rho_s$ is the thermal diffusivity of the substrate, and the F - function is given by

$$F(z,t)=\frac{1}{2}e^{\alpha^2\chi t}\left\{e^{\alpha z}erfc\left[\alpha\sqrt{\chi t}+\frac{z}{2\sqrt{\chi t}}\right]+e^{-\alpha z}erfc\left[\alpha\sqrt{\chi t}-\frac{z}{2\sqrt{\chi t}}\right]\right\}. \tag{5}$$

The smooth pulse shape, $I_0(t)$, typical for excimer laser, can be described by (Bäuerle, 2000):

$$I_0(t)=\frac{\Phi t}{t_\ell^2}\exp\left[-\frac{t}{t_\ell}\right], \tag{6}$$

where $t_\ell = 0.409\, t_{FWHM}$ (t_{FWHM} is the duration of the pulse defined as the full width at half maximum), and Φ is the (homogeneous) laser fluence incident on the sample.

The total temperature T then can be presented as a sum of three distributions, $T = T_{1D} + T_{3D} - T_{sh}$. All values are calculated from the 3-D expression (4). For the 1-D temperature, T_{1D} , one should put $r_0 = \infty$ and $S_0 = 1$. The effect of the shadow, T_{sh}, is modeled by (4) with radius of shadow r_s and $S_0 \to S_1$. It is clear that the maximal surface temperature in the POS-model is higher than that in 1-D model (some examples are given in Luk`yanchuk, 2002 d, 2003).

The next step is related to the calculation of the surface displacement $z_s(r,t)$, caused by inhomogeneous heating of the substrate. In a general case one should solve the problem of thermal elasticity (Sokolnikoff, 1956; Nowacki, 1962). To find the epicentral surface displacement, $z_s(t)=z_s(0,t)$, we shall use the solution of the 3-D thermal elasticity problem, which relates the *x-y* Fourier transforms (denoted by tilde) with wave-vector $\boldsymbol{k}$ of surface thermal expansion and temperature rise distribution (Arnold, 2003)

$$\tilde{z}_s = 2\alpha_T(1+\sigma)\int_0^\infty e^{-kz}\tilde{T}(z)dz\,. \tag{7}$$

Here α_T is the linear thermal expansion coefficient, and σ is the Poisson coefficient, both for the substrate.

Substituting the solution of 3-D heat equation (4) and performing the direct and inverse Fourier transforms (i.e., zero order Hankel transforms for axial symmetry, inverse transform for $r = 0$ only), one obtains:

$$z_s(t)=S_0\frac{\beta_3 A_0}{c_s\rho_s}\int_0^t\frac{I(t-t_1)dt_1}{1+4\chi t_1/r_0^2}\int_0^\infty\alpha\,dz\left[1-\sqrt{\pi}\varsigma\, e^{\varsigma^2}erfc(\varsigma)\right]F(z,t_1), \tag{8}$$

where r_0 is the radius of Gaussian beam, $\varsigma = \dfrac{z/r_0}{\sqrt{1+4\chi t_1/r_0^2}}$, and the F - function is given by formula (5).

This surface deformation is considered to be the driving force, which produces the particle acceleration. It is convenient to describe particle dynamics in terms of deformation parameter, $\delta(t)$

$$\delta(t) = z_s(t) + \Delta a(t) - f(t) + \delta_0, \tag{9}$$

where $f(t)$ is the displacement of the particle center, and the term $\Delta a(t) = \alpha_T^{(p)} a T_s(t)$ describes the effect of the particle heating due to the thermal contact. When we assume ideal thermal contact the temperature of the particle is the same as the substrate temperature. $\alpha_T^{(p)}$ is the linear thermal expansion coefficient for the particle. Parameter δ_0 describes the initial deformation under the action of the adhesion force. This parameter is expressed in the DMT theory (Lu, 2000b) as

$$\delta_0 = \frac{1}{8}\left[\frac{9a\langle\hbar\omega\rangle^2}{2\pi^2 h^4 E^{*2}}\right]^{1/3}, \quad \frac{1}{E^*} = \left(\frac{1-\sigma_p^2}{E_p} + \frac{1-\sigma_s^2}{E_s}\right), \tag{10}$$

where σ_p, σ_s and E_p, E_s are the Poisson coefficients and Young's moduli for the particle (p) and substrate (s), $\langle\hbar\omega\rangle$ presents the Lifshitz constant and h~0.4 nm is the separation between the surfaces in contact.

The acceleration of the particle due to the combined elastic and adhesive force can be expressed as (Lu, 2000b)

$$\frac{4}{3}\pi a^3 \rho_p \frac{d^2 f(t)}{dt^2} = \frac{4}{3}\sqrt{a}\, E^*[\delta(t)^{\frac{3}{2}} - \delta_0^{\frac{3}{2}}], \tag{11}$$

where ρ_p is the density of the particle. The last (constant) term here reflects the fact that the Van der Waals component of the adhesion force is independent of deformation. The initial conditions for eqn. (11) are

$$\frac{df}{dt}\Big|_{t=0} = \frac{d\Delta a}{dt}\Big|_{t=0} = f\big|_{t=0} = 0. \tag{12}$$

Equation (11) can be rewritten in equivalent form for the deformation parameter $\delta(t)$, which "moves" in macroscopic elastic-adhesion potential, U, under the action of "inertial" force F_i (Arnold, 2003):

$$\frac{4}{3}\pi a^3 \rho_p \frac{d^2\delta(t)}{dt^2} = F_i - \frac{\partial U}{\partial \delta}, \; F_i = \frac{4}{3}\pi a^3 \rho_p \left(\frac{d^2 z_s(t)}{dt^2} + \frac{d^2 \Delta a(t)}{dt^2} \right), \tag{13}$$

where the adhesion potential $U(\delta)$ is given by

$$U(\delta) = U_0 \left[\frac{2}{5}\left(\frac{\delta}{\delta_0}\right)^{5/2} - \frac{\delta}{\delta_0} \right], \; U_0 = \frac{4}{3}\sqrt{a}\, E^* \delta_0^{5/2}. \tag{14}$$

Numerical integration of the eqns. (11) or (13) shows that under the action of a ns-laser pulse the particle may perform oscillations with a period of a few nanoseconds (Luk`yanchuk, 2001, 2002 d; Arnold, 2002 a). This period decreases almost linearly with the particle size. These oscillations are due to the existence of natural resonance frequency for the particle on surface system. Corresponding approximate expressions are given in (Arnold, 2002 a, b, 2003).

Different conditions for the particle removal can be written from the force and energetic considerations (Bäuerle, 2000, Arnold, 2002 a). Force criterion considers a "static" condition for the particle removal (inertial force exceeds the Van der Waals force of attraction). The energy criterion assumes, that the particle is "thrown out" from a potential hole either at the expense of critical deformation, $\delta \le 0$, or by gaining sufficient kinetic energy to overcome the potential barrier (Lu, 1999).

These detailed calculations were performed for SiO_2 particles of different sizes on different substrates (Luk`yanchuk, 2003, 2004). It is interesting to compare the results of calculations for the threshold fluence with experimental data, see in Fig. 5.

One can see four curves in each picture in Fig. 5. The upper curve is calculated from the 1-D theory, and the curves below are calculated from the 3-D theory. The curve annotated "Mie" utilizes the parameters S_0 and r_0 found from the Mie theory, and the two lowest "POS" curves utilize parameters S_0 and r_0 calculated from the POS-theory for two different refractive indexes $n = 1.5$ (SiO_2 - particle) and $n = 1.6$ (Polystyrene particle). Solid stars present experimental data. It is clear from Fig. 5 that the POS-theory is much closer to the experiment than the 1-D model. Nevertheless, some problems still exist. In the range of 1 μm particles the theory is quite close to the experiment. For bigger particles e.g. 2.5 μm the experiment shows a threshold several times smaller.

In Fig. 6 we present calculations of the maximal surface temperature at threshold fluences for cleaning of SiO_2 particles from different substrates.

We were looking for resources of the theory to predict a diminishing laser cleaning threshold. One way is related to a more precise description of laser cleaning within the frame of 3-D model (detailed description of the intensity

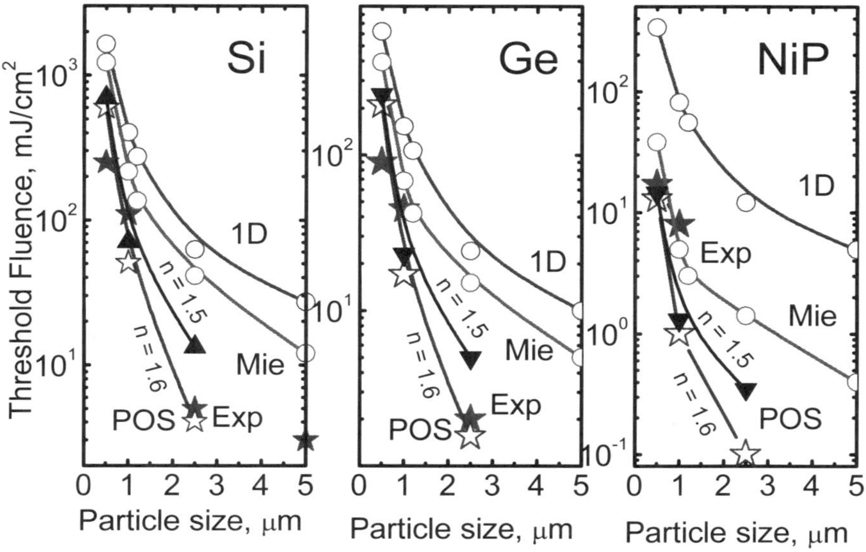

Fig. 5. Theoretical and experimental results of the threshold laser fluences for SiO_2 particles on Si, Ge and NiP substrates. Excimer laser 248 nm, pulse duration 23 ns.

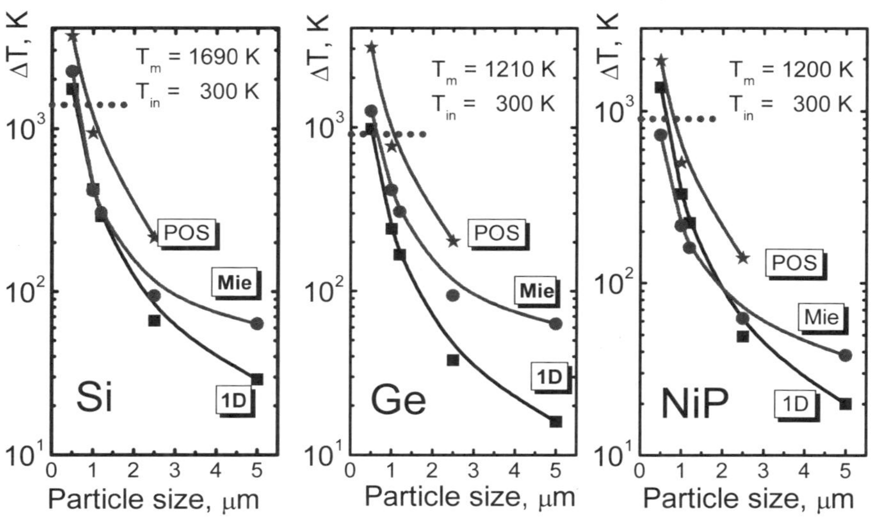

Fig. 6. Maximal surface temperature at threshold fluences, which were presented in Fig. 5.

distribution, solution of non-linear heat equation, etc.). Another way is to look for additional mechanisms which can enhance dry cleaning.

We started with a thorough examination of the scattering problem. It is clear that distributions (3) are simplified. In reality, moving from the center to the particle periphery one can see many oscillations in the field intensity, see Fig. 4. These oscillations are quite general, they exsist even within the frame of the Mie theory. The only difference is that with the Mie theory peaks of intensity do not exceed the input intensity (enhancement factor is smaller than 1). In contrast, it follows from the POS-theory that in some places the enhancement factor can be higher than 1. In Fig. 4 one can see that it arises at the distance $r = 1.9\ a$. This intensity produces a ring of enhanced intensity, which can be seen in distribution shown in Fig. 7.

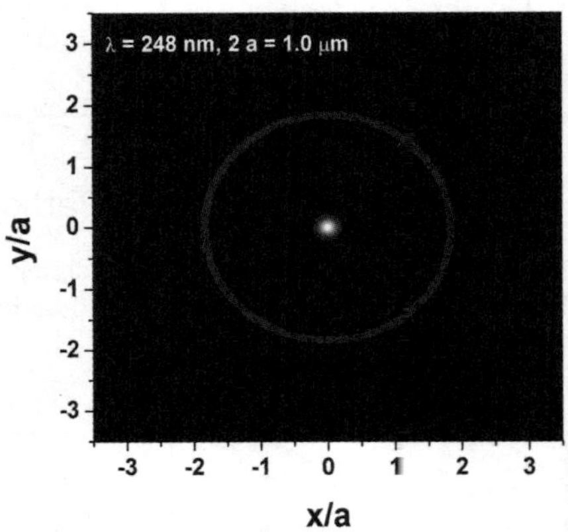

Fig. 7. Calculated topography of rings that correspond to regions with enhancement factor bigger than 1 for a 1.0 μm PS particle on GeSbTe (GST) substrate. The laser light is unpolarized and is incident normally on the substrate.

We decided to check theoretical predictions experimentally by "writing" the intensity distributions on the surface of phase change material (Wang, 2004 a, b). By controlling the laser fluence, we were able to choose the condition when intensity within the ring exceeds the threshold necessary for melting (it is a conventional way to write CDs and DVDs on phase-change material). With increasing laser intensity we can see two rings (second ring at $r = 2.5\ a$, see in Fig. 4). For a particle with a different size the position of the rings (and even the number of rings) may change. This was a direct way to check the predictions of the POS-theory. The details of these experiments are given in (Wang, 2004 a, b).

In Fig. 8 the pattern which is formed around of 1 μm Polystyrene particle is shown. The position of this ring coincides with that shown in Fig. 4.

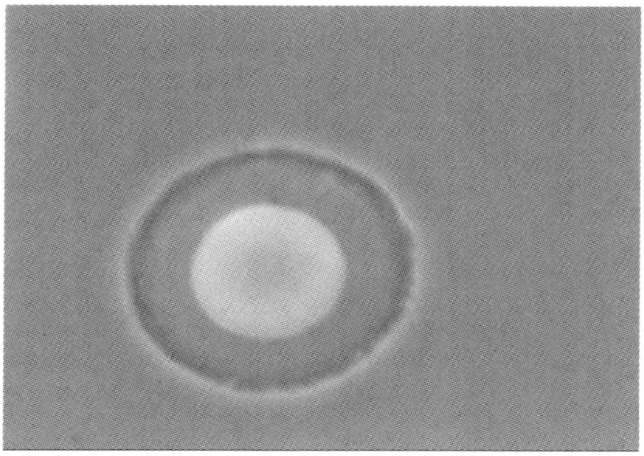

Fig. 8. The ring formed around an un-removed 1.0 μm polystyrene particle. Laser light is incident normally on the substrate. The ring is concentric for an isolated particle. The laser fluence used is below the film damage threshold.

Another example of intensity distribution is shown in Fig. 9 for a 2 μm polystyrene particle. For this case two rings at $r/a = 1.3$ and $r/a = 1.55$ have enhancement bigger than 1, thus they can be seen without background melting of the film. Another two rings with $r/a = 2.05$ and $r/a = 2.35$ have enhancement slightly below 1, nevertheless these rings can also be written because of the high contrast with the surrounds in the intensity distribution. Positions of all the rings for 2 μm Polystyrene particle on GeSbTe (GST) substrate are shown in Fig. 10.

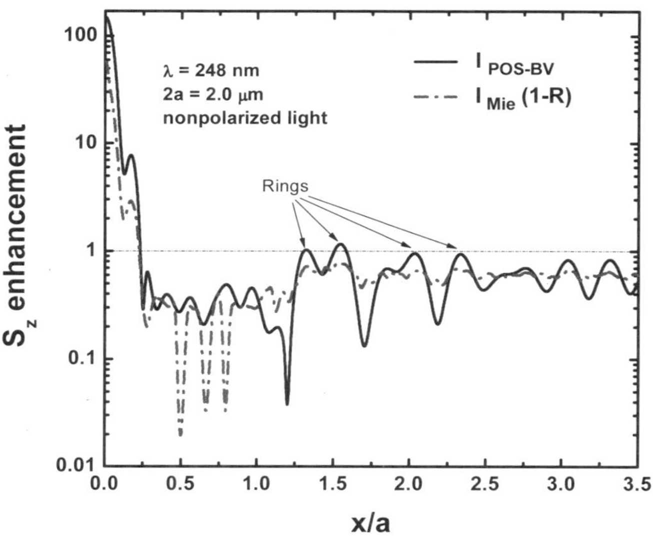

Fig. 9. Calcula-ted S_z intensity fields on substra-te GST surface for a 2.0 μm PS particle. The laser light is nonpolarized and the incident is normally on the substrate.

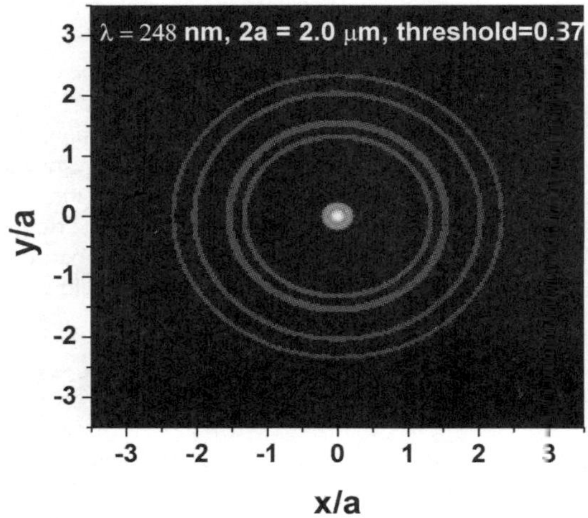

Fig. 10. Multiple rings formed around a single PS particle illuminated by a 248 nm laser pulse. Topography shows rings with enhancement higher than 0.87.

The results of our experiments are presented in Fig. 11. Laser fluence was slightly above the threshold, thus on the whole substrate one can see the background melting. Nevertheless, the rings are clearly detected. Their numbers and positions coincide well with those shown in Figs. 9 and 10.

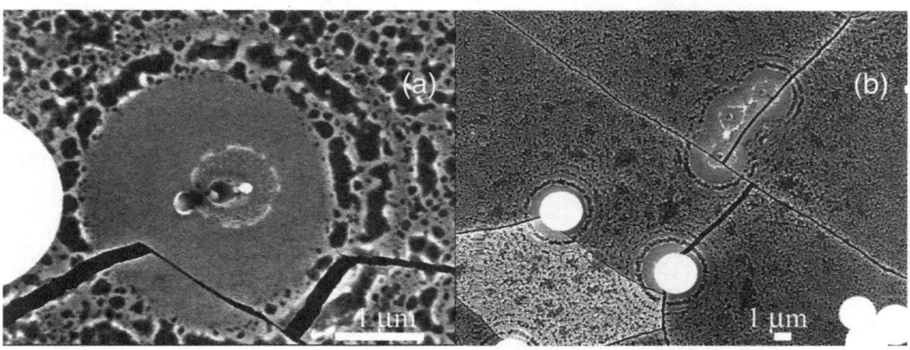

Fig. 11. Multiple rings formed around a single (a) and array-arranged (b) 2.0 μm PS particles illuminated by a 248 nm laser pulse at a laser fluence of 14 mJ/cm^2. The scale bar is 1.0 μm.

Another ability to verify POS-theory arises from the investigation of effects with incident radiation. One effect is trivial; the central nanodent under the particle (area with highest intensity) moves to the periphery with increase of the incidence angle, see in Fig. 12. This displacement follows simple geometrical relation, see Fig. 13.

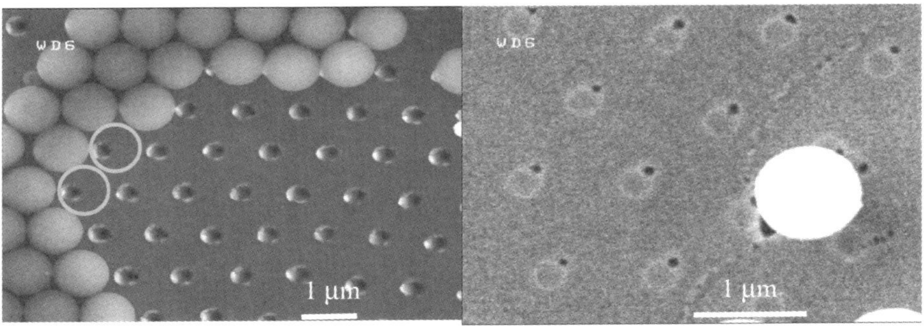

Fig. 12. SEM images of nanodent structures formed on the GST films under the removed particles after irradiation with one laser pulse at different angles of incidence (a) $\alpha = 30°$, (b) $\alpha = 60°$, respectively. Laser fluence is 7.5 mJ/cm^2. Scale bar is 1.0 µm. Circles in (a) illustrate particle position before removal. White shadows in (b) are around the contact area of removed particles.

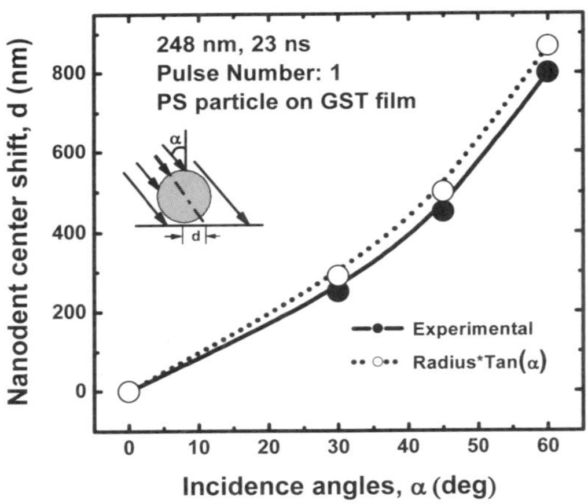

Fig. 13. Displacement of the nanodent center as a function of incidence angle. The solid line presents experimental data and the dotted line presents the value of $a\,Tan(\alpha)$ from conventional geometrical optics.

As for the distribution of the intensity in the periphery following from POS-theory, one can see two effects: the first, with incident radiation "rings" becomes "elliptical" and the second – laser intensity along these "ellipses" becomes inhomogeneous. While the ring line presents a homogeneous "canal embankment", the elliptical line presents inhomogeneous "chain of hills". This "chain of hills" structure is the most pronounced with close to threshold intensity and less pronounced with higher intensity, see the example in Fig. 14.

The "chain of hills" structure was found in experiments (Wang, 2004, 2005), see an example in Fig. 15. From the given examples one can see that POS-theory is

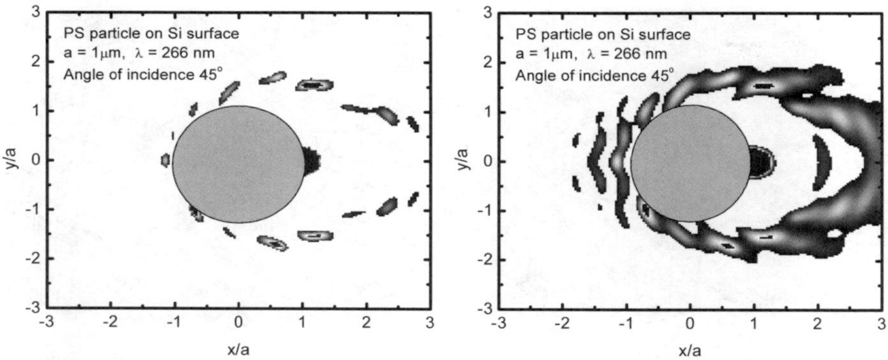

Fig. 14. Contour plot of enhanced intensity the value of which is higher than critical. The particle contour is also presented in the central part. On the border of the particle one can see the main maximum in the intensity distribution. "Chain of hills" structure for threshold intensity is shown in the left picture. Contour plot of intensity distribution with a slightly smaller enhancement is shown in the right picture.

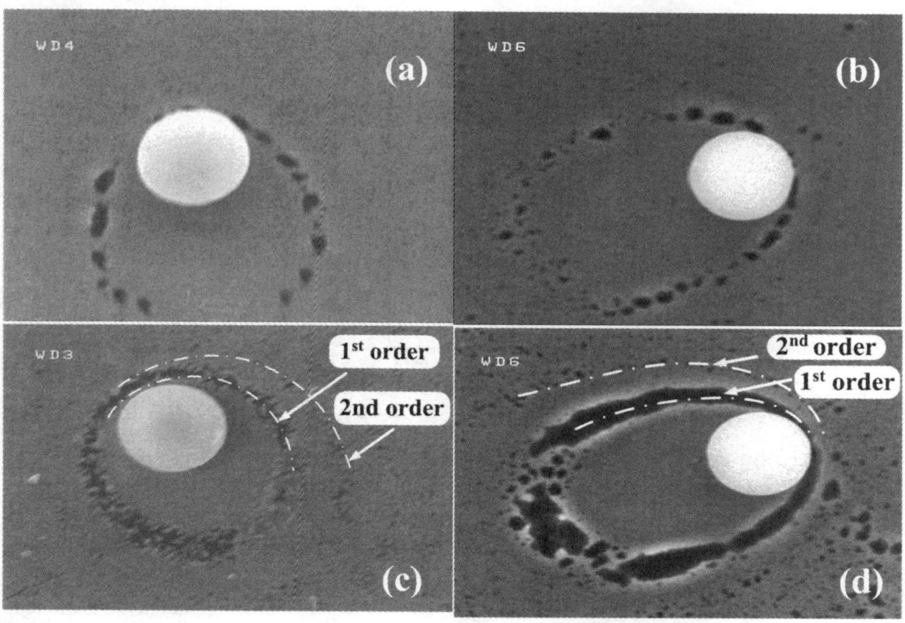

Fig. 15. SEM images of "chain of hills" patterns formed on GST film around un-removed particles at different angles and laser fluences, (a) $\alpha = 45^o$, 7.5 mJ/cm^2, (b) $\alpha = 60^o$, 7.5 mJ/cm^2, (c) $\alpha = 45^o$, 10.5 mJ/cm^2 and (d) $\alpha = 60^o$, 10.5 mJ/cm^2. The dashed lines show the different orders of the rings.

in good agreement with the experiment both near the central peak and on the periphery. The central peak effect and total energy conservation effect was taken into account with simplified profiles (3). It means that the influence of the periphery structure onto the maximal temperature under the particle is small.

At the same time thresholds in Fig. 5 were calculated on the basis of solution (4) of the linear heat equation. In reality thermal properties of materials depend on temperature, thus one should solve the nonlinear heat equation. An example of this solution (Luk`yanchuk, 2002 d) shows that the "nonlinear" temperature for Si can be typically 25% higher than the "linear" one. It leads to a reduction in the laser cleaning threshold fluence. It means that nonlinear effects are more pronounced for small particles, which need a higher temperature at threshold fluences, eg. see in Fig. 6. Meanwhile the problem with Si and Ge substrates exists also for bigger sizes, which demonstrate lower experimental values for the threshold fluence than theory predicts In the next Section we want to demonstrate that necessary corrections may follow from acoustic effects.

3. Acoustic Effects in Laser Cleaning

Acoustic effects in laser cleaning have been discussed previously in a few papers (Kolomenskii, 1991 a, b, 1993 a, b, 1995, 1998, Pleasants, 2004). The acoustic effects in the 1-D case, which were discussed in the paper of Pleasants, are probably important for weakly absorbing materials. There are more important effects related to formation of surface acoustic wave (SAW), for our purpose, which were discussed in the papers of Kolomenskii.

The generation of SAWs can be modeled by a pressure pulse, $p(r,t)$, acting on the free surface of the solid. Assuming a Gaussian distribution

$$p(r,t) = p_0 \exp\left[-\frac{r^2}{r_a^2}\right] \exp\left[-\frac{t^2}{t_0^2}\right], \tag{15}$$

the normal velocity component $v(r,t)$ in the excited SAW pulse can be described in a linear approximation by the expression (Kolomenskii, 1998)

$$v(r,t) = \frac{c_R}{4}\left(\frac{r_a^2}{2br}\right)^m \frac{r_a c_R t_0}{b^2} \frac{p_0}{\rho c_t^2} \Gamma \gamma \Phi_m\left(\frac{c_R t - r}{b}\right), \tag{16}$$

where the shape of the SAW is given by the function

$$\Phi_m(\xi) = \int_0^\infty k^{1+m} \exp\left(-\frac{k^2}{4}\right) \cos\left(k\xi + \frac{\pi m}{2}\right) dk\,, \tag{17}$$

and the characteristic wavelength of the SAW pulse is given by parameter

$$b = \sqrt{r_a^2 + c_R^2 t_0^2}\ . \tag{18}$$

The exponent $m = 0$ is used for a line shaped source (near acoustic field), and $m = 1/2$ for a point like source. The dimensionless combination of the elastic constants in (16) were introduced

$$\gamma = \left[\left(1 - c_R^2/c_l^2\right)/\left(1 - c_R^2/c_t^2\right)\right]^{1/4}, \tag{19}$$

$$\Gamma = \left\{\left(c_t^2 - c_R^2/2\right)\left[\left(c_t^2 - c_R^2\right)^{-1} + \left(c_l^2 - c_R^2\right)^{-1}\right] - 2\right\}^{-1}, \tag{20}$$

where $c_{t,l,R}$ are the propagation velocities of the transverse, longitudinal, and Rayleigh waves. For different solids the product $\Gamma \times \gamma$ ranges from 0.1 to 1. For the Si substrate the typical values are $c_R = 3.4 \cdot 10^5$ cm/s, $c_t/c_R =$ 1.10, $c_l/c_R =$ 1.59, thus $\Gamma\gamma$ = 0.74. Kolomenskii estimated the SAW intensity for a line-shaped source with $m = 1/2$ and pulse pressure $p_0 = 10^8$ Pa. Normal velocity and acceleration for a SAW with $r_0 = 7\ \mu\text{m}$ and $t_0 = 10\ \text{ns}$ are shown in Fig. 16(a). These are quite close to the results published by Kolomenskii.

For a line-shaped source the SAW profiles (in the linear approximation) do not depend on the distance. Values of acceleration are sufficient to remove 0.4 μm Al_2O_3 particles from the Si substrate, in agreement with the experiment. Remember that Kolomenskii discussed the situation when a SAW was created by a line focused laser pulse.

However if we discuss the point source, then one can see from (16) the amplitude of SAW is reduced with distance as $1/\sqrt{r}$. It follows from the total conservation of the energy flux. The flux density varies with distance as $\dfrac{1}{2\pi r}$. As the acoustic energy is proportional to v^2 it means that velocity (and acceleration) amplitudes vary as $1/\sqrt{r}$. These quantities are shown in Fig. 16(b) (other parameters are the same as in Fig. 16(a). One can see that for a distance bigger than 7 μm acceleration amplitude is insufficient for the removal of the same particles. With a point source the SAW shape becomes asymmetrical. The amplitude of the velocity for the leading edge is higher than for the falling edge. However with higher value of p_0 one can consider the following situation when a sufficiently intensive SAW is created by the neighboring particle.

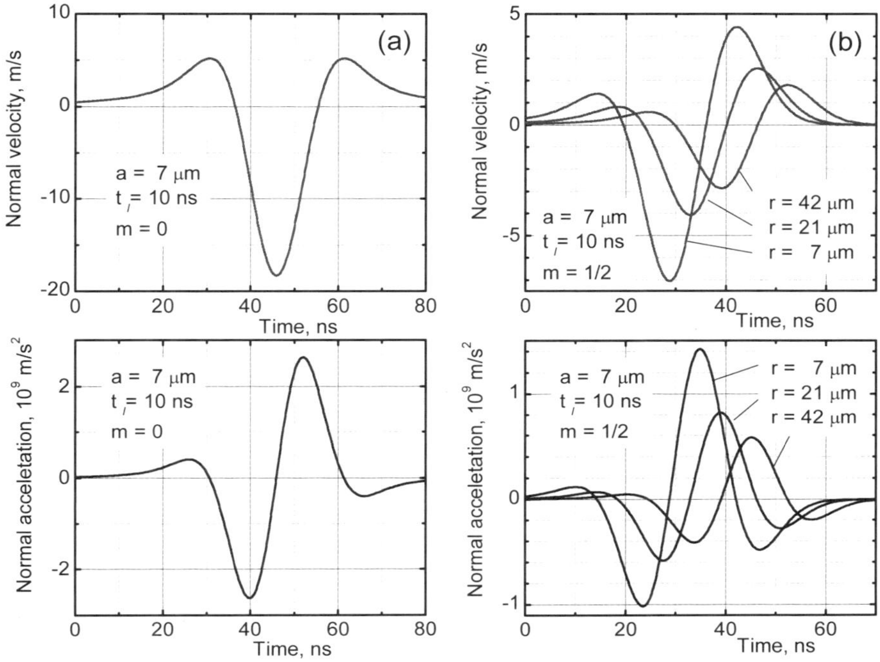

Fig. 16. Normal velocity and acceleration for a line-shaped source (a) and for a point source at different distances (b).

We can roughly estimate the effect of convergent acoustic wave, replacing in (15) the pressure profile by

$$p(r,t) = p_1\left\{\exp\left[-\frac{r^2}{r_a^2}\right] - \exp\left[-\frac{r^2}{(r_a-\delta_r)^2}\right]\right\}\exp\left[-\frac{t^2}{t_0^2}\right], \qquad (21)$$

where δ_r is the width of the ring. It does not practically change the pulse shape if $\delta_r << c_R t_0$ but the amplitude strongly depends on the position *r*, near the center it can be sufficiently big, at least comparable with quasistatic thermal expansion.

To illustrate the basic effect let us discuss distribution (21) with a narrow ring, $\delta_r << r_a$, and the same total energy as in (15), i.e. $p_1 \approx p_0 r_a / 2\delta_r$. Examples are shown in Fig. 17.

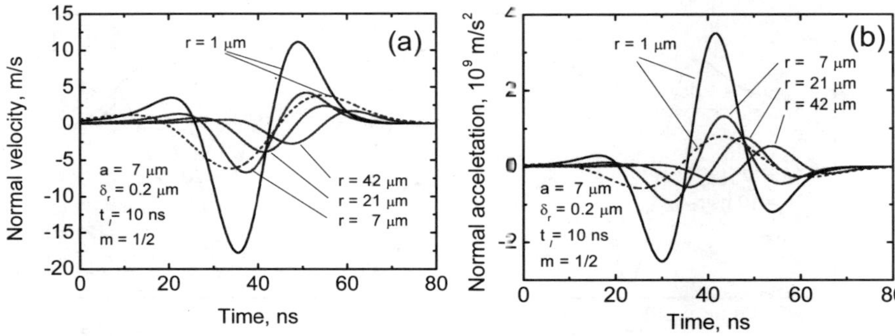

Fig. 17. Normal velocity (a) and acceleration (b) for a ring-shaped source with radius $a = 7$ μm and width $\delta_r = 0.2$ μm. Other parameters are the same as in Fig. 16. Dot lines present the acoustic wave which came from the region of homogeneous heating. For this case $a = 42$ μm and $\delta_r = 41$ μm. Total energy is the same as in Fig. 16.

One can see that the divergence wave out of ring is practically the same as in Fig. 16, while the convergent wave produces enhanced velocities and accelerations. Other acoustic waves can appear from the area of homogeneous heating out of the "shadow region", see distribution of intensity in (3). This case also can be estimated from the ring-shaped profile (21) if one puts $r_a >> c_R t_\ell$ and $\delta_r = r_a - r_{sh}$, where r_{sh} is the radius of the shadow region. Example of this pulse is also presented in Fig. 17.

Formula (16) was written for a particular case of the far field, thus we cannot apply this formula for small r values. It is clear from the energy conservation, that the limiting increase in the amplitude of acceleration under the particle can be estimated as $(ka)^{1/2}$, where $k = 2\pi/\lambda_R$ is the wave vector of the Rayleigh wave. This estimation is applicable just for the "big ring" with $a >> c_R t_\ell$. The effects of a convergent cylindrical wave for this limited case were analyzed by Cielo, 1985, Günalp, 1989 and recently by Kolomenskii, 2005. However, in laser cleaning we are within the acoustic near-field, where all characteristic sizes are smaller than $c_R t_\ell$. A careful examination of the near-field problem is beyond the scope of the present study. We just want to draw attention that in the near field region the radius of the focal spot for a convergent wave can be significantly smaller than the wavelength of the SAW. Experimentally a focal spot size of less than 1/14 of one wavelength was recorded (De Rosny & Fink, 2002). Thus, one should expect that acoustic effects give a contribution to removal for the particles with a size above $\approx c_R t_\ell / 15\pi$. For a laser pulse of 23 ns it means the size (diameter) is bigger than

1.7 μm. It is also clear that acoustic effects are more pronounced for bigger particles.

4. Plasmonic Effect in Laser Cleaning of Small Metal Particles

As it was discussed in Sect. 2 it is not possible, by the theory presented, to clean small transparent particles by the dry laser cleaning method. The necessary temperature exceeds the melting and boiling temperatures, which leads to a change of removal mechanism to ablative cleaning. Thus, laser cleaning of transparent nanoparticles is performed by other methods (Song, 2004, Graf, 2004).

One of the reasons for difficulties with small particles is related to small optical enhancement. For 100 nm particles the enhancement is equal to one (see in Fig. 18). Sub 50 nm transparent particles generally produce a "shadowing effect" instead of an enhancement effect, under most kinds of laser irradiation.

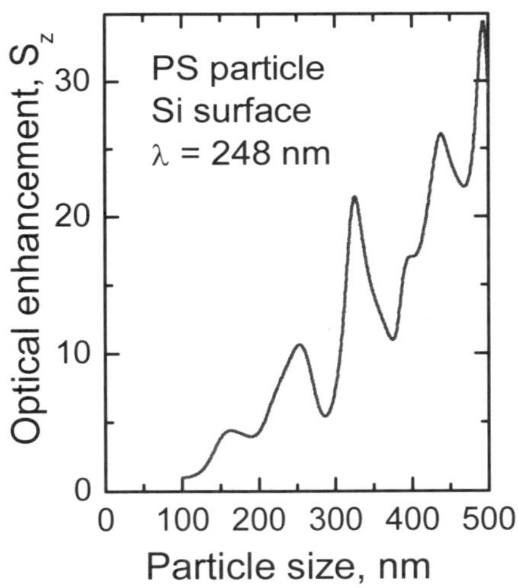

Fig. 18. Optical enhancement under Polysterene particle on Si substrate for radiation of 248 nm.

However, one can think about significant field enhancement for metallic nanoparticles, using radiation to excite a localized surface plasmon (Zayats, 2003). Typical enhancement of the field in the vicinity of the particle varied from several times to several tens, depending on the particle properties. There are three effects which one can expect in the discussed problem: 1) absorption of the particle by itself may play an important role; 2) enhancement of intensity on the substrate surface due to coupling to a surface plasmon can enhance substrate heating; and 3) pronounced dependence on the angle of incidence of the radiation may arise.

In the Mie theory the extinction, absorption and scattering cross sections are given by $\sigma_{ext} = \pi a^2 Q_{ext}$, $\sigma_{abs} = \pi a^2 Q_{abs}$, $\sigma_{sca} = \pi a^2 Q_{sca}$, where related efficiencies Q are presented by (Born & Wolf, 1999):

$$Q_{ext} = \frac{2}{q^2}\sum_{\ell=1}^{\infty}(2\ell+1)\mathrm{Re}(a_\ell + b_\ell),\ \ Q_{sca} = \frac{2}{q^2}\sum_{\ell=1}^{\infty}(2\ell+1)\left\{|a_\ell|^2 + |b_\ell|^2\right\},\tag{22}$$

$$Q_{abs} = Q_{ext} - Q_{sca},$$

where coefficients a_ℓ and b_ℓ are defined as

$$a_\ell = \frac{q_p\psi'_\ell(q_m)\psi_\ell(q_p) - q_m\psi_\ell(q_m)\psi'_\ell(q_p)}{q_p\varsigma'_\ell(q_m)\psi_\ell(q_p) - q_m\psi'_\ell(q_p)\varsigma_\ell(q_m)},\ \ q_m = k_m a,$$

$$b_\ell = \frac{q_p\psi'_\ell(q_p)\psi_\ell(q_m) - q_m\psi_\ell(q_p)\psi'_\ell(q_m)}{q_p\psi'_\ell(q_p)\varsigma_\ell(q_m) - q_m\psi_\ell(q_p)\varsigma'_\ell(q_m)},\ \ q_p = k_p a.\tag{23}$$

The wave vector for vacuum is indicated by $k_0 = 2\pi/\lambda$ and $q = k_0 a$. The wave vector of the radiation in the media is indicated by $k_m = 2\pi\sqrt{\varepsilon_m}/\lambda$, and for the particle $k_p = 2\pi\sqrt{\varepsilon_p}/\lambda$. The functions ψ_ℓ and ψ'_ℓ are expressed using the Bessel function (regular at $\rho = 0$), where prime indicates differentiation

$$\psi_\ell(\rho) = \sqrt{\frac{\pi\rho}{2}}\, J_{\ell+\frac{1}{2}}(\rho),\ \ \psi'_\ell(\rho) = \frac{\partial\psi_\ell(\rho)}{\partial\rho}.\tag{24}$$

The functions ς_ℓ and ς'_ℓ are expressed in a similar way through the Hankel function

$$\zeta_\ell(\rho) = \sqrt{\frac{\pi\rho}{2}}\, H_{\ell+\frac{1}{2}}(\rho),\ \ \zeta'_\ell(\rho) = \frac{\partial\zeta_\ell(\rho)}{\partial\rho}.\tag{25}$$

Plasmon resonances for small particle arise when

$$\mathrm{Re}\,\varepsilon_p + \frac{\ell+1}{\ell} = 0,\tag{26}$$

where $\ell = 1$ corresponds to dipole resonance, $\ell = 2$ to quadrupole, $\ell = 3$ to octopole, etc.

We shall discuss a particular case of gold particles, for which we performed experimental investigations. The optical properties of Au (bulk material from Palik, 1985-1998) are shown in Fig. 19. Using these values we can find the Drude's parameters ω_p and γ for each frequency ω. The Drude formula for dielectric permittivity is used as interpolation formula:

$$\varepsilon(\omega) = \varepsilon' + i\varepsilon'' = 1 - \frac{\omega_p^2}{\omega^2 + \gamma^2} + i\frac{\gamma}{\omega}\frac{\omega_p^2}{\omega^2 + \gamma^2}. \qquad (27)$$

The variation of collision frequency versus particle size is performed by standard renormalization $\gamma \to \gamma_\infty + v_F/a$ (Kreibig & Vollmer, 1995), where γ_∞ is the collision frequency for the bulk material, and v_F is Fermi velocity.

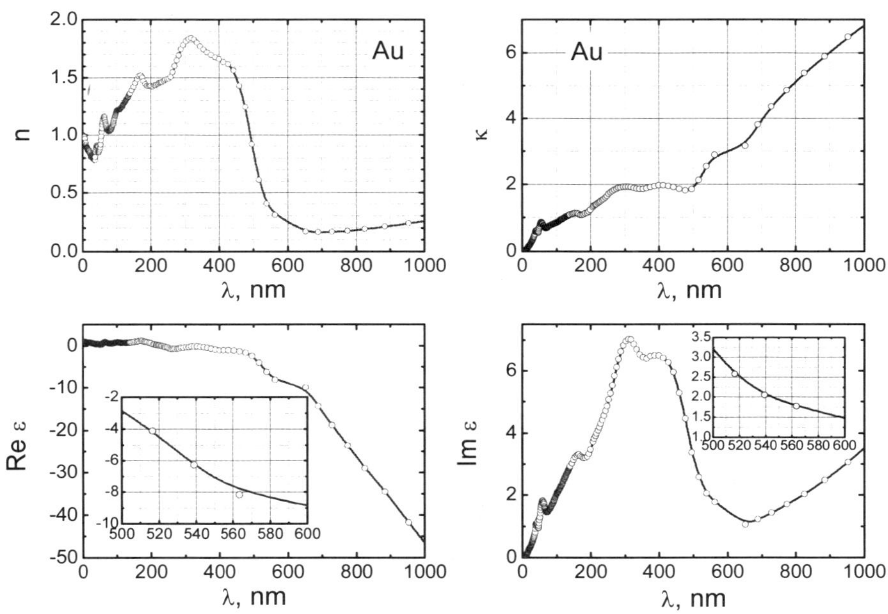

Fig. 19. Optical properties of bulk Au according to Palik, 1985-1998.

From Fig. 19 it follows that the necessary condition for plasmon resonance, $\mathrm{Re}\,\varepsilon < 0$, is fulfilled everywhere with $\lambda > 227$ nm. Renormalized optical constants were used to calculate the extinction, scattering and absorption cross-sections for spherical Au particles of 20 nm radius as a function of the wavelength λ, see in

Fig. 20. One can see a plasmon resonance at $\lambda = 498$ nm. This resonance corresponds to dipole excitation that can be seen clearly from the field distribution in Fig. 21. The left peak at $\lambda = 207$ nm is also a dipole Mie resonance but without the formation of localized plasmon, because of the condition $\mathrm{Re}\,\varepsilon > 0$. For the case of particles with weak dissipation one can see quadrupole and octopole plasmon resonances (Luk'yanchuk & Tribelsky, 2005).

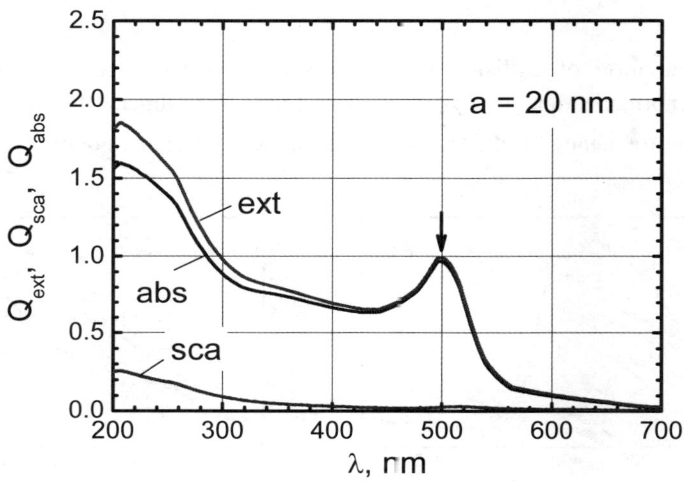

Fig. 20. The extinction, scattering and absorption cross-sections for gold particle of 20 nm radius sphere as a function of the wavelength λ.

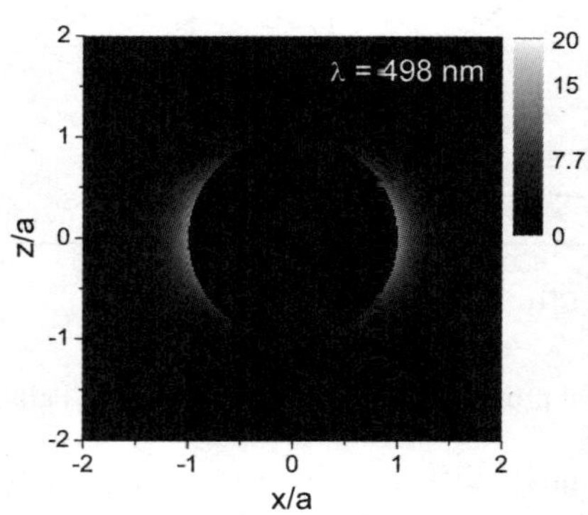

Fig. 21. The distributions of field E^2 around the Au particle at exact dipole resonance with $\lambda = 498$ nm.

From Fig. 21 one can see that the highest field enhancement arises on the particle "equator", while for laser cleaning one needs enhancement under the particle. The natural way to increase enhancement under the particle is to change the angle of incidence of the incident radiation. To illustrate this effect we performed calculations with the particle-on-surface theory. The results of these calculations are shown in Fig. 22.

One can see from Fig. 20 that the particle is mainly absorbing for radiation of 532 nm ($Q_{abs} = 0.43$). It means that the free particle can be heated efficiently. The corresponding temperature rise can be estimated from the energy balance

$$T = \frac{3}{4}\frac{Q_{abs}\Phi}{c\rho a}, \qquad (28)$$

where Φ is the (homogeneous) laser fluence. For example for $\Phi = 50$ mJ/cm^2 formula (28) yields a very high temperature above 6000 K for a 40 nm ($a = 20$ nm) gold particle. Although this temperature cannot be reached for the particle on the surface due to heat conductivity of the substrate, it is clear that effects related to the direct heating of the particle may play an important role. Heating of the substrate arises due to radiation enhancement under the particle and due to thermal contact of the particle and substrate.

We performed an experimental study of laser cleaning of sub-50 nm gold particles from Si substrate by 7 ns laser pulse of 532 nm radiation and found that these nanoparticles can be efficiently removed.

In the experiment an n-type polished silicon wafer was used as the substrate. The sample was cleaned with acetone in an ultrasonic bath for 5 min. After that, the sample was rinsed with DI water and dried with N_2. A suspension of 40 nm gold spherical particle with 5% size deviation was used. The particles were applied to the Si surface by a small dispenser. The solvent was dried due to evaporation and consequently the self-assembled particles were left on the surface. Q-switched 2rd harmonic of Nd:YAG laser (BMI industry Series 500) was used as the laser source. The wavelength was 532 nm with a pulse duration of 7 ns. The laser spot size is about 6 mm. The repetition rate varied from 1 to 10 Hz. The output laser beam was linearly polarized.

Fig. 23 shows the SEM images of 40-nm gold nanoparticles left on the Si substrate surface before (a) and after (b) 300 laser pulses irradiation at a laser fluence of 50 mJ/cm^2 and at an incident angle of 45°. It can be seen from Fig. 23 (a) that the as-deposited mask of gold nanoparticles on Si surface have arranged in different forms: region 1 is free from the particles; in region 2 particles form a monolayer, in region 3 particles form a multiplayer, and in region 4 one can see isolated particles.

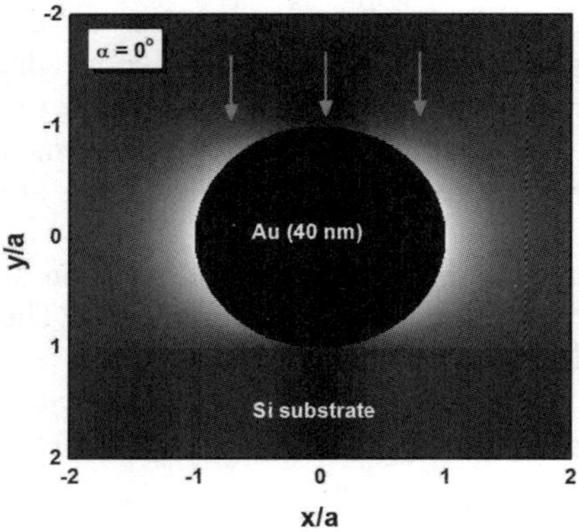

(a) Intensity distribution in xz-plane $\alpha = 0^{\circ}$

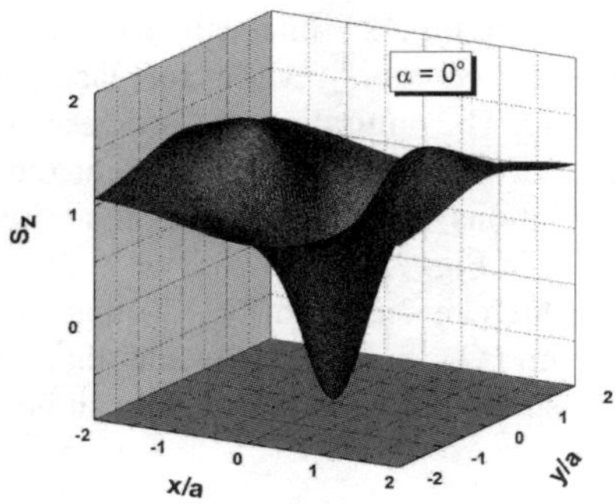

(b) Intensity distribution under particle in *xy*-plane $\alpha = 0^{\circ}$

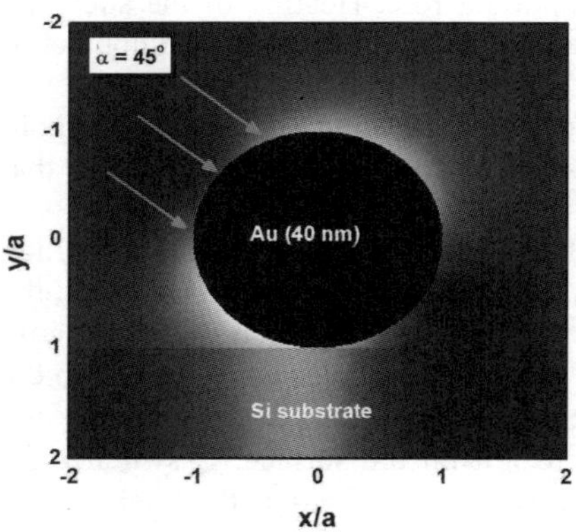

(c) Intensity distribution in *xz*-plane $\alpha = 45^{\circ}$

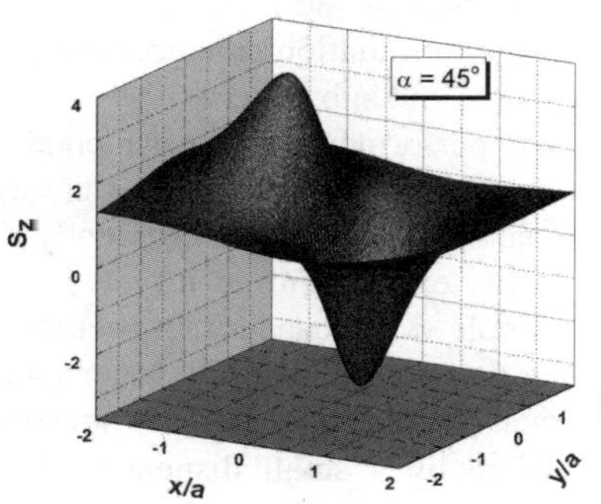

(d) Intensity distribution under particle in *xy*-plane $\alpha = 45^{\circ}$

Fig. 22. Contour plots for intensity distribution in *xz*-plane (a, c) and normalized intensity (z-component of the Poynting vector) under the 40-nm gold particle on n-Si surface (b, d) at different incidence angles: $\alpha = 0^{\mathrm{O}}$ (a, b) and $\alpha = 45^{\mathrm{O}}$ (c, d). One can see that the field enhancement under the particle with $\alpha = 45^{\mathrm{O}}$ is about two times higher than that for $\alpha = 0^{\mathrm{O}}$ due to a more efficient coupling.

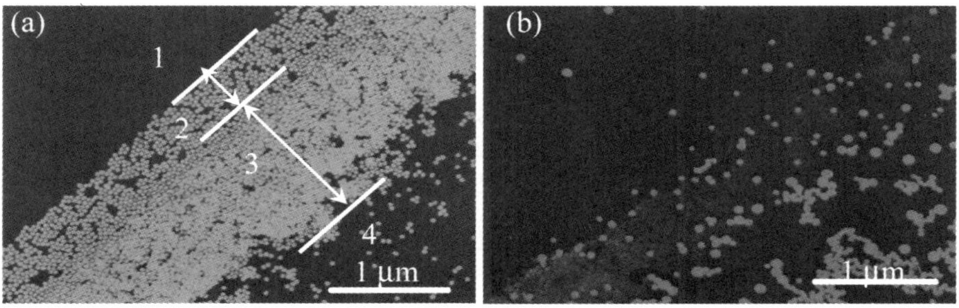

Fig. 23. SEM images of 40 nm gold nanoparticles on n-Si substrate surface before (a) and after (b) 300 pulses (532 nm, 7 ns) at a laser fluence of 50 mJ/cm^2 and an incidence angle of 45^0.

After laser cleaning, one can see from Fig. 23 (b) that the majority of gold nanoparticles were removed from the surface (The total cleaning efficiency was estimated to be ~80% in this case). Some particles removed from regions 2 and 3 were found to re-deposit in regions 1 and 4, more in region 4 than 1.

As it was shown in Fig. 22, using a large, non-normal angle of incidence for the light, the peak position in the intensity field is not at the contact point between the particle and surface, but is at a point offset up the side of the particle. Thus, the substrate thermal deformation force repels the particle away from the surface at some angle. This could explain why most of the removed particles were redeposited in region 4.

Another phenomenon seen in Fig. 23(b) is the appearance of big gold nanoparticles after laser irradiation. An upper limit in size of a ~200 nm nanoparticle (an aggregation of ~5 individual nanoparticles) has been observed in our experiments. As we mentioned above it could be due to the considerable heating of the particles up to the melting temperature. One should remember also that nanoparticles have a lower melting temperature as compared to the bulk material. The lower melting temperature of nanoparticles is due to the large ratio of surface atoms to inner atoms, in which case the surface energy of the surface atoms is reduced relative to the bulk material. The melting point of bulk gold is about 1064 °C while it is 600 – 800 °C for several-tens-of-nanometer sized gold nanoparticles (Buffat & Borel, 1975). In experiments, it was found that "multilayered" gold nanoparticles (particles in region 3) are more readily melted to form larger sized nanoparticles than the particles in the "monolayer" region (particles in region 2).

Fig. 24 shows the SEM image of the damage patterns under multilayer particles (region 3) after 300 pulses of laser irradiation at a laser fluence of 50 mJ/cm^2, and at an incidence angle of 45°. The damage sites are in irregular shapes with a lateral size in a range from 20 to 40 nm. In contrast, no damage was observed under monolayer particles in region 2. The reason could be attributed to the electromagnetic (EM) energy coupling between different layers of gold nanoparticles which lead to a more efficient energy coupling with the substrate. From the viewpoint of nanofabrication, this small gold nanoparticle aggregation permits the production of an array of nanostructures with feature size smaller than 40 nm by single/multiple laser shots. Compared to other techniques such as E-beam processing and near-field scanning optical microscope (NSOM), nanopatterning with small particles is a rather cheap and easymethod to implement in a parallel manner. In order to obtain sub-50 nm pattern features by this method, it should be emphasized that a metal plasmon particle should be used. Meanwhile, an appropriate laser source should be employed whose frequency must be close to the plasmon resonance frequency of the metal particles.

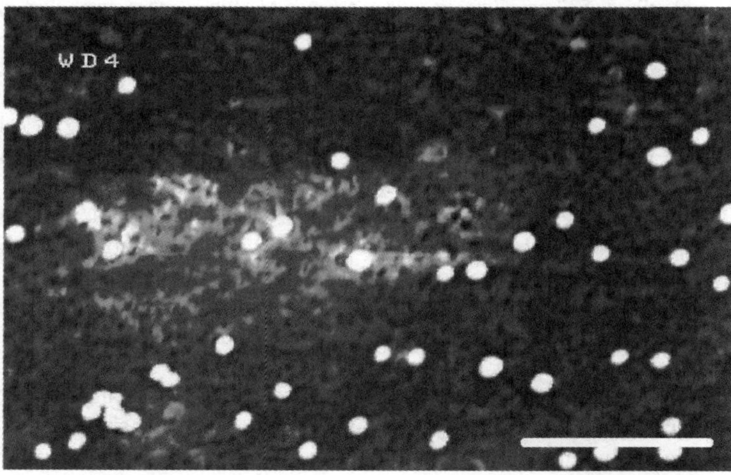

Fig. 24. SEM image of damage patterns under 40-nm gold nanoparticles after 532 nm laser processing. A total of 300 laser pulses were used and the laser beam is incident on sample at 45-degree. The used laser fluence is 50 mJ/cm^2.

Fig. 25 presents the cleaning efficiency as a function of angle of incidence for the laser irradiation, for gold particles. The cleaning efficiency increases smoothly with incidence angle. This tendency is quite different from that in laser cleaning of transparent particles on the surface, in which a steep decline of cleaning efficiency appears with increasing incidence angles (Zheng, 2001).

Effective removal of metallic particles (Cu) using large-angle-of-incidence laser cleaning was previously discussed in the paper of Lee, 2000. However the authors demonstrated this effect for "large" particles (diameter about 10 μm) and they also suggested a different mechanism for the increased efficiency – a shadowing effect.

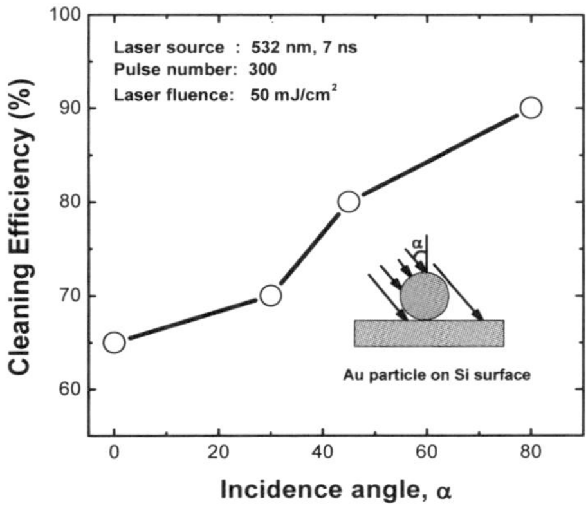

Fig. 25. Cleaning efficiency as a function of incidence angle.

5. Conclusion

The present work was devoted to the discussions of possible acoustic and plasmonics effects in dry laser cleaning. The results can be summarized as follows:

1) Calculations of the intensity distribution were performed on the basis of an exact solution of Maxwell equations for the problem "particle on surface". These calculations were verified experimentally by writing the real intensity distribution on the surface of a phase-change material. The theoretical results are in a good agreement with experiment.

2) Field enhancement and its consequences play a major role in the dry laser cleaning process. However some effects related to intensity distribution on the periphery of the particle were not taken into account in previous calculations. For example in the case of a transparent particle and normal incidence of the radiation, the intensity has near-field focusing near the center and additional ring(s) symmetrically around the basic maximum.

3) We consider that the ring of intensity produces a cylindrical convergent surface acoustic wave, which can enhance the particle removal for sufficiently "big" particles (above 2 μm).

4) Another effect is discussed for metallic particles. The intensity on the surface under a metallic particle typically diminishes due to the screening effect, in contrast to transparent particles, which act as a near-field lens. Nevertheless, if one uses radiation frequencies near a surface plasmon resonance, the conditions for the efficient coupling of the radiation with a metallic surface can be

provided. This plasmonic effect permits "laser" cleaning of metallic nanoparticles from a metallic surface. We demonstrated experimentally the ability to clean 40 nm gold particles from a Si substrate.

Acknowledgement

We wish to thank Prof. S. I. Anisimov, Dr. N. Arnold and Dr. Al. Kolomenskii for discussions. We are also thankful to the Editor of this book, Prof. D. M. Kane for her great help and patience. B. L. is thankful to Russian Basic Research Foundation (grants 04-02-17225, 04-02-16972) and discussions with Prof. V. G. Mikhalevich.

References

Arnold N., *Dry laser cleaning of particles by nanosecond pulses: Theory*, Chapter 2 in "Laser Cleaning", Ed. by B. Luk`yanchuk (World Scientific, New Jersey, London, Singapore 2002a), pp. 51-102.

Arnold N., Schrems G., Mühlberger T., Bertsch M., Mosbacher M., Leiderer P., Bäuerle D., *Proc. SPIE,* vol. 4426, pp. 340-346 (2002b)

Arnold N., *Theoretical description of dry laser cleaning*, Appl. Surf. Sci., vol. **208–209**, 15 (2003)

Arnold N., Schrems G., Bauerle D., *Ablative thresholds in laser cleaning of substrates from particulates*, Appl. Phys. A 79, 729 (2004)

Barber P. W., Hill S. C., *Light Scattering by Particles: Computational Methods*, (World Scientific, Singapore 1990)

Bäuerle D., *Laser Processing and Chemistry*, 3^{d} ed., (Springer, Berlin, 2000)

Bobbert P. A., Vlieger J., *Light scattering by a sphere on a substrate*, Physica A, vol. **137**, 209 (1986 a)

Bobbert P. A., Vlieger J., Greef R., *Light reflection from a substrate sparsely seeded with spheres – comparison with an ellipsometric experiment*, Physica A, vol. **137**, 243 (1986 b)

Bohren C. E., Huffman D. R., *Absorption and Scattering of Light by Small Particles*, (John Willey & Sons, 1983)

Born M., Wolf E., *Principles of Optics*, 7-th Edition, (Cambridge University Press, 1999)

Buffat P., Borel J. P., *Size effect on the melting temperature of gold particles*, Phys. Rev. A **13**, 2287 (1975)

Cielo P., Nadeau F., Lamontagne M., *Laser generation of convergent acoustic waves for materials inspection*, Ultrasonics, 55, March 1985

Denk R., Piglmayer K., Bäuerle D., *Laser-induced etching and deposition of W using a-SiO_2 microspheres*, Appl. Phys. A **76**, 549 (2003)

De Rosny J., Fink M., *Overcoming the diffraction limit in wave physics using a time-reversal mirror and a novel acoustic sink*, Phys. Rev. Lett., **89**, 124301 (2002)

Graf J., Luk`yanchuk B.S., Mosbacher M., Hong M. H., Chong T.C., Leiderer P., *Optimization of the energy transfer medium in laser cleaning: a new strategy*, Abstracts of 4th International Workshop on Laser Cleaning, Sydney, Australia, 14-17 December 2004 http://www.physics.mq.edu.au/IWLC4/program.htm

Günalp N., Atalar A., *Response of acoustic imaging systems using convergent leaky waves to cylindrical flaws*, IEEE Trans. On Ultrasonics, vol. **36**, 507 (1989)

Hong M.H., Huang S.M., Luk'yanchuk B.S., Chong T.C., *Laser assisted surface nanopatterning,* Sensors and Actuators A-Physical, vol. **108**, 69 (2003)

Huang S. M., Hong M. H., Luk'yanchuk B. S., Lu Y. F., *Laser assisted nanofabrications on metal surfaces with optical near field effects*, Proc. SPIE, vol. **4760** (2002a)

Huang S. M., Hong M. H., Luk`yanchuk B. S., Zheng Y. W., Song W. D., Lu Y.F., Chong T. C., *Pulsed laser-assisted surface structuring with optical near-field enhanced effects,* J. Appl. Phys., (2002b)

Huang S. M., Hong M. H., Luk`yanchuk B., Wang Z. B., *Nanostructures fabricated on metal surfaces assisted by laser with optical near-field effects*, Appl. Phys. **A 77**, 293 (2003)

Kelley J. D., Hovis F. E., *A thermal detachment mechanism for particle removal from surfaces by pulsed laser irradiation,* Microelectronic Engineering **20**, 159 (1993)

Kerker M., *The scattering of Light*, (Academic Press, New York & London, 1969)

Kerker M., *Selected Papers on Light Scattering*, Proc. SPIE, vol. **951** (Part One), (1989), see Section 4 "*Optical Resonances*"

Kolomenskii A. A., Maznev A. A., *Observation of phonon focusing with pulsed laser excitation of surface acoustic waves in silicon*, JETP Letters, vol. **53**, pp. 423-426 (1991a)

Kolomenskii A. A., Maznev A. A., *Shaking the mechanical micro-particles off the silicon surface induced by surface acoustic-wave excited by laser-pulses,* Sov. Tech. Phys. Lett., vol. **17**, No. 13, pp. 62-66 (1991b)

Kolomenskii A. A., Maznev A. A., *Laser induced acoustic cleaning of surfaces from mechanical microparticles*, Bull. Acad. Sci. USSR, Phys. Ser., vol. **57**, No. 2, pp. 180-189 (1993a)

Kolomenskii A. A., Maznev A. A., *Phonon-focusing effect with laser-generated ultrasonic surface-waves*, Phys. Rev. B, vol. **48**, No. 19, pp. 14502-14508 (1993b)

Kolomenskii A. A., Maznev A. A., *Propagation of laser-generated surface acoustic-waves visualized by shake-off of fine particles,* J. Appl. Phys., vol. **77**, pp. 6052-6054 (1995)

Kolomenskii A. A., Schuessler H. A., Mikhalevich V. G., Maznev A. A., *Interaction of laser-generated surface acoustic pulses with fine particles: Surface cleaning and adhesion studies*, J. Appl. Phys., vol. **84**, No. 5, pp. 2404-2410 (1998)

Kolomenskii A. A., Jerebtsov S.N., Schuessler H. A., *Focal transformation of converging one-cycle surface acoustic waves excited by femtosecond laser pulses*, Optics Letters, **30**, 2005.

Kreibig U., Vollmer M., *Optical Properties of Metal Clusters* (Springer Verlag, Berlin, Heidelberg 1995)

Lee J. M., Watkins K. G., Steen W. M., *Angular laser cleaning for effective removal of particles from a solid surface*, Appl. Phys. A **71**, 671 (2000)

Lu Y. F., Song W. D., Ang B. W., Chan D. S. H., Low T. S., *A Theoretical Model for Laser Removal of Particles from Solid Surface,* Appl. Phys. A **65**, 9 (1997)

Lu Y. F., Zheng Y. W., Song W. D., *An energy approach to the modelling of particle removal by pulsed laser irradiation,* Appl. Phys. A **68**, 569 (1999)

Lu Y. F., Zhang L., Song W. D., Zheng Y. W., Luk'yanchuk B. S., *Laser writing of sub-wavelength structure on silicon (100) surfaces with particle enhanced optical irradiation,* JETP Letters, vol. **72**, 457 (2000a)

Lu Y. F., Zheng Y. W., Song W. D., *Laser induced removal of spherical particles from silicon wafers,* J. Appl. Phys. **87**, 1534 (2000b)

Luk'yanchuk B. S., Zheng Y. W., Lu Y. F., *Laser Cleaning of the surface: Optical resonance and near-field effects*, Proc. SPIE, vol. **4065**, 576 (2000)

Luk'yanchuk B. S., Zheng Y. W., Lu Y. F., *A new mechanism of laser dry cleaning*, Proc. SPIE, vol. **4423**, 115 (2001)

Luk'yanchuk B. S., Zheng Y. W., Lu Y. F., *Basic physical problems related to dry laser cleaning*, RIKEN Review, No. **43**, 37 (2002a)

Luk'yanchuk B. S., Zheng Y. W., Lu Y. F., *Particle on the surface: Basic physical problems related to laser cleaning,* Proc. SPIE, vol. **4426**, 284 (2002b)

Luk'yanchuk B. S., Huang S. M., Hong M. H., *3D effects in dry laser cleaning,* Proc. SPIE, vol. **4760**, 204 (2002c)

Luk'yanchuk B. S., Mosbacher M., Zheng Y. W., Münzer H. - J., Huang S. M., Bertsch M., Song W. D., Wang Z. B., Lu Y. F., Dubbers O., Boneberg J., Leiderer P., Hong M. H., Chong T. C., *Optical Resonance and Near-Field Effects in Dry Laser Cleaning*, Chapter 3 in "Laser Cleaning", Ed. by B. Luk`yanchuk (World Scientific, New Jersey, London, Singapore 2002d), pp. 103-178.

Luk`yanchuk B. S., Arnold N., Huang S. M., Wang Z. B., Hong M. H., *Three-dimensional effects in dry laser cleaning*, Appl. Phys. **A 77**, pp. 209-215 (2003)

Luk`yanchuk B. S., Wang Z. B., Song W. D., Hong M. H., *Particle on surface: 3D - effects in dry laser cleaning*, Appl. Phys. A **79**, pp. 747–751 (2004)

Luk'yanchuk B. S., Tribelsky M. I., *Anomalous Light Scattering by Small Particles and inverse hierarchy of optical resonances*, Collection of papers devoted to

memory of Prof. M. N. Libenson. The St.-Petersburg Union of the Scientists, Russia, pp. 101-117 (2005)
Morse P. M., Feshbach H., *Methods of Theoretical Physics*, vol. 2 (McGraw-Hill, 1953)
Mosbacher M., Münzer H. -J., Zimmermann J., Solis J., Boneberg J., Leiderer P., *Optical field enhancement effects in laser-assisted particle removal*, Appl. Phys. **A 72**, 41 (2001)
Münzer H.-J., Mosbacher M., Bertsch M., Zimmermann J., Leiderer P., Boneberg J., *Local field enhancement effects for nanostructuring of surfaces*, Journal of Microscopy, vol. **2002**, 129 (2001)
Münzer H.-J., Mosbacher M., Bertsch M., Dubbers O., Burmeister F., Pack A., Wannemacher R., Runge B.-U., Bäuerle D., Boneberg J., Leiderer P., *Optical near-field effects in surface nanostructuring and laser cleaning,* Proc. SPIE, vol. **4426**, 180 (2002)
Nowacki W., *Thermoelasticity* (Pergamon, Oxford 1962)
Palik E. D., *Handbook of optical constants of solids*, (Academic Press, Orlando, 1985-1998)
Pleasants S., Arnold N., Kane D. M., *Acoustic substrate expansion in modelling dry laser cleaning of low absorbing substrates*, Appl. Phys. **A 79**, 507 (2004)
Sokolnikoff I. S., *Mathematical Theory of Elasticity*, 2nd Ed., (McGraw-Hill, 1956)
Song W. D., Hong M. H., Luk'yanchuk B. S., *A method and apparatus for cleaning surfaces*, US patent No. 6,777,642 B2, August 2004
Stratton J. A., *Electromagnetic Theory*, (McGraw-Hill, New York & London, 1941)
Van de Hulst H. C., *Light Scattering by Small Particles*, (Dower Publ., New York, 1981)
Wang Z. B., Hong M. H., Luk'yanchuk B. S., Lin Y., Wang Q.F., Chong T. C., *Angle effect in laser nanopatterning with particle-mask*, J. Appl. Phys., vol. **96**, 6845 (2004 a)
Wang Z. B., Hong M. H., Luk`yanchuk B. S., Huang S. M., Wang O.F., Shi L.P., Chong T. C., *Parallel nanostructuring of GeSbTe films with particle-mask*, Appl. Phys. A **79**, 1603 (2004 b)
Wang Z. B., *Optical Resonance and Near-Field Effects: Small Particles under Laser Irradiation*, PhD Thesis, National University of Singapore, 2005, 234 pp.
Zayats A.V., Smolyaninov I. I., *Near-field photonics: surface plasmon polaritons and localized surface plasmons*, J. Opt. A: Pure Appl. Opt. **5**, S16 (2003)
Zheng Y. W., Luk'yanchuk B. S., Lu Y. F., Song W. D., Mai Z. H., *Dry laser cleaning of Particles from Solid Substrates: Experiments and Theory,* J. Appl. Phys. **90** (5), 2135 (2001)

Chapter 4

AXIALLY SYMMETRIC FOCUSING OF LIGHT IN DRY LASER CLEANING AND NANOPATTERNING

J KOFLER AND N ARNOLD

Institute for Applied Physics, Johannes Kepler University Linz, Altenbergerstrasse 69, 4040 Linz, Austria
johannes.kofler@jku.at and nikita.arnold@jku.at

The field enhancement by spherical particles plays a fundamental role in nanopatterning and in dry laser cleaning where such particles are used as model contaminants. We describe the general vectorial axially symmetric and strongly aberrated focusing by matching the solution of geometrical optics with a wave field built by the integral canonical for this topology, i.e., the Bessoid integral. For the focusing by transparent spheres with a few wavelengths in diameter, the results are much more compact and intuitive than the Mie theory. The results of the latter are reproduced with good agreement down to Mie parameters of about 30. Analytical expressions for the intensity on the axis and the position of the diffraction focus are derived.

1. Introduction

In this work we describe theoretically the focusing of light by transparent spheres with diameters of several wavelengths. Such focusing plays an important role in dry laser cleaning [1, 2] and has been used lately for different types of high-throughput laser material processing [3].

Strong spherical aberration makes the focusing non-trivial. Usually, the exact solution is obtained using the Mie theory [4], which does not give much of a physical insight as it is based on a multipole expansion. At the same time, the main focusing properties of transparent dielectric spheres originate rather from the picture of geometrical optics.

In the lowest approximation the sphere acts as an ideal lens. In many cases this picture does not even provide a description which is qualitatively correct. Also classical formulas for *weak* spherical aberration [5] do not yield useful results for the field behind a sphere: The maximum intensity is kept unchanged and its position does not depend on the wavelength.

Our approach — following the method of uniform caustic asymptotics in Ref. [6] — is based on the canonical integral for the cuspoid ray topology of strong spherical aberration. Though this Bessoid integral appears naturally in small angles approximation, it can be used to describe *arbitrary* axially symmetric strong spherical aberration by appropriate coordinate and amplitude transformations. For angularly dependent vectorial amplitudes the formalism uses higher-order Bessoid integrals.

The Bessoid integral is the axially symmetric generalization of the Pearcey integral [7], which plays an important role in many short wavelength phenomena [8].

2. The Bessoid Integral

2.1. *Definition*

We first consider the diffraction of a *scalar* spherically aberrated wave on a circular aperture with radius a in the plane $z = -f$ around the z-axis, where f is the focal distance. The origin of the coordinate system is put into the focus F. In cylindrical coordinates (ρ, z) and small angle approximation, the Fresnel–Kirchhoff diffraction integral [5] yields the field amplitude

$$U(\rho, z) = -\frac{\mathrm{i}\,k\,U_0}{f}\,\mathrm{e}^{\mathrm{i}\,k\,z}\int_0^a J_0\left(k\,\frac{\rho\,\widetilde{\rho}_1}{f}\right)\mathrm{e}^{-\mathrm{i}\,k\,\frac{z\,\widetilde{\rho}_1^2}{2\,f^2}\,-\,\mathrm{i}\,k\,B\,\widetilde{\rho}_1^4}\,\widetilde{\rho}_1\,\mathrm{d}\widetilde{\rho}_1. \quad (1)$$

Here U_0 is the amplitude of the incident wave in the center of the aperture, k is the wavenumber ($k = 2\pi/\lambda$, where λ is the wavelength) and $\widetilde{\rho}_1$ is the distance from the axis on the aperture. The Bessel function J_0 comes from the integration over the polar angle φ. The parameter B in the exponent determines the strength of the spherical aberration. For $B > 0$ the diffraction focus shifts towards the aperture, while $B = 0$ corresponds to ideal focusing [5].

We introduce dimensionless coordinates $\rho_1 \equiv \sqrt[4]{4\,k\,B}\,\widetilde{\rho}_1$, $R \equiv \sqrt[4]{k^3/4\,B}\,\rho/f$ and $Z \equiv \sqrt{k/4\,B}\,z/f^2$ and consider an infinitely large aperture. Then the field (1) becomes proportional to the *Bessoid integral* [9]

$$I(R, Z) = \int_0^{\infty} \rho_1\,J_0(R\,\rho_1)\,\mathrm{e}^{-\mathrm{i}\left(Z\,\frac{\rho_1^2}{2} + \frac{\rho_1^4}{4}\right)}\mathrm{d}\rho_1$$
$$= \frac{1}{2\,\pi}\int_{-\infty}^{\infty}\int_{-\infty}^{\infty}\mathrm{e}^{-\mathrm{i}\left(R\,x_1 + Z\,\frac{x_1^2+y_1^2}{2} + \frac{(x_1^2+y_1^2)^2}{4}\right)}\mathrm{d}x_1\mathrm{d}y_1. \quad (2)$$

Its square is shown in figure 1. In the Cartesian representation $x_1 = \rho_1\cos\varphi$ and $y_1 = \rho_1\sin\varphi$ are dimensionless coordinates in the plane of integration.

Expression (2) is the axially symmetric generalization of the Pearcey integral [7] $I_P(X, Z) = (2\pi)^{-1/2} \int_{-\infty}^{\infty} \exp\left[-\mathrm{i}\left(X x_1 + Z x_1^2/2 + x_1^4/4\right)\right] \mathrm{d}x_1$, which is also shown in figure 1.

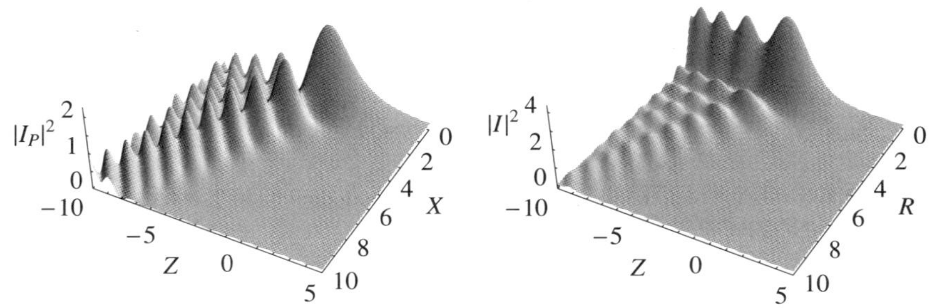

Figure 1. Absolute square of the Pearcey integral I_P (left) and the Bessoid integral I (right). The latter is proportional to the field of a spherically aberrated wave within small angles approximation

Both integrals correspond to so-called diffraction catastrophes [6]. Their field distribution contains caustic zones where the intensity predicted by geometrical optics goes to infinity. The Pearcey integral corresponds to a cusp caustic, i.e., a single one-dimensional curve in a two-dimensional space, and does not reveal a high intensity along the axis, while the Bessoid integral corresponds to a cuspoid caustic, i.e., to a surface of revolution of the cusp in three dimensions, as well as the caustic line up to the focus F at $z = Z = 0$. The equation of the cusp[a] is given by the semicubic parabola $27\,R^2 + 4\,Z^3 = 0$.

The cusp is the envelope of the family of rays. They correspond to the points of stationary phase in the Bessoid integral, i.e., those points where the two first partial derivatives with respect to R and Z of the phase

$$\phi \equiv -R\,x_1 - Z\,\frac{x_1^2 + y_1^2}{2} - \frac{(x_1^2 + y_1^2)^2}{4} \tag{3}$$

in the exponent in (2) vanish. Inside the cusp, for $27\,R^2 + 4\,Z^3 < 0$, three rays (tangents to the cusp) arrive at each point of observation $P \equiv (\rho, z)$, and outside, for $27\,R^2 + 4\,Z^3 > 0$, there is only one real ray (figure 2). Thus, the cusp forms the border between the lit region and the (partial) geometrical shadow.

[a] Normally, the term cusp refers to its vertex only, but we use it to denote the two branches. Besides, henceforth we will apply the term cusp also for the cuspoid.

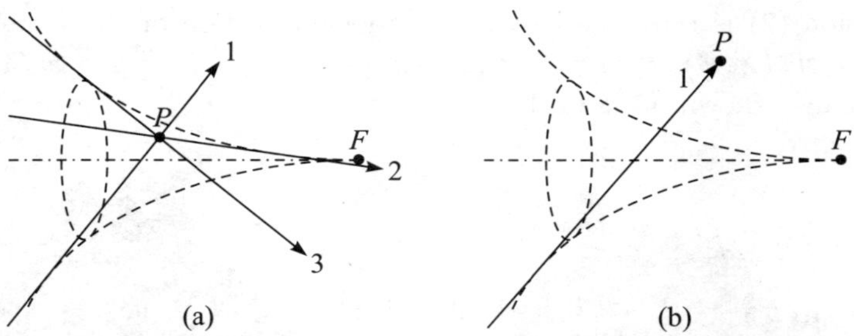

Figure 2. (a) 3-ray region inside the cuspoid (dashed line). (b) 1-ray region outside. The z-axis is represented by a dashed-dotted line

Without loss of generality, we assume that all rays lie in the meridional plane $\varphi = 0$ ($y_1 = 0$) and hence correspond to the roots $x_{1,j}$ ($j = 1, 2, 3$) of the cubic equation

$$R + Z\,x_1 + x_1^3 = 0\,, \tag{4}$$

which are given by Cardan's formulas [10]. On the axis, $R = 0$, a cone of an infinite number of rays converges. These rays originate from the circle $x_1^2 + y_1^2 = -Z$ on the aperture. They are all in phase and produce a high intensity along the axis (compare the two pictures in figure 1). The oscillations occur due to interference with the ray propagating along the z-axis.

2.2. *Asymptotic expressions*

Off the caustic — away from the cusp and the focal line — the Bessoid integral can be approximated by the method of stationary phase [11]:

$$I(R, Z) \approx \sum_{j=1}^{m} \frac{\mathrm{e}^{\,\mathrm{i}\,\phi_j + \mathrm{i}\,\frac{\pi}{4}\,\mathrm{sign}\,\mathbf{H}_j}}{\sqrt{|\det \mathbf{H}_j|}}\,, \tag{5}$$

where the summation runs over all real rays, i.e., $m = 1$ (lit) or 3 (shadow). ϕ_j is obtained by inserting the j-th stationary point ($x_{1,j}, y_{1,j} = 0$) into the phase (3). The determinant and signature of the Hessian are given by

$$\det \mathbf{H}_j = Z^2 + 4\,x_{1,j}^2\,Z + 3\,x_{1,j}^4\,, \tag{6}$$

$$\mathrm{sign}\,\mathbf{H}_j = \mathrm{sgn}(-Z - 3\,x_{1,j}^2) + \mathrm{sgn}(-Z - x_{1,j}^2)\,. \tag{7}$$

Near the cusp the Bessoid integral shows an Airy-type behavior typical for caustics where two real rays disappear and become complex.

Near the axis (small R) one can derive the approximation

$$I(R,Z) \approx \frac{\sqrt{\pi}}{2}\, J_0(R\sqrt{-Z})\, \mathrm{e}^{\,\mathrm{i}\,\frac{Z^2-\pi}{4}}\, \mathrm{erfc}\Big(\frac{Z}{2}\,\mathrm{e}^{\,\mathrm{i}\,\frac{\pi}{4}}\Big), \tag{8}$$

where erfc is the complementary error function,[b] which can also be written in terms of Fresnel sine and cosine functions [9]. Expression (8) becomes exact for $R = 0$, where $J_0(0) = 1$. It shows that near the axis the Bessoid integral is virtually a Bessel-beam with a variable cross section.

2.3. *Numerical evaluation*

As the Bessoid integrand is highly oscillatory, its evaluation for the whole range of coordinates R and Z is non-trivial and of large practical importance. Direct numerical integration along the real axis and the method of steepest descent in the complex plane both have their disadvantages. By far the fastest technique is based on the numerical solution of the *ordinary* differential equation [12]

$$L_R - Z\, I_R + \mathrm{i}\, R\, I = 0\,. \tag{9}$$

Indices denote (partial) derivatives and $L \equiv I_{RR} + I_R/R$ is an abbreviation for the radial Laplacian applied onto I. The three initial conditions at $R = 0$ are (i) $I(0,Z)$ calculated from (8), (ii) $I_R(0,Z) = 0$ due to symmetry and (iii) $L(0,Z) = Z\, I(0,Z) + \mathrm{i}$, arising from the fact that the Bessoid integral satisfies the paraxial Helmholtz equation $2\,\mathrm{i}\, I_Z + L = 0$.

In the literature the Pearcey integral was calculated by solving differential equations [13], by a series representation [14] and by the first terms of its asymptotic expansion [15]. The Bessoid integral was expressed in terms of parabolic cylinder functions [16] and as a series [9]. The latter work gives reference to an unpublished work of Pearcey [17], stating that differential equations for the Bessoid integral were employed there. No other indication of the existence of such equations is known to us.

2.4. *Geometrical optics for the cuspoid*

In geometrical optics, rays carry the information of amplitude and phase. The total field in a point P is given by the sum of all ray fields there. A ray's field at P is determined by

$$U(P) = U_0\, \frac{\mathrm{e}^{\,\mathrm{i}\, k\, \psi}}{\sqrt{J}}\,, \tag{10}$$

---

[b] http://mathworld.wolfram.com/Erfc.html

where U_0 is the amplitude at some initial wavefront, ψ is the eikonal (optical path along the ray from this wavefront to P) and J is the generalized geometrical divergence, which can be calculated from flux conservation along the ray [18]. For a homogeneous medium with constant refractive index $J = R_m R_s/(R_{m0} R_{s0})$, where R_m, R_s are the main radii of curvature at the point P and R_{m0}, R_{s0} on the initial wavefront, where $U = U_0$. Caustics correspond to an intersection of infinitesimally close rays ($R_m R_s \to 0$, $|U| \to \infty$). They are the envelopes of ray families and the rays are tangent to the caustic. When one crosses the caustic, an even number of rays appears or disappears.

When a ray touches the caustic, its radius of curvature changes the sign and the ray undergoes a phase delay of $-\pi/2$, which is taken into account by the proper choice of the square root in (10). When a ray touches several caustics, these delays must be added. The total caustic phase shift, denoted as $\Delta\varphi$, can be explicitly written in the phase. For the cuspoid topology and ray numbering ($j = 1, 2, 3$) according to figure 2, we obtain:

$$U(P) = U_0 \frac{\mathrm{e}^{\mathrm{i}k\psi}}{\sqrt{J}} = U_0 \frac{\mathrm{e}^{\mathrm{i}k\psi + \mathrm{i}\Delta\varphi}}{\sqrt{|J|}} \quad \text{with} \quad \Delta\varphi_j = \begin{cases} -\pi & \text{for } j = 1, \\ 0 & \text{for } j = 2, \\ -\pi/2 & \text{for } j = 3. \end{cases} \qquad (11)$$

Ray 1 touched the cuspoid and the focal line, ray 2 is not shifted, and ray 3 touched the cuspoid.

3. Relation between Geometrical and Wave Optics

3.1. *Matching with the Bessoid integral*

If we have found the phases $\varphi \equiv k\psi$ and divergences J of the rays, the (scalar) geometrical optics solution with an axially symmetric 3-ray cuspoid topology can be written as

$$U(\mathbf{r}) = \sum_{j=1}^{3} \frac{U_{0,j}\, \mathrm{e}^{\mathrm{i}\varphi_j(\mathbf{r})}}{\sqrt{J_j(\mathbf{r})}} . \qquad (12)$$

Here $\mathbf{r} \equiv (\rho, z)$ are the real-space coordinates and we have allowed for different initial amplitudes $U_{0,j}$ of the rays. This field shows singularities at the caustic, especially on the axis, which is the most interesting region for applications. On the other hand, the paraxial spherically aberrated wave resulted in the Bessoid integral (2), which has no singularities.

We want to describe arbitrary axially symmetric focusing by matching the solution of geometrical optics (where it is correct) with a wave field constructed from the Bessoid integral and its partial derivatives I_R and I_Z (*method of uniform caustic asymptotics*). We make the Ansatz [6]

$$U(\mathbf{r}) = \left(A(\mathbf{r})\, I(\mathbf{R}) + \frac{1}{\mathrm{i}}\, A_R(\mathbf{r})\, I_R(\mathbf{R}) + \frac{1}{\mathrm{i}}\, A_Z(\mathbf{r})\, I_Z(\mathbf{R}) \right) \mathrm{e}^{\,\mathrm{i}\,\chi(\mathbf{r})} . \tag{13}$$

Here $\mathbf{R} \equiv (R, Z)$ are the yet unknown coordinates of the Bessoid integral, A, A_R and A_Z are three amplitude factors[c] and χ is a phase function. Now the stationary phase approximation of (13) is matched with the geometrical optics solution (12) by equating the amplitudes and phases:

$$\left(A(\mathbf{r}) + A_R(\mathbf{r})\, \phi_R(\mathbf{R}, \mathbf{t}_j) + A_Z(\mathbf{r})\, \phi_Z(\mathbf{R}, \mathbf{t}_j)\right) \frac{1}{\sqrt{H_j}} = \frac{U_{0,j}}{\sqrt{J_j}} , \tag{14}$$

$$\chi(\mathbf{r}) + \phi(\mathbf{R}, \mathbf{t}_j) = \varphi_j(\mathbf{r}) , \tag{15}$$

where ϕ_R and ϕ_Z are the partial derivatives of (3) and $1/\sqrt{H_j} \equiv \mathrm{e}^{\,\mathrm{i}\,\frac{\pi}{4}\,\mathrm{sign}\,\mathbf{H}_j} / \sqrt{|\det \mathbf{H}_j|}$.[d]

The points of stationary phase were denoted as $\mathbf{t}_j \equiv (t_j, 0)$, where the t_j are given by Cardan's solutions of $R + Z\,t + t^3 = 0$. Note that stationary points[e] $\mathbf{t}_j = \mathbf{t}_j(\mathbf{R})$ and coordinates $\mathbf{R} = \mathbf{R}(\mathbf{r})$. The conditions (14) and (15) give 6 equations for the 6 unknowns R, Z, χ, A, A_R, and A_Z.

It is convenient to solve (15) using quantities that are permutationally invariant with respect to the roots t_j [8, 19]. This yields

$$\begin{aligned} R &= \sqrt{\frac{Z^3}{54} - \frac{4\,b_2}{9\,Z}} , \\ Z &= \pm \sqrt[4]{\frac{2}{3}} \sqrt{-2\,\mathrm{sgn}(b_3)\,\sqrt{b_2 + q} + 2\sqrt{2\,b_2 - q + 2\sqrt{b_2^2 - b_2\,q + q^2}}} , \\ \chi &= b_1 - \frac{1}{6}\, Z^2 , \end{aligned} \tag{16}$$

where $\mathrm{sgn}(Z) = \mathrm{sgn}(Z^4 - 24\,b_2)$. The b_l $(l = 1, 2, 3)$ are given by $b_1 \equiv (1/3) \sum_{j=1}^{3} \varphi_j$, $b_2 \equiv \sum_{j=1}^{3} (\varphi_j - b_1)^2$ and $b_3 \equiv \sum_{j=1}^{3} (\varphi_j - b_1)^3$. The quantity q can be expressed in different ways:

$$\begin{aligned} q^3 \equiv 6\,b_3^2 - b_2^3 &= \frac{1}{2^{11}}\, R^2\, (27\,R^2 + 4\,Z^3)^3 \\ &= -2\,(\varphi_1 - \varphi_2)^2\, (\varphi_2 - \varphi_3)^2\, (\varphi_3 - \varphi_1)^2 . \end{aligned} \tag{17}$$

---

[c] The indices R and Z in the amplitude factors do not indicate derivatives.

[d] Outside the cusp, the rays 2 and 3 are complex and the definition of H_j is more subtle.

[e] The partial derivatives with respect to R and Z in (14) must be evaluated in such a way as the t_j were held constant, although they are functions of $\mathbf{R}$ themselves.

Hence, it vanishes exactly at the caustic where two phases are equal.[f]

The solutions of (14) are

$$A = -\,U_{0,1}\,\frac{\sqrt{H_1}}{\sqrt{J_1}}\,\frac{t_2\,t_3}{(t_3-t_1)\,(t_1-t_2)} - \ldots - \ldots \text{ (cyclic)}\,,$$
$$A_R = U_{0,1}\,\frac{\sqrt{H_1}}{\sqrt{J_1}}\,\frac{t_1}{(t_3-t_1)\,(t_1-t_2)} + \ldots + \ldots \text{ (cyclic)}\,, \qquad (18)$$
$$A_Z = 2\,U_{0,1}\,\frac{\sqrt{H_1}}{\sqrt{J_1}}\,\frac{1}{(t_3-t_1)\,(t_1-t_2)} + \ldots + \ldots \text{ (cyclic)}\,,$$

where the cyclic terms permutate the numbering of rays: $(1,2,3) \to (2,3,1) \to (3,1,2)$. The Bessoid matching solution (13) does not show the divergences of geometrical optics.

3.2. *General expressions on and near the axis*

Inside the cusp ($z < f$), the formulas can be strongly simplified on and near the axis (small ρ). From figure 3 one can see that up to the first order in ρ the phases can be written as

$$\varphi_1 = \varphi_{\rm np} + k\,\rho\,\sin\beta\,, \quad \varphi_2 = \varphi_{\rm p}\,, \quad \varphi_3 = \varphi_{\rm np} - k\,\rho\,\sin\beta\,. \qquad (19)$$

Here $\varphi_{\rm np}$ and $\varphi_{\rm p}$ denote the phases of the non-paraxial rays and the (par)axial ray (with $\rho = 0$).

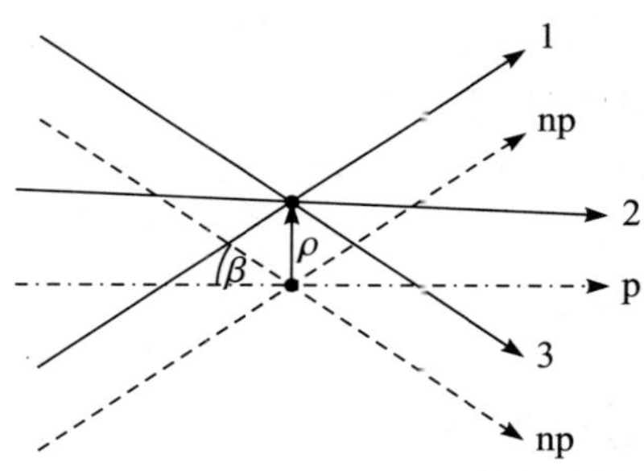

Figure 3. Near the axis, the phases of the rays 1 and 3 differ from the phase of the non-paraxial rays by $\pm\,k\,\rho\,\sin\beta$, whereas the phases of ray 2 and the (par)axial ray are the same in first order. The non-paraxial ray crosses the axis at the angle $\beta > 0$

We insert these phases into the exact expressions in (16), Taylor expand with respect to ρ and resubstitute $\varphi_{\rm np} = (\varphi_1 + \varphi_3)/2$, $\varphi_{\rm p} = \varphi_2$ and

[f] At the cuspoid $\varphi_2 = \varphi_3$ ($27\,R^2 + 4\,Z^3 = 0$) and on the axis $\varphi_1 = \varphi_3$ ($R = 0$).

$k\,\rho\,\sin\beta = (\varphi_1 - \varphi_3)/2$ from (19). This yields

$$R = \frac{(\varphi_1 - \varphi_3)/2}{\sqrt{2}\,\sqrt[4]{(\varphi_1 + \varphi_3)/2 - \varphi_2}} = \frac{k\,\rho\,\sin\beta}{\sqrt{-Z}}\,, \tag{20}$$

$$Z = -2\sqrt{\frac{\varphi_1 + \varphi_3}{2} - \varphi_2} = -2\sqrt{\varphi_{\mathrm{np}} - \varphi_{\mathrm{p}}}\,. \tag{21}$$

On the axis $\rho = 0$, we obtain $R = 0$ and $Z = -2\sqrt{\varphi_1 - \varphi_2}$. The Bessoid integral has its global maximum at $Z_{\mathrm{m}} \approx -3.051$. This corresponds to the phase difference

$$\varphi_1 - \varphi_2 = Z_{\mathrm{m}}^2/4 \approx 2.327\,. \tag{22}$$

The geometrical meaning of this result is that ray 1 and 3 are shifted by $-\pi/2$ as they touch the cusp. In addition, they acquire a further shift of $-\pi/2$ when crossing the focal line. But exactly *on the axis* only half of this delay has occurred yet, which yields $\varphi_1 - \varphi_2 \approx 3\,\pi/4 \approx 2.356$, which is close to 2.327. This is the condition for the (first) constructive interference of the axial and the non-paraxial rays.

The width of the focal line caustic, ρ_w, is defined by the first zero $w_0 \approx 2.405$ of the Bessel function in (8). Hence, with (20),

$$\rho_w \equiv \frac{w_0}{k\sin\beta} \approx 0.383\,\frac{\lambda}{\sin\beta}\,. \tag{23}$$

In the geometrical optics picture the first minimum occurs when ray 1 and 3 interfere destructively, i.e., when their phase difference becomes π. This results in $\varphi_1 - \varphi_3 = \pi + \pi/2$, where the term $\pi/2$ takes into account the caustic phase shift of ray 1: $\rho_w \approx (\varphi_1 - \varphi_3)/2\,k\sin\beta = 0.375\,\lambda/\sin\beta$.

Finally, we present an expression for the field (13) on the axis. The equations for the amplitudes (18) simplify tremendously. Furthermore, we use the linear relationship between the Bessoid and its Z-derivative, $\mathrm{i}\,Z\,I(0,Z) - 2\,I_Z(0,Z) = 1$. Finally, we end up with [12]

$$U = \left[\frac{U_{0,1}\sqrt{2\,k\,\rho\,\sin\beta}}{\sqrt{J_1}}\left(\mathrm{i}\,I - \frac{1}{Z}\right) + \frac{U_{0,2}}{\sqrt{J_2}}\right]\mathrm{e}^{\mathrm{i}\,\varphi_2}\,. \tag{24}$$

Note that (inside the cusp) on the axis ($\rho \to 0$) both $1/\sqrt{J_2}$ and the ratio $\sqrt{\rho}/\sqrt{J_1}$ remain finite.

3.3. *Angular dependences and vectorial problems: Higher-order Bessoid matching*

Often — especially in vectorial problems (see section 4) — there exists axial symmetry with respect to the wavefronts, ray phases and generalized divergences, but not with respect to the amplitudes. In this case, new functions are required to represent arbitrary angular dependence of the field. The natural generalization of (2) are the *higher-order Bessoid integrals* [16] with the non-negative integer m:

$$I_m(R, Z) = \int_0^\infty \rho_1^{m+1}\, J_m(R\,\rho_1)\, \mathrm{e}^{-\mathrm{i}\left(Z\,\frac{\rho_1^2}{2} + \frac{\rho_1^4}{4}\right)}\, \mathrm{d}\rho_1, \tag{25}$$

where $I_0 \equiv I$. They obey the recurrence relation [12] $I_{m+1} = -I_{m,R} + m\, I_m/R$. The integral I_m is canonical for angular dependent geometrical field components $U^{(m)}(\rho, z)\, \sin m\,\varphi$ or $U^{(m)}(\rho, z)\, \cos m\,\varphi$. In matching similar to (13), the angular dependence cancels. Since higher-order Bessoid integrals can be written in terms of I_0, it can be shown that the points of stationary phase, the matching of phases and thus the higher-order coordinates and phases are identical with the original ones. From the physical point of view, this reflects the conservation of the wavefront and thus the ray phases and divergences.

The equations for the amplitudes have to be generalized. The higher-order amplitudes A_m, A_{mR} and A_{mZ} have the same form as (18) but with an additional factor $(\mathrm{i}\, t_j)^m$ in each denominator [12], e.g.,

$$A_m = -U_{0,1}\, \frac{\sqrt{H_1}}{\sqrt{J_1}}\, \frac{t_2\, t_3}{(\mathrm{i}\, t_1)^m\, (t_3 - t_1)\, (t_1 - t_2)} - \ldots - \ldots \text{ (cyclic)}\,. \tag{26}$$

4. The Sphere

4.1. *Geometrical optics solution*

Consider a plane wave falling on a transparent sphere in vacuum. Figure 4 illustrates the refraction of a single ray in the meridional plane, containing the point of observation P and the axis. Within the frame of geometrical optics the cuspoid is formed behind the sphere in analogy to figure 2.

Let a be the sphere radius and $n > 1$ its refractive index. The center M is located at the origin of an axially symmetric cylindrical coordinate system (ρ, z). The incident plane wave propagates parallel to the z-axis. The geometrical optics focus, formed by the paraxial rays, is located at $F \equiv (0, f)$ with $f \equiv a\, n/[2\,(n-1)]$ [20].

A ray passes the point Q, is first refracted at Q_1, a second time at Q_2 and propagates to P. The incident and transmitted angle, θ_i and θ_t, are related by Snell's law, $\sin\theta_i = n \sin\theta_t$. Writing the position of $P \equiv (\rho, z)$ in polar coordinates, $\rho = l \sin\theta$ and $z = l \cos\theta$, one can find the following

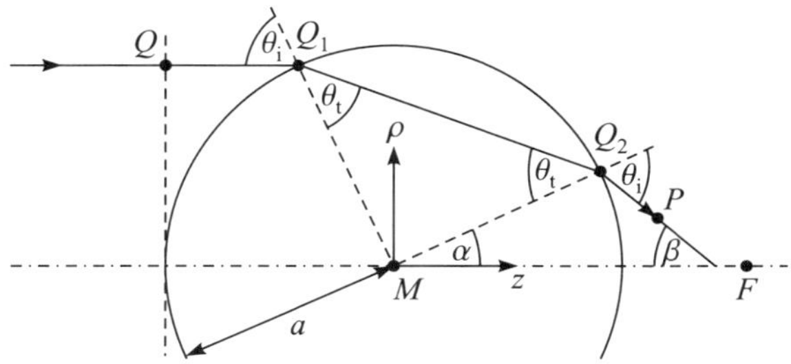

Figure 4. Refraction of a ray — propagating from Q to P — by a sphere with radius a and refractive index n. The picture is drawn in the meridional plane and all indicated angles are positive

expression, determining the three rays that arrive at P:

$$l \sin(\theta + 2\,\theta_i - 2\,\theta_t) = a \sin\theta_i, \tag{27}$$

where one has to substitute $\theta_t = \arcsin[(\sin\theta_i)/n]$. This is a transcendental cubic-like equation which has three roots,[g] either all real or one real and two complex conjugate. We denote them as $\theta_{i,j}$ $(j = 1, 2, 3)$ and choose their order consistently with the previous notations.[h] When the $\theta_{i,j}$ are known, we find the $\theta_{t,j}$ from Snell's law and the α_j and β_j from $\alpha_j = 2\,\theta_{t,j} - \theta_{i,j}$ and $\beta_j = 2\,\theta_{i,j} - 2\,\theta_{t,j}$. Omitting the index j, the three ray coordinates can be written as $s \equiv \overline{Q_2P} = (l\cos\theta - a\cos\alpha)/\cos\beta$. The eikonal is the optical path accumulated from Q to P (on the dashed vertical line in figure 4 all rays are still in phase):

$$\psi = \overline{QQ_1} + n\,\overline{Q_1Q_2} + \overline{Q_2P} - a = a\,(2\,n\cos\theta_t - \cos\theta_i) + s\,. \tag{28}$$

We have subtracted the sphere radius a from the path contributions to make the eikonal zero in the center M, if there were no sphere.

Next we calculate the geometrical optics amplitudes by determining the meridional and sagittal radii of curvature, R_m and R_s, and their changes due to refraction. Formulas for the refraction on an arbitrary surface with arbitrary orientation of the main radii are given in Refs. [18, 21]. A simple derivation for the sphere can be found in Ref. [12]. It yields the dependence of the actual radii of curvature R_m and R_s (right after the refraction) on

[g] For $n \geq \sqrt{2}$ this is true for $z \geq a$. If $n < \sqrt{2}$, the 3-ray region does not start until some distance behind the sphere.

[h] Therefore, $\theta_{i,1}$ is always real and negative, whereas $\theta_{i,2}$ and $\theta_{i,3}$ are either real and positive (lit region) with $\theta_{i,2} < \theta_{i,3}$ or complex conjugate (geometrical shadow).

the initial radii R_{m0} and R_{s0} (just before the refraction):

$$R_m = \frac{n\,a\,R_{m0}\cos^2\theta_t}{a\cos^2\theta_i + R_{m0}\,(\cos\theta_i - n\cos\theta_t)}, \quad R_s = \frac{n\,a\,R_{s0}}{a + R_{s0}\,(\cos\theta_i - n\cos\theta_t)}. \tag{29}$$

For a plane wave, $R_{m0}, R_{s0} \to \infty$, the radii of curvature in the points Q_1 (inside the sphere) and Q_2 (outside the sphere) have the compact form

$$\begin{aligned} R_{m,Q_1} &= -a\,\frac{\sin\theta_i\cos^2\theta_t}{\sin(\theta_i - \theta_t)}, \quad & R_{m,Q_2} &= -a\,\frac{\cos\theta_i}{2}\left(\frac{\cos\theta_i\sin\theta_t}{\sin(\theta_i-\theta_t)} - 1\right), \\ R_{s,Q_1} &= -a\,\frac{\sin\theta_i}{\sin(\theta_i - \theta_t)}, \quad & R_{s,Q_2} &= -a\,\frac{\sin(2\,\theta_t - \theta_i)}{\sin(2\,\theta_i - 2\,\theta_t)}. \end{aligned} \tag{30}$$

The overall geometrical generalized divergence after both refractions reads (index j omitted)

$$\frac{1}{\sqrt{J}} = \frac{\sqrt{R_{m,Q_1}\,R_{s,Q_1}}}{\sqrt{(R_{m,Q_1} + d)\,(R_{s,Q_1} + d)}}\,\frac{\sqrt{R_{m,Q_2}\,R_{s,Q_2}}}{\sqrt{(R_{m,Q_2} + s)\,(R_{s,Q_2} + s)}}, \tag{31}$$

where $d \equiv 2\,a\cos\theta_t$ is the distance of propagation within the sphere. Note that ray 1 has a negative angle θ_i. Besides, double caustic phase shift should be inserted manually for this ray (minus sign) as in (11). The caustic shifts of the rays 2 and 3 are taken into account automatically by the standard definition of the square root.

Finally, the geometrical optics solution for the sphere is given by (12),[i] where the eikonal ψ and divergence J are given by (28) and (31). The equation determining the three rays is (27).

To incorporate Fresnel transmission coefficients, we assume that the incident light is linearly polarized in x-direction, i.e., the incident electric field vector is $\mathbf{E}_0 = E_0\,\mathbf{e}_x$, with $\mathbf{e}_x$ the unit vector in x-direction and $E_0 \equiv U_0$. Since axial symmetry is broken, we introduce the polar angle φ which is measured from x to y. The point of observation $P \equiv (\rho, \varphi, z)$ will be reached by three rays (two may be complex) and their angles $\theta_{i,j}$ are still determined by (27), for all three rays lie in the meridional plane, containing P and the z-axis (figure 5a).

The initial π- and σ-polarized components depend on φ: $E_{0,\pi} = E_0\cos\varphi$ and $E_{0,\sigma} = E_0\sin\varphi$ (figure 5b). We define the overall transmission coefficients $T_\pi \equiv t_{12,\pi}\,t_{21,\pi} = 1 - r^2_{12,\pi}$, and $T_\sigma \equiv t_{12,\sigma}\,t_{21,\sigma} = 1 - r^2_{12,\sigma}$. Here the t_{12} (r_{12}) are the standard Fresnel transmission (reflection) coefficients [5] from the medium 1, i.e., vacuum, into the medium 2, i.e., the sphere.

[i] In the geometrical shadow the sum becomes only the term with $j = 1$.

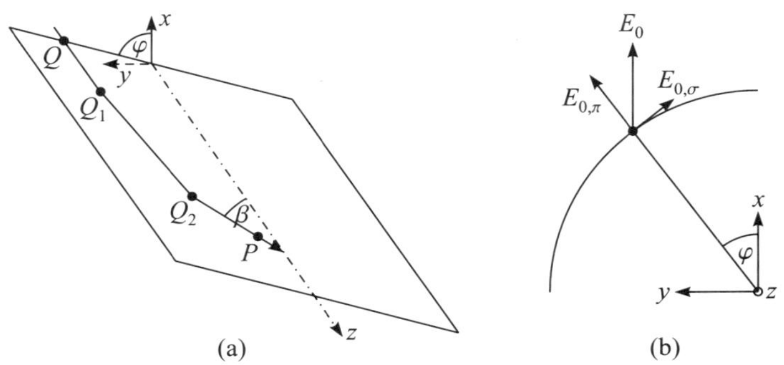

Figure 5. (a) A ray propagates from Q to P in the meridional plane (plane of incidence). (b) Decomposition of the initial electric field vector with length E_0 into its π- and σ-component parallel and perpendicular to the meridional plane

The ray field behind the sphere is found by the projection onto the original Cartesian system (x, y, z). We write the components of the transmission vector $\mathbf{T} \equiv (T_x, T_y, T_z)$ and show the ray index $j = 1, 2, 3$ explicitly. The φ-dependence is indicated with the superscript (m):

$$\begin{aligned} T_{x,j} &= T_j^{(0)} + T_j^{(2)} \cos 2\varphi\,, & & T_j^{(0)} \equiv \frac{T_{\pi,j}\cos\beta_j + T_{\sigma,j}}{2}\,, \\ T_{y,j} &= T_j^{(2)} \sin 2\varphi\,, & \text{where}\quad & T_j^{(1)} \equiv T_{\pi,j}\sin\beta_j\,, \\ T_{z,j} &= T_j^{(1)} \cos\varphi\,, & & T_j^{(2)} \equiv \frac{T_{\pi,j}\cos\beta_j - T_{\sigma,j}}{2}\,. \end{aligned} \tag{32}$$

Hence, the geometrical optics solution for the electric field $\mathbf{E} \equiv (E_x, E_y, E_z)$ — including the eikonal ψ (28) and divergence J (31) — reads

$$\begin{aligned} E_{x,j} &= E_j^{(0)} + E_j^{(2)} \cos 2\varphi\,, & & \\ E_{y,j} &= E_j^{(2)} \sin 2\varphi\,, & \text{where}\quad & E_j^{(m)} \equiv E_0\,\frac{T_j^{(m)}\,\mathrm{e}^{\mathrm{i}\,k\,\psi_j}}{\sqrt{J_j}}\,. \\ E_{z,j} &= E_j^{(1)} \cos\varphi\,, & & \end{aligned} \tag{33}$$

4.2. *The Bessoid matching solution*

Matching each term $E^{(m)} = \sum_{j=1}^{3} E_j^{(m)}$ by the Ansatz (13) in its higher-order formulation with the corresponding integral I_m, we obtain the vectorial electric field in the form

$$\mathbf{E}(\rho, \varphi, z) = \begin{pmatrix} E_x \\ E_y \\ E_z \end{pmatrix} = E^{(0)} \begin{pmatrix} 1 \\ 0 \\ 0 \end{pmatrix} + E^{(1)} \begin{pmatrix} 0 \\ 0 \\ \cos\varphi \end{pmatrix} + E^{(2)} \begin{pmatrix} \cos 2\varphi \\ \sin 2\varphi \\ 0 \end{pmatrix}. \tag{34}$$

Figure 6 illustrates the intensity, i.e., the absolute square of the electric field $|E|^2 \equiv \mathbf{E}\,\mathbf{E}^*$, for $\varphi = 0$ (x,z-plane) and $\varphi = \pi/2$ (y,z-plane).

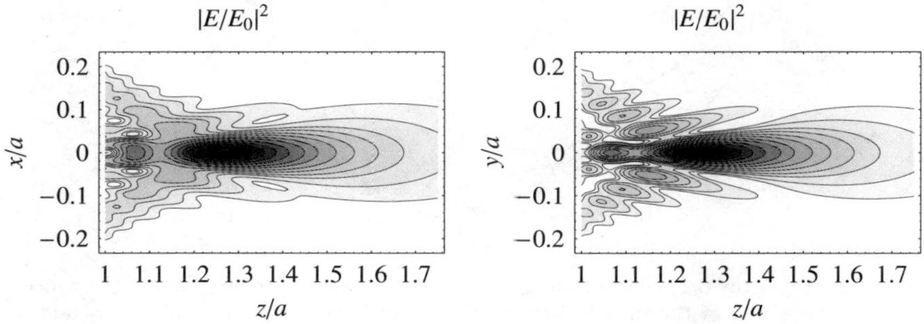

Figure 6. Normalized intensity $|E/E_0|^2$ in the normalized x,z-plane (left) and in the y,z-plane (right). Contour shadings go from white (zero) to black (≈ 700). Parameters: refractive index $n = 1.5$, dimensionless wavenumber $k\,a = 100$. The initial electric field amplitude is $\mathbf{E}_0 = E_0\,\mathbf{e}_x$. The sphere with radius a is situated in the origin. The focus of geometrical optics is located at $z = f = 1.5\,a$, whereas the diffraction focus — the point of maximum intensity — is significantly shifted towards the sphere: $f_d \approx 1.25\,a$. In dimensional units, for a wavelength of $\lambda = 0.248\,\mu$m the sphere radius is $a \approx 4\,\mu$m

The magnetic field $\mathbf{H}$ can be calculated similarly (incident magnetic field $\mathbf{H}_0 = H_0\,\mathbf{e}_y$, $H_0 = E_0$) and the (normalized) Poynting vector is given by $\mathbf{S} \equiv \mathrm{Re}(\mathbf{E}\times\mathbf{H}^*)$.

4.3. *On the axis*

On the axis the electric field is given by its x-component only (direction of polarization) due to averaging over φ in (34). For $z < f$ (inside the cusp) it is given by the analytical expression (24). After several simplifications [12] it can be written as

$$E = E_0 \left[T_1\, D_1 \left(\mathrm{i}\,I - \frac{1}{Z} \right) + \frac{T_2}{1 - z/f} \right] \mathrm{e}^{\,\mathrm{i}\,\varphi_2} \tag{35}$$

where the transmission factors $T_j \equiv T_j^{(0)}$ are given in (32) and for dielectric spheres have the form:

$$T_1 = \frac{n\,(1 + 3\cos\beta_1)\cos\theta_{i,1}\cos\theta_{t,1}}{(n\cos\theta_{i,1} + \cos\theta_{t,1})^2}, \quad T_2 = \frac{4\,n}{1 + n^2}. \tag{36}$$

The phases in the coordinate $Z = -2\sqrt{\varphi_1 - \varphi_2}$ are given by

$$\varphi_1 = \varphi_3 = k\,a\left(2\,n\cos\theta_{t,1} - \cos\theta_{i,1} + \frac{\sin\alpha_1}{\sin\beta_1}\right),\ \varphi_2 = k\,a\,(2\,n-2) + k\,z\,, \tag{37}$$

and $D_1 \equiv \sqrt{\varphi_1 - \varphi_3}/\sqrt{J_1}$ is the first ray's compensated divergence:[j]

$$\begin{aligned} D_1 &= -\frac{\sqrt{(R_{m,Q_1})_1\,(R_{s,Q_1})_1}}{\sqrt{[(R_{m,Q_1})_1 + d_1]\,[(R_{s,Q_1})_1 + d_1]}}\,\frac{\sqrt{(R_{m,Q_2})_1\,(R_{s,Q_2})_1}}{\sqrt{(R_{m,Q_2})_1 + s_1}}\,\sqrt{2\,k}\,\sin\beta, \\ &= -2\sqrt{2\,k\,a}\,\cos(\beta_1/2)\sqrt{\frac{\cot\theta_{i,1}\cos\theta_{t,1}\sin(\beta_1/2)}{1 + 1/n^2 - 3\sin^2(\beta_1/2)/\sin^2\theta_{i,1}}}. \end{aligned} \tag{38}$$

which manifestly has no singularity until the geometrical focus where $(R_{m,Q_2})_1 + s_1 \to 0$. Equation (35) is valid even near the focus, since the diverging terms D_1/Z and $(1 - z/f)^{-1}$ almost cancel. For $z \to f$, however, the divergence of D_1 itself becomes important, as the non-paraxial ray 1 becomes axial.

In figure 7 we show the position and the value of the maximum of $|E|^2$ as a function of the refractive index and the dimensionless product $k\,a$, calculated from (35). The z-coordinate of this global maximum is denoted with f_d (diffraction focus) and the intensity there is $|E(f_d)|^2$.

The main contribution in (35) stems from the Bessoid integral, that is from the term $\propto T_1\,D_1\,I$. Thus, the position of the maximum can be estimated from condition (22), i.e., $\varphi_1 - \varphi_2 \approx 3\,\pi/4$. If the phase difference $\varphi_1 - \varphi_2$ from (37) is expressed as a function of $\theta_{i,1}$, Taylor expanded and equated to $3\,\pi/4$, then we get in the lowest non-trivial order of the inverse product $k\,a$:[k]

$$f_d \approx \frac{a}{2}\,\frac{n}{n-1}\left(1 - \sqrt{\frac{3\,\pi}{4\,k\,a}\,\frac{n\,(3-n) - 1}{n\,(n-1)}}\right). \tag{39}$$

---

[j] The minus sign on the right hand side comes from the manually inserted phase shift of the first ray. The structure of this equation (first line only) is general and is valid for arbitrary axially symmetric systems. D_1 is always finite on the axis, since both the sagittal radius of curvature and the phase difference $\varphi_1 - \varphi_3$ are proportional to the distance ρ.

[k] The factor $3\,\pi/4 \approx 2.356$ can be replaced by the more exact Bessoid value 2.327. Expression (39) approximates the position of the maximum within an error of about $< 5\,\%$ for $k\,a > 100$ and values of the refractive index near $1.4 < n < 1.6$. The transcendental phase difference condition itself, which holds for large angles, naturally has a wider range of applicability.

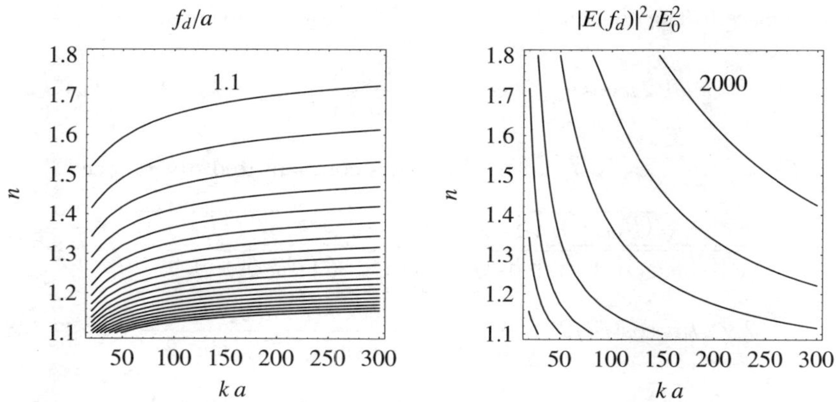

Figure 7. Left: Diffraction focus in units of the sphere radius as a function of n and $k\,a$ (contour lines from top to bottom go from 1.1 to 3.0 in steps of 0.1). Right: Intensity enhancement at f_d (contour lines from bottom left to top right are 20, 50, 100, 200, 500, 1000 and 2000)

Hence, in the limit of small wavelengths or large spheres the relative difference between the diffraction and the geometrical focus decreases proportionally to the inverse square root of $k\,a$.

4.4. *Comparison with the theory of Mie*

We presented a general way to match geometrical optics solutions with the Bessoid integrals. It can be applied to any axially symmetric system with the cuspoid topology of spherical aberration.

For the sphere we can compare our approximate results with the theory of Mie [4]. A main quantity characterizing the sphere is the dimensionless Mie parameter $q \equiv k\,a$. Figure 8 compares the intensity on the axis obtained from the Mie theory with the Bessoid approximation. The parameters are as in figure 6 and the Mie parameter is $q = 300$, 100, 30 and 10.

We see very good agreement down to $q \approx 30$ ($a/\lambda \approx 4.8$). For $q = 10$ ($a/\lambda \approx 1.6$) the asymptotic behavior far from the sphere is still correct. However, for small q the characteristic scale a is no longer large compared to the wavelength λ and geometrical optics becomes invalid.

Next, we compare the off-axis electric and magnetic field as well as the z-component of the Poynting vector, S_z. Right behind the sphere ($z = a$) the agreement is not perfect (see again figure 8), most probably due to evanescent contributions, which are taken into account in the Mie theory. If we assume that they disappear at a distance of about $\lambda/2$ from the

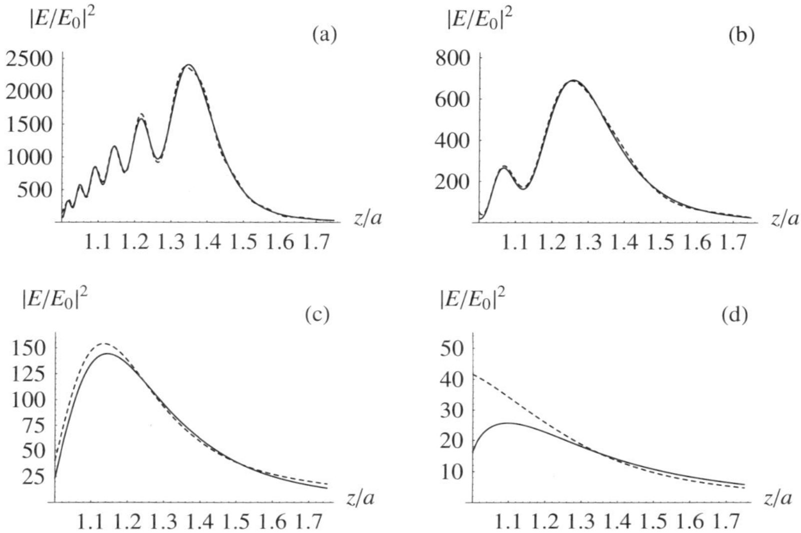

Figure 8. $|E/E_0|^2$ on the axis. Dashed lines represent the Mie theory, solid lines are the results of Bessoid matching. The cases (a), (b), (c) and (d) correspond to $q = 300$, 100, 30 and 10, respectively. In dimensional units, for $\lambda = 0.248\,\mu$m this corresponds to sphere radii of $a \approx 12\,\mu$m, $4\,\mu$m, $1.2\,\mu$m and $0.4\,\mu$m

sphere surface, good correspondence is expected for distances $z \geq a + \lambda/2$, i.e., $z/a \geq 1 + \pi/q$. For $q = 100$ this results in $z/a \gtrsim 1.03$. The following sections are made at $z = 1.02\,a$, showing good agreement (figure 9). For $z \gtrsim 1.05\,a$ the pictures become indistinguishable and even at $z = a$ all qualitative features are preserved.

The double-peak structure of $|E|^2$ along the direction of polarization has a purely geometrical origin. It is mainly related to the axial field component and is unrelated to near field effects. This is explained in detail in Ref. [12], where the estimation for the distance of the maxima from the axis is also given.

5. Conclusions

We described theoretically general axially symmetric aberrated focusing and studied light focusing by microspheres as an example. Following the method of uniform caustic asymptotics in Ref. [6], we introduced a canonical integral describing the field for the given cuspoid ray topology. This Bessoid integral appears naturally in the approximation of small angles. In some

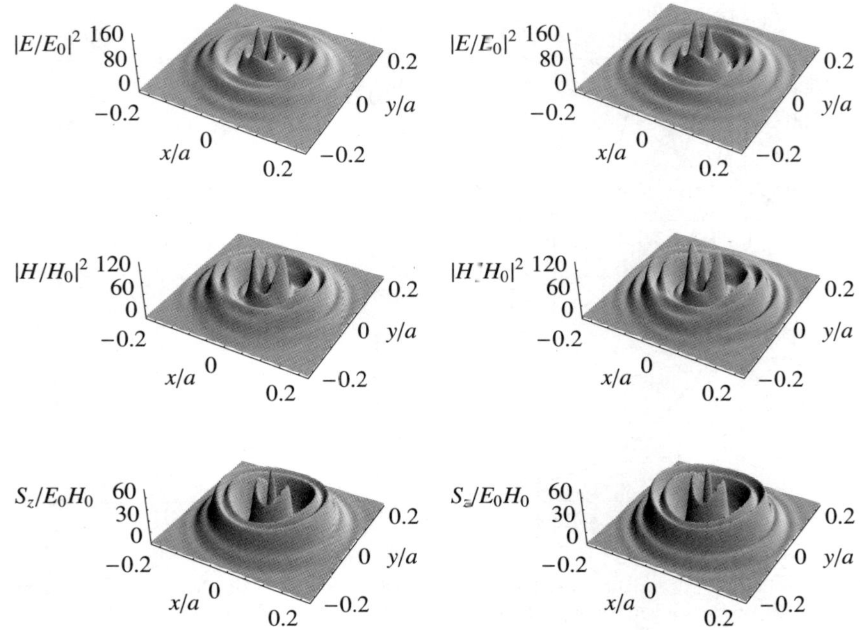

Figure 9. Normalized $|E|^2$, $|H|^2$ and S_z in the normalized x,y-plane for $z = 1.02\,a$ calculated with Bessoid matching (left) and with the Mie theory (right). The parameters are the same as in figure 6.

regions (off the caustic or exactly on the axis) it reduces to simple analytical expressions. In other regions we efficiently computed this highly oscillatory integral using a single ordinary differential equation.

For arbitrary axially symmetric and strongly spherically aberrated focusing, coordinate and amplitude transformations match the Bessoid wave field and the solution of geometrical optics. Caustic divergences of the latter are removed thereby. For vectorial problems with angular dependent field components, higher-order Bessoid integrals are used for the matching. The formulas significantly simplify on and near the axis. An approximate universal condition for the diffraction focus can be given in terms of phase differences. Here, the concept of caustic phase shifts is of main importance.

The central part of the Bessoid integral is essentially a Bessel beam with a variable cross section due to the variable angle of the non-paraxial rays. Its local diameter is always smaller than in the focus of an ideal lens with the same numerical aperture. Besides, the largest possible apertures can be physically realized, which is hardly possible with lenses. All this is achieved at the expense of longitudinal confinement.

As an example the focusing of a linearly polarized plane wave by a transparent sphere is studied in detail. We calculate the geometrical optics eikonals and divergences, incorporate Fresnel transmission coefficients and perform Bessoid matching. Using the general theory, simple expressions for the light field on the axis and for the diffraction focus are derived. The two strong maxima in the intensity observed immediately behind the sphere can be explained as well.

Finally, the results of the Bessoid matching procedure are compared with the theory of Mie. The agreement is good for Mie parameters $k\,a > 30$. Near the sphere the correspondences is worse due to unaccounted evanescent contributions. With small radius to wavelength ratio geometrical optics breaks down and deviations increase everywhere.

The whole approach and the developed formulas can be applied in other areas of physics where axially symmetric focusing is of importance, e.g., acoustics, semiclassical quantum mechanics, radio wave propagation, scattering theory, etc.

Acknowledgments

The authors want to thank Prof. D. Bäuerle (Applied Physics, University of Linz) for stimulating discussions on microsphere patterning experiments, which initiated this study. The program used for the Mie calculations was provided by Prof. B. Luk'yanchuk and Dr. Z. B. Wang (both at the Data Storage Institute, Singapore). N. A. thanks Prof. V. Palamodov (Tel Aviv University) for illuminating mathematical suggestions, and the FWF (Austrian Science Fund) as well as the Christian Doppler Laboratory of Surface Optics (University of Linz) for financial support. We are indebted to Prof. D. Kane (Macquarie University Sydney) for the invitation to contribute to this book.

References

1. Luk'yanchuk, B. (Editor), *Laser Cleaning*, World Scientific Publishing (2002)
2. Bäuerle, D., *Laser Processing and Chemistry*, Springer–Verlag, 3rd Edition (2000)
3. Bäuerle, D., Landström, L., Kofler, J., Arnold, N., and Piglmayer, K., Laser-processing with colloid monolayers, Proc. SPIE **5339** (2004), 20–26

4. Mie, G., *Beiträge zur Optik trüber Medien speziell kolloidaler Matallösungen*, Ann. d. Physik (4), **25** (1908), 377
5. Born, M. and Wolf, E., *Principles of Optics*, Cambridge University Press, 7th Edition (2002)
6. Kravtsov, Y. A. and Orlov, Y. I., *Caustics, Catastrophes and Wave Fields*, Springer Series on Wave Phenomena (Vol. 15), Springer–Verlag, 2nd Edition (1999)
7. Pearcey, T., *The structure of an electromagnetic field in the neighbourhood of a cusp of a caustic*, Lond. Edinb. Dubl. Phil. Mag. **37** (1946), 311–317
8. Connor, J. N. and Farrelly, D., *Theory of cusped rainbows in elastic scattering: uniform semiclassical calculations using Pearcey's Integral*, J. Chem. Phys. **75**(6) (1981), 2831–2846
9. Kirk, N. P., Connor, J. N., Curtis, P. R., and Hobbs, C. A., *Theory of axially symmetric cusped focusing: numerical evaluation of a Bessoid integral by an adaptive contour algorithm*, J. Phys. A: Math. Gen. **33** (2000), 4797–4808
10. Bronstein, I. N. and Semendjajew, K. A., *Handbook of Mathematics*, Springer–Verlag, 4th Edition (2004)
11. Focke, J., *Wellenoptische Untersuchungen zum Öffnungsfehler*, Optica Acta **3** (1956), 110
12. Kofler, J., *Focusing of Light in Axially Symmetric Systems within the Wave Optics Approximation*, Master Thesis, University of Linz, Austria (2004)
13. Connor, J. N. and Curtis, P. R., *Differential equations for the cuspoid canonical integrals*, J. Math. Phys. **25**(10) (1984), 2895–2902
14. Connor, J. N., *Semiclassical theory of molecular collisions: Three nearly coincident classical trajectories*, Mol. Phys. **26** (1973), 1217–1231
15. Stamnes, J. J. and Spjelkavik, B., *Evaluation of the field near a cusp of a caustic*, Optica Acta **30** (1983), 1331–1358
16. Janssen, A. J., *On the asymptotics of some Pearcey-type integrals*, J. Phys. A: Math. Gen. **25** (1992), L823–L831
17. Pearcey, T. and Hill, G. W., *Spherical aberrations of second order: the effect of aberrations upon the optical focus*, Melbourne: Commonwealth Scientific and Industrial Research Organization (1963), 71–95
18. Kravtsov, Y. A. and Orlov, Y. I., *Geometrical Optics of Inhomogeneous Media*, Springer Series on Wave Phenomena (Vol. 6), Springer–Verlag (1990)
19. Brekhovskikh, L. M. and Godin, O. A., *Acoustics of Layered Media II*, Springer Series on Wave phenomena (Vol. 10), Springer–Verlag (1992)
20. Bergmann, L. and Schäfer, C., *Lehrbuch der Experimentalphysik* (Band 3), Optik (Hrsg: Gobrecht, H.), Walter de Gruyter, 7. Auflage (1978)
21. Červený, V., *Seismic Ray Theory*, Cambridge University Press (2001)

Chapter 5

LIQUID-ASSISTED LASER SHOCK CLEANING FOR NANOSCALE PARTICLE REMOVAL

DEOKSUK JANG, BUKUK OH AND DONGSIK KIM†

Department of Mechanical Engineering, POSTECH, Pohang Pohang, 790-784, Korea

The laser-shock cleaning (LSC) method based on the photomechanical effect of a high-power laser pulse has been shown to be effective for removal of nanoscale contaminants from solid surfaces. In the dry LSC technique, laser-induced breakdown of the ambient gas creates a strong shock wave, and impingement of this shock wave onto a solid surface removes submicron-sized particles from the surface. However, removal of nanoscale contaminants is still a challenging task and it is important to develop cleaning methods effective for particulates less than 100 nm in size. In this work, a novel method entitled "liquid-assisted laser shock cleaning (LLSC)" is proposed for removal of nanoscale particles that cannot be efficiently removed by conventional laser cleaning techniques. The LLSC process utilizes the combined effect of the laser-induced shock impingement and explosive liquid-film evaporation. The key to the technique is synchronizing the onset of bubble nucleation and the timing of shock-wave collision with the solid surface. Both the pressure force created in the explosive vaporization of the liquid film and the synchronized impact by the shock wave in the ambient air increase the cleaning force to detach the bound particles which could not be removed otherwise. It is shown that the novel process leads to substantial enhancement of particle-removal efficiency, compared with the conventional laser cleaning methods.

1. Introduction

As the feature sizes of semiconductor and microelectronic device shrink to the nanoscale, surface cleaning technology, particularly removal of nanoparticles, has become a critical issue in the semiconductor and microelectronics industries. Nanoscale contaminants on a surface affect device performance and reliability of integrated circuit (IC) manufacture. Therefore, various novel cleaning methods have recently been suggested for the contamination control of silicon wafers. Among them, a number of laser cleaning processes including the dry

† To whom correspondence should be addressed. Phone: +82 (54) 279-2179. Fax: +82 (54) 279-3199. E-mail: dskim87@postech.ac.kr.

laser cleaning (DLC), steam laser cleaning (SLC) and laser shock cleaning (LSC) have been shown to be effective for submicrometer particle removal [1-12]. In the LSC method, the surface is not directly exposed to laser pulse irradiation, which precludes the chance for potential damage to photo-sensitive parts, and a relatively large area can be cleaned by a single laser pulse [9]. Since the LSC is a dry process, there is no need to control the liquid purity. However, despite these advantages, the adoption of LSC process is difficult because this method requires large pulse energy for removal of submicrometer particles in comparison with the SLC process and becomes ineffective as the particle size is below 100 nm. In recent studies of the SLC process, it has been shown that the SLC process can detach particles as small as a few tens of nanometers from solid surfaces [7, for example]. However, even with the SLC process, removal of nanoscale particles less than 100 nm in diameter is still a challenging task, and the effectiveness of the conventional cleaning methods is not yet thoroughly revealed in the nanoscale regime. Especially, the cleaning efficiency data show increasingly large scattering as the particle dimension gets small, strongly depending on the characteristics of the contaminant particles that adhere to solid surfaces.

In this work, a novel hybrid method to remove nanoscale particles from a solid surface is developed using the combined effect of the laser-generated shock wave and the liquid-film vaporization as in the steam laser cleaning. The key to this novel method, entitled "liquid-assisted laser shock cleaning (LLSC)" technique, is synchronizing the onset of explosive vaporization of liquid film and the timing of shock-wave collision with the solid surface. In the LLSC process, the pressure force created in the explosive vaporization of the liquid film, and the synchronized impact by the shock wave in the ambient air, increase the combined cleaning force to detach many of the particles that could not be removed by either technique independently. Cleaning tests using a Q-switched Nd:YAG laser and an KrF excimer laser were carried out to verify this concept. Accordingly, the surfaces cleaned by the LSC and SLC processes sequentially are compared with those cleaned through the equivalent number of LLSC cycles.

2. Principles of the Cleaning Method

2.1. *Laser shock cleaning (LSC) method*

In the LSC method, a high-power laser pulse is focused onto a small spot in the ambient gas above the sample to be cleaned. The laser-induced breakdown (LIB) of the gas then produces a strong shock wave, which impinges onto the surface. It has been reported that the LSC (laser shock cleaning) method, first proposed by Lee *et al.*[10], is effective for particles as small as a few hundred nanometers in diameter [9,10]. Also, the laser shock cleaning technique has been combined with a liquid-phase chemical cleaning process for oxide scale removal [13,14]. In the modified process, laser-induced breakdown (LIB) occurs in a liquid solution and the shock wave generated in the liquid enhances the cleaning process significantly. In our recent work, the dynamics of the laser-induced shock waves in various ambient gases was characterized using the optical beam deflection technique and the laser flash photography [15]. Experimental analysis of the shock wave propagation revealed that optical breakdown of air, N_2, He, and Ar by a Q-switched Nd:YAG pulse could produce shock waves of speed of order 1000 m/s and intensity of order 1 MPa, in a typical operation range of the LSC process.

The process of shock wave generation by LIB can be calculated theoretically and the preliminary results of our two-dimensional numerical simulation show that the theoretical prediction of the process is in reasonable agreement with the experimental observation. The shock waves are generated by multiphoton ionization [16] and by cascade ionization [17] in the LIB process. In the model, these ionization mechanisms were considered simultaneously in a similar fashion to Ireland *et al.*[18]. The ionization rate equation used in the numerical simulation is

$$\frac{dn_e}{dt} = n_e N\left[\frac{377q}{\omega^2}\left(\frac{\nu_m}{N}\right)^2\right]I(t) + \frac{NA}{k^{3/2}}I^k(t) \tag{1}$$

(n_e: electron number density [#·cm^{-3}], t: time [s], $N \sim 3.56\times10^{16}\cdot p$: density of neutral atom, p: ambient pressure [Torr], ω: laser beam frequency [Hz], ν_m: electron-atom collision frequency, q=10^{21} : a constant for noble gas [cm^{-1}·s^{-1}·V^{-2}], k: number of absorbed photons, A: multiphoton ionization coefficient [cm^{2k}·W^{-k}·s^{-1}], $I(t)$: laser irradiance [W·cm^{-2}]). Once the electron number density is obtained by Eq. (1), the Euler equations for gas dynamics are solved with the heat source term coming from the laser beam absorption [19]. The electron and ion transport equations are coupled with the Eq. (1) and the Euler equations [20]. Figure 1 shows an example of the numerical prediction of the shockwave propagation (2D) as compared to the experimental observations. Though the predictions are different from the experimental results quantitatively, the numerical calculation properly reflects the overall characteristics of the shock wave dynamics, yielding the same order of magnitude of the physical variables. Development of three dimensional computation code and parametric studies of the LSC process are currently under way.

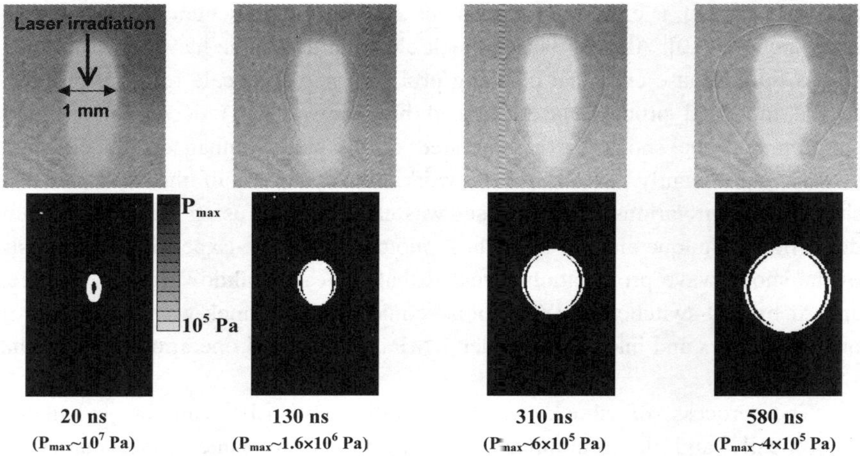

Figure 1. Time resolved shadowgraph images of shockwave propagation (upper row) and theoretically computed pressure contours. The maximum pressure P_{max} is displayed at each moment (laser pulse energy = 31 mJ, irradiance = 3.81×10^{12} W/cm^2).

2.2. *Steam laser cleaning (SLC) method*

Many investigations have shown that the explosive vaporization in a transparent liquid film in contact with an opaque solid surface generates a strong pressure pulse at the liquid/solid interface [21]. The SLC process utilizes the pressure pulse, which can be as high as the thermodynamic critical pressure ~P_c. Its effectiveness has been demonstrated for removing small particles from a solid surface [1-3,5,7, for example]. In the typical SLC process, saturated vapor is ejected onto the sample prior to the laser pulse irradiation and a thin liquid film is formed by condensation of the vapor. The SLC based on a KrF excimer laser or a Nd:YAG laser is capable of cleaning submicrometer-sized particles. Experiments demonstrated that both near-IR and UV laser pulses could remove 0.3 μm particles from NiP substrates at relatively low laser fluences without inflicting damage [5]. Test cleaning results with several different contaminant-substrate sample systems, including alumina particles and epoxy film on NiP, Si, and Cr surfaces, show that the liquid-assisted cleaning with appropriate choice of operation parameters can be very effective in removing a variety of contaminants from solid substrates, especially in removing particles larger than 200~300 nm.

2.3. *Liquid-assisted laser shock cleaning (LLSC) method*

Figure 2 depicts the concept of the novel LLSC process. A thin liquid film is first deposited by impingement of the saturated vapor of water/isopropanol mixture. The Nd:YAG laser pulse then induces LIB of air and a spherical shock wave propagates from the center of the plasma. When the shock wave arrives at the surface after the travel time d/c (d: focal point height, c: shock wave speed), an excimer laser pulse is fired which induces explosive vaporization of the liquid film. As the liquid vaporization occurs over ~1 μs and the bubble expansion occurs on a time scale of the order of 100 ns [21], a second laser pulse is triggered right before the shock wave arrival for synchronizing the impacts. However, when cleaning a finite area, the shock wave arrival time varies depending on the position, i.e., R in Fig. 2 and the sample surface needs to be scanned to provide the combined cleaning force uniformly. The excimer laser pulse is fired at the moment when the shock wave touches the center of the cleaning zone (delay time between laser pulses $t_0=d/c$), with the sample moving periodically on a translation stage under multiple-number-of-laser-pulse irradiation.

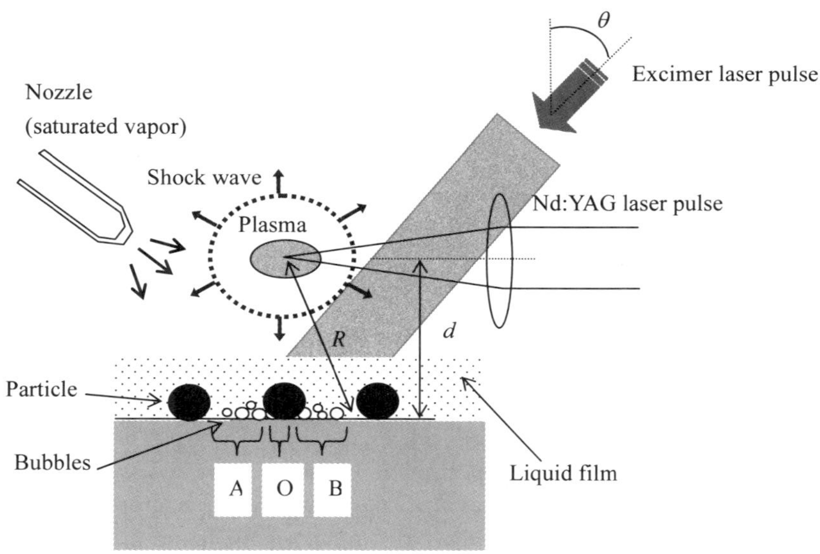

Figure 2. Concept of the novel liquid-assisted laser shock cleaning (LLSC) method.

The performance of the LLSC technique depends on the location of the particle on the surface. In Fig. 2, three different regions are marked with labels A, O, and B. In the stagnation region O, the pressure pulse acts downwards, pushing the particle back into the surface. Accordingly, neither the LSC nor the LLSC process works in this relatively small region. Our recent analysis of the flow field in the LSC process indicates that the shock wave intensity on the surface, i.e., shock wave speed, is weaker in the region A than B [15]. Consequently, the best cleaning performance is obtained by scanning the region B with multiple laser pulses.

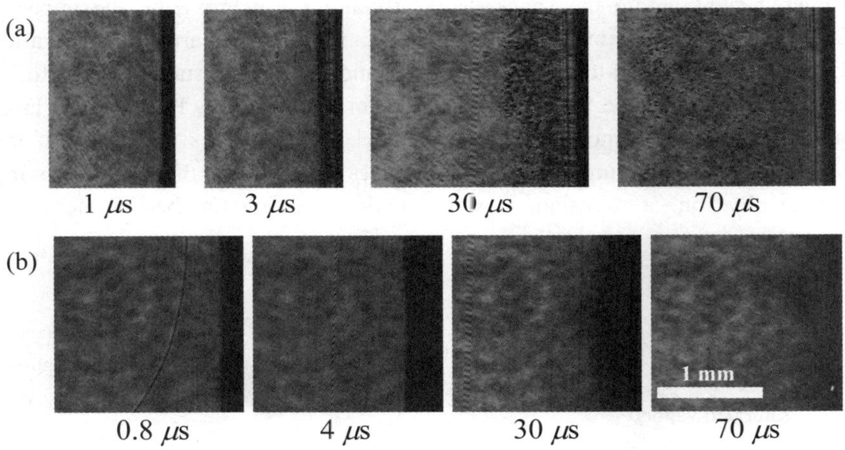

Figure 3. Shadowgraphs of liquid-film vaporization in the (a) laser steam cleaning and (b) liquid-assisted laser shock cleaning processes. The delay time between Nd:YAG laser and excimer laser irradiation is 1 μs, in the LLSC method. The Nd:YAG laser pulse energy and excimer laser fluence are 520 mJ and 240 mJ/cm^2, respectively.

In Fig. 3, the typical LLSC process is depicted by the laser flash photography technique, as compared with the SLC process. The details of the visualization technique can be found in other published wotk, in which similar methods have been employed, eg. [5]. Figure 3(a) shows the sequence of the laser-induced thin liquid film vaporization. Collision of the shock wave with the surface and subsequent evaporation of the liquid (water/isopropanol mixture) film is visualized in Fig. 3(b) while Fig. 3(a) shows the liquid film evaporation by the explosive vaporization only. The shock wave produced by the Nd;YAG laser pulse touches the Si surface at 1 μs. It is evident that the impingement of the shock wave complicates liquid film ablation. Especially, relatively slow propagation of the vapor/droplet plume and gas flow in the direction parallel to the surface are observed in the case of the LLSC process. These results suggest that further investigations on the explosive vaporization of liquid film under the shock impingement condition should be carried out for optimizing the LLSC process. However, detailed analysis of the shock wave interaction with vaporization kinetics is beyond the scope of the present study.

3. Experimental

3.1. *Sample preparation*

The contaminated samples used in this cleaning experiment were prepared by following steps. Alumina (Al_2O_3, nominal average diameter 50 and 100 nm) and copper-oxide (CuO, nominal average diameter 50 nm) particles were dispersed in deionized water and the suspension was sonicated for 30 min in an ultrasonic cleaner bath. The suspension was then agitated for 30 min using a magnetic stirrer to form a uniform suspension. Once the suspension was prepared, Si-wafer samples were immersed in the suspension and dried for 24 hr before the cleaning operation. A typical example can be seen in Fig. 4 where 50 nm sized CuO nanoparticles were deposited onto a Si wafer. There are some agglomerated particles, however, most of particles are well dispersed on the Si wafer surface.

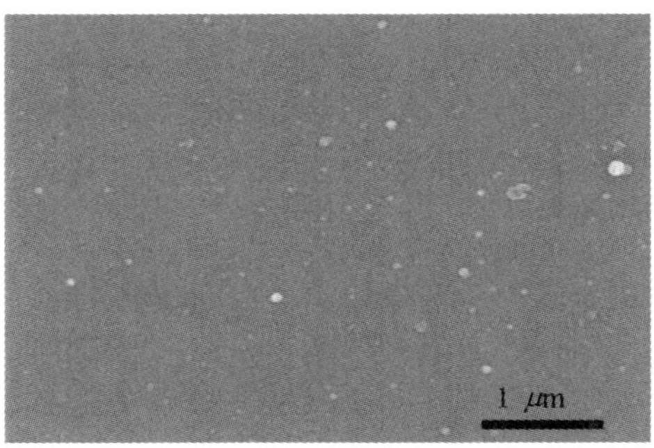

Figure 4. Typical image of a contaminated sample surface used in the LLSC experiments. CuO (50 nm nominal diameter) nanoparticles were dispersed on Si wafer.

3.2. *Cleaning experiment*

A schematic diagram of the experimental setup for LLSC is shown in Fig. 5. By focusing of a Q-switched Nd:YAG laser (wavelength: λ = 1064 nm; FWHM: τ = 6 ns; pulse energy: E=520 mJ) beam, a strong shock wave is generated. A

KrF excimer laser (λ = 248nm; τ = 25 ns) is employed to heat the substrate and vaporize the thin liquid film. For attaining a uniform spatial distribution of laser energy, the excimer laser beam having Gaussian intensity distribution was cut off by an aperture and only the core part with a relatively flat intensity distribution. The Nd:YAG laser beam is incident on the sample at a grazing angle ($\theta \approx 0°$). The gap between the sample and the shock wave center is d = 2 mm and the angle of excimer laser incidence is θ = 30°. A thin liquid film (from a mixture of isopropanol and water, 1:10 by volume) is deposited by a nozzle driven by a high-pressure purified air line. A heater is used to maintain the liquid inside the source container at an elevated temperature around 60 °C. The puffing duration for the liquid film deposition is about 50 - 100 ms. After a delay of 100 ms, Nd:YAG laser is fired to generate the shock wave and finally the excimer laser is fired when the shock wave contacts the cleaning spot on the sample surface (approximately 1~2 μs after the Nd:YAG laser pulse). After the cleaning test, the sample surfaces were analyzed using an optical microscope (×1000) and a scanning electron microscope (SEM).

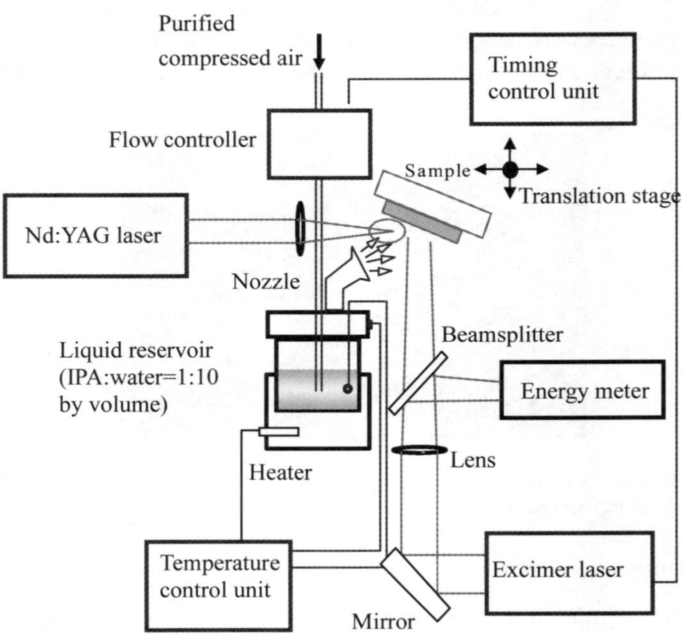

Figure 5. Experimental setup for LLSC.

4. Results and Discussion

4.1. *Combined effect of explosive vaporization and shock wave impingement*

Before performing the experiment using nanoscale contaminant particles, cleaning tests were carried out using relatively large (mean diameter 5 μm) alumina particles to verify the combined effect of the two laser pulses. The excimer laser fluence and Nd:YAG laser pulse energy were adjusted to be about 220 mJ/cm^2 and 140 mJ, respectively, neither of which was enough for the SLC or the LSC process alone to remove the particles completely. The image displayed in Fig. 6(c) demonstrates that the combined effect indeed increases the cleaning force when the two laser pulses are properly synchronized. It should be noted that three SLC cycles followed by three LSC cycles could not remove the particles completely. Figure 6 confirms that the synchronization of the two laser pulses is critical for this method to be effective.

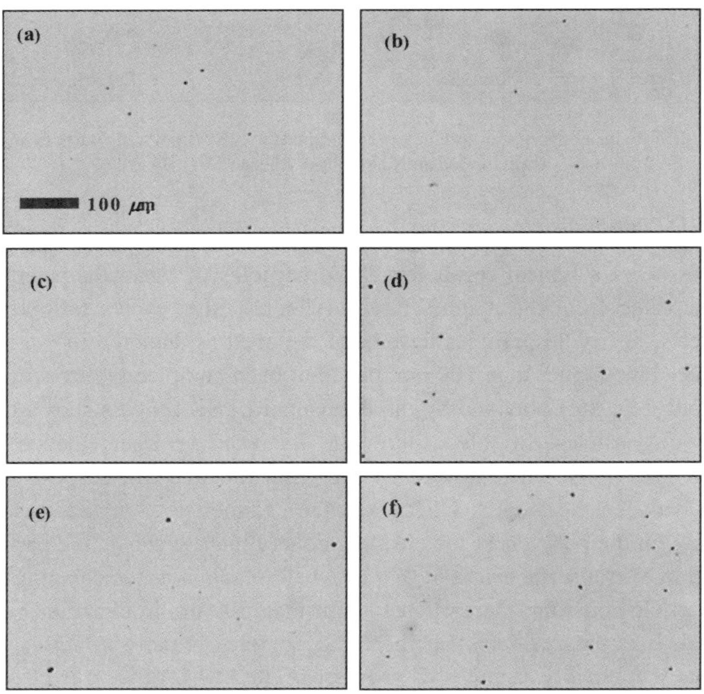

Figure 6. Optical micrographs of Si-wafer surfaces after three LLSC cleaning cycles (d=2 mm) for various delay times: (a) t_0=0 s, (b) 0.4 μs, (c) 0.8 μs, (d) 1.2 μs, (e) 1.6 μs, and (f) 2.0 μs.

4.2. CuO *nanoparticle cleaning*

The cleaning performance of LLSC method was tested using CuO (mean diameter 50 nm) particles. After 100 LLSC cycles carried out, the contaminated surface was analyzed using SEM. The experimental result confirms that almost all particles have been removed as shown in Fig. 7. The delay t_0 between the Nd:YAG and the excimer pulses was fixed to 1 μs, and the sample was moved during the cleaning experiment for the combined force to scan all cleaning area uniformly. Though the SEM images displayed in Fig. 7 are not for the same spot, it has been verified that similar results are obtained regardless of the choice of the test spot, within the cleaned area of several millimeters in length.

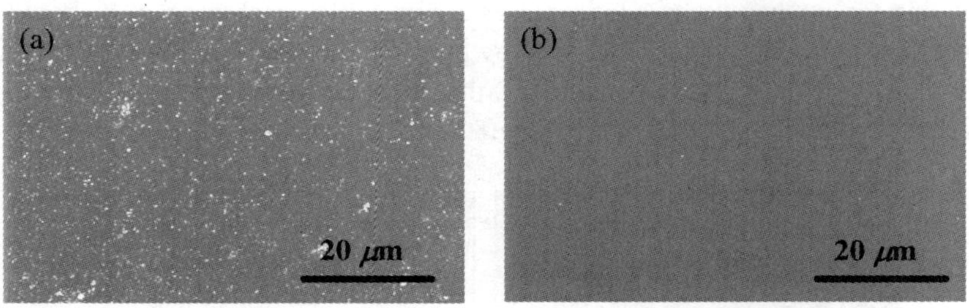

Figure 7. SEM images of Si wafer surface (a) contaminated with 50 nm CuO particle and (b) after cleaning. 100 LLSC cycles is performed at excimer laser fluence 170 mJ/cm^2.

4.3. Al_2O_3 *nanoparticle cleaning*

Figure 8 shows a typical result for Al_2O_3 particles of mean diameter 100 nm (nominal value from the manufacturer). After 20 SLC cycles followed by 20 LSC cycles, nearly all particles have been removed as depicted in Fig. 8(b) but some particles, smaller than 100 nm, have not been completely removed. On the other hand, Fig. 8(c) obtained by the equivalent LLSC process shows complete removal of particles. In this experiment, the excimer laser fluence and the Nd:YAG laser pulse energy were 270 mJ/cm^2 and 520 mJ, respectively. The delay t_0 between the Nd:YAG laser and the excimer laser pulses was 1~2 μs, depending on the position of the cleaning zone. Figure 9 shows the particle size distribution over a surface area of 726 μm^2 before and after the cleaning process. In the particle counting, the particles larger than 500 nm in diameter have been neglected. It is thus evident that the LLSC process is more effective than the sequentially applied SLC and LSC processes. The LLSC process is effective for particles greater than 100 nm approximately but the efficiency decreases rapidly as the particle size is reduced below 50 nm.

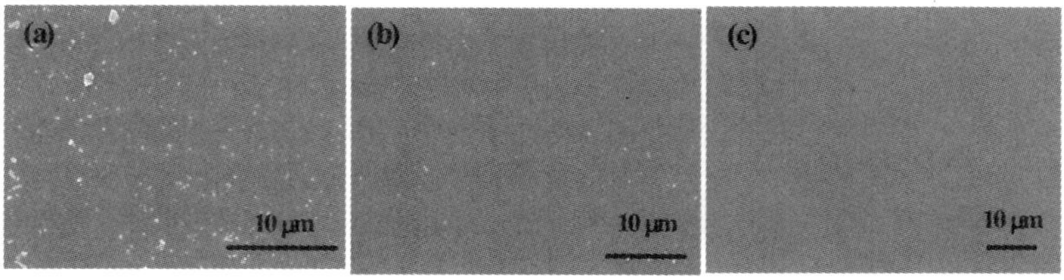

Figure 8. SEM images of a silicon substrate with Al_2O_3 particles 100 nm in nominal mean diameter: (a) before cleaning, (b) after 20 SLC cycles followed by 20 LSC cycles, and (c) after 20 LLSC cleaning cycles.

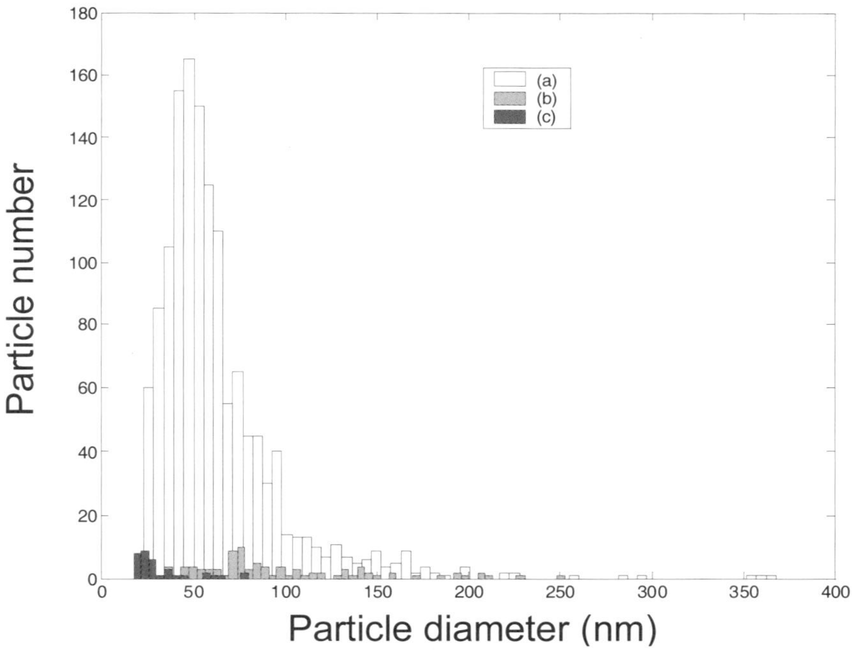

Figure 9. Particle size distribution (100 nm nominal mean diameter): (a) before cleaning, (b) after 20 SLC cycles followed by 20 LSC cycles, and (c) after 20 LLSC cleaning cycles.

Cleaning tests were performed for Al_2O_3 particles of 50 nm in nominal mean diameter, with the same operation parameters. The results are shown in Fig. 10. In this experiment, the number distribution was obtained for a surface area of 1425 μm^2, leading to the results displayed in Fig. 11. These results are largely similar to those in Fig. 9. In particular, no particle larger than 80±10 nm remains while the particles less than this size are only partially removed in both cases.

Figure 10. SEM images of a silicon substrate with Al_2O_3 particles 50 nm in nominal mean diameter: (a) before cleaning, (b) after 20 SLC cycles followed by 20 LSC cycles, and (c) after 20 LLSC cleaning cycles.

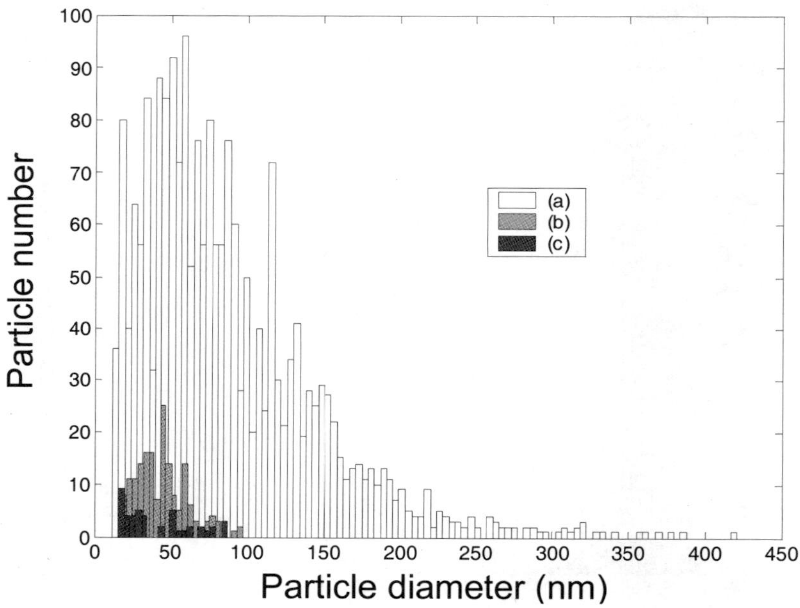

Figure 11. Particle size distribution (50 nm nominal mean diameter): (a) before cleaning, (b) after 20 SLC cycles followed by 20 LSC cycles, and (c) after 20 LLSC cleaning cycles.

In Fig. 12, the cleaning efficiency is plotted for the case of 50 nm alumina particles. The cleaning efficiency is defined as the ratio of the number of removed particles to that of the originally deposited particles. Perfect, i.e., 100 %, removal efficiency has been obtained for particles larger than 100 nm in diameter. However, the efficiency diminishes as the particle size becomes small. It is clear from Fig. 12 that the LLSC process is effective for removal of nanoscale particles.

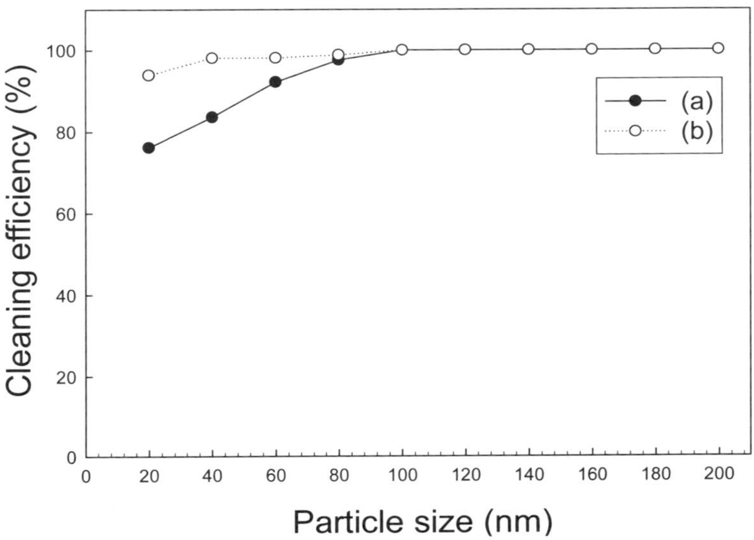

Figure 12. Cleaning efficiency (50 nm nominal mean diameter) of (a) after 20 SLC cycles followed by 20 LSC cycles, and (b) after 20 LLSC cleaning cycles.

5. Conclusions

In this study, a novel hybrid laser cleaning method based on the combined effect of liquid film evaporation and laser-induced shock impingement has been introduced. The results demonstrated that the cleaning force to remove particles from a solid surface can be enhanced significantly by the combined effect. Compared with the conventional laser shock cleaning (LSC) and steam laser cleaning (SLC) processes, the LLSC process shows superior cleaning performances for Al_2O_3 and CuO nanoparticles. Also, the results of the cleaning experiment reveal that such process parameters as the synchronization of the two laser pulses, the liquid film thickness, and the method of deposition of the contaminant particles have a strong effect on the cleaning performance. Consequently, this work suggests that this novel cleaning method can be employed with process optimization for removal of nanoscale contaminant particles which cannot be removed by other conventional cleaning methods. As the dynamics of liquid film evaporation in the LLSC process is considerably different from that in the SLC process, further investigations are required for analyzing the interaction between the shock wave and the explosively vaporizing liquid film.

Acknowledgments

This work was supported by The Korean Government National R&D Project for Nano Science and Technology.

References

1. A. C. Tam, W. P. Leung, W. Zapka and W. Ziemlich, *J. Appl. Phys.* **71**, 3515 (1992).
2. W. Zapka, W. Ziemlich, W.P. Leung and A.C. Tam, *Microelectronic Eng.* **20**, 171 (1993).
3. S. J. Lee, K. Imen and S. D. Allen, *J. Appl. Phys.* **74**, 7044 (1993).
4. Y. W. Zheng, B. S. Luk'yanchuk, Y. F. Lu, W. D. Song, Z. H. Mai, *J. Appl. Phys.* **90**, 2135 (2001)
5. M. She, D. Kim and C. Grigoropoulos, *J. Appl. Phys.* **86**, 6519 (1999).
6. Y. F. Lu, Y. Zhang, Y. H. Wan and W. D. Song, *Appl. Surface Sci.* **138-139**, 140 (1999).
7. M. Mosbacher, V. Dobler, J. Bonebnerg and P. Leiderer, *Appl. Phys. A* **70**, 669 (2000).
8. G. Vereecke, E. Rohr and M. M. Heyns, *Appl. Surf. Sci.* **157**, 67 (2000)
9. J. Lee and K. Watkins, *J. Appl. Phys.* **89**, 6496 (2001).
10. J. M. Lee, S. H. Cho, J. G. Park, S. H. Lee, Y. P. Han and S. Y. Kim, *Proc.3rd Int. Symp.on Laser Precision Microfab.*, Osaka, Japan, 287 (2002).
11. C. Cetinkaya, R. Vanderwood and M. Rowell, *J. Adhesion Sci. Technol.* **16**, 1201 (2002)
12. R. Vanderwood and C. Cetinkaya, *J. Adhesion Sci. Technol.* **17**, 129 (2003)
13. D. Kim and H. Lim, *The Iron and Steel Institute of Japan International* **43**, 1289 (2003).
14. H. Lim and D. Kim, *J. Laser Appl.* **16,** 25 (2004).
15. H. Lim and D. Kim, *Appl. Phys. A* **79**, 965 (2004).
16. P. Agostini, G. Barjot, G. Mainfray, C. Manus and J. Thebault, *IEEE J. Quant. Electron.* **QE6**, 782 (1970)
17. F. Morgan, L.R. Evans and C.G. Morgan, *J. Phys. D: Appl. Phys.* **4**, 225 (1971)
18. C.L.M. Ireland and C.G. Morgan, *J. Phys. D: Appl. Phys.* **6**, 720 (1973)
19. J.R. Ho, C.P. Grigoroupoulos and J.A.C. Humphrey, *J. Appl. Phys.* **79**, 7205 (1996)
20. H. K. Park, D. Kim, C. P. Grigoropoulos and A. C. Tam, *J. Appl. Phys.* **80**, 4072 (1996).
21. D. Kim, H. K. Park and C. P. Grigoropoulos, *Int. J. Heat Mass Transfer* **44**, 3843 (2001).

Chapter 6

UV LASER-INDUCED DEHYDROXYLATION OF UV FUSED SILICA SURFACES

A J FERNANDES, D M KANE
Department of Physics, Macquarie University, Sydney, NSW 2109, Australia

B GONG, R N LAMB
Surface Science and Technology, The University of New South Wales, Sydney, NSW 2052, Australia

The 'clean' surface of silica glass is usually covered with a quasi-layer of hydroxyl groups. These groups are significant as their concentration on a surface affects surface adhesion and chemical reactivity. Removal of hydroxyl groups from the surface by a UV pulsed laser treatment has been demonstrated to be an alternative technique to the dehydroxylation of glass by the traditional oven heat treatment. Silica so treated has improved resistance to particulate adhesion. Dehydroxylation using this UV laser treatment has key advantages of being: a much faster process; largely limited to heating the surface not the bulk of the silica; and which allows selective spatial patterning of the dehydroxylation of the silica surface. This work outlines a technique developed to allow systematic, quantitative measurements of the dehydroxylation of UV fused silica. The removal of hydroxyl groups using laser irradiation is shown to be a thermal process.

1. An Introduction to the Surface Structure and Dehydroxylation of Silica Glass

There have been several interesting spin-offs of research into using short pulsed lasers to remove particles from surfaces such as particle mediated ablation and patterning[1-8], and the dehydroxylation of silica surfaces using laser irradiation[9-11]. The work presented here reports on the dehydroxylation of the surface of ultraviolet (UV) fused silica slides, by irradiation with a KrF excimer laser, with a detailed investigation into how the process can be quantified and controlled. A number of experimental issues of surface masking by normal open lab environment contaminants (moisture and hydrocarbons) have had to be uncovered and solved.

The surface of fused silica (SiO_2) is characterized by various hydroxyl groups (OH) which are attached to the silicon atoms[12-15], and are formed when water (H_2O) reacts with the residual valencies on the siloxane (Si-O-Si) surface. These surface hydroxyl groups are significant, as they are key sites in adhesion and bonding, and their concentration affects the physical performance of silica in photonics, microelectronics, catalysis and chromatography[15-17]. As silica is amorphous, the silicon atoms on the surface are not in a 'regular geometrical arrangement', and hence the hydroxyl concentration across the surface varies.

The term 'dehydroxylation' refers to the removal of hydroxyl groups from the silica surface and is normally achieved by heating the bulk material to high temperatures (over 1000°C) over several hours[13, 15-18]. Figure 1 depicts the most commonly observed surface hydroxyl groups. At room temperature, physisorbed water occurs on the surface (see Figure 1a), and must be totally removed before dehydroxylation can occur. This occurs at ~200°C[15], leaving only single silanol, geminal, vicinal and siloxane groups on the surface. The vicinal groups (Figure 1b) are removed at temperatures of ~450-500°C, while geminal (Figure 1c) and the majority of single silanol groups (Figure 1d) are removed between 600-900°C. Once the silica is heated to between 1000-1100°C, the surface is predominantly covered with siloxane groups (Figure 1e)[15].

A silica surface is said to be *hydroxylated* when the surface is dominated by single silanol groups. The surface is also *hydrophilic* as the polar OH groups hydrogen-bond water molecules to the surface. In contrast, if siloxane groups are dominant on the surface, the silica is said to be *dehydroxylated* and *hydrophobic*.

Irradiation with a pulsed, frequency-doubled, copper vapour laser (UV-CVL, 255nm) was first found to cause dehydroxylation of a silica surface[9, 10]. During laser cleaning experiments, Halfpenny observed that the surface of a silica slide which had been irradiated with several hundred UV-CVL laser pulses was resistant to particles adhering post process. Further examination found that the surface was hydrophobic. The native and laser irradiated silica surface were investigated using X-ray Photoelectron Spectroscopy (XPS) and Time-of-Flight Secondary Ion Mass Spectrometry (ToF-SIMS) from which it was discovered that hydroxyl groups had been removed from the laser-irradiated surface.

The laser-induced dehydroxylation of silica is a relatively new field, and the study by Halfpenny is the only known prior work to the KrF laser dehydroxylation described here and in Fernandes *et al.*[11] From a laser cleaning perspective the laser induced dehydroxylation is a useful processing tool, making it possible to treat a surface to make it hydrophobic and resistant to much contamination in a clean environment.

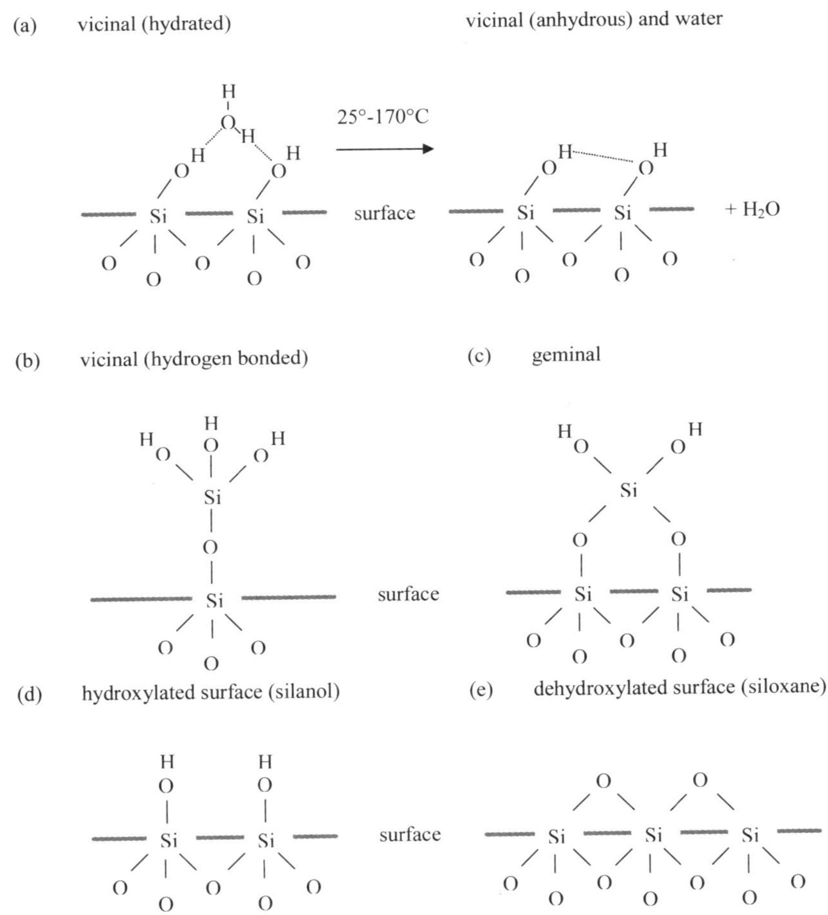

Figure 1: Commonly observed hydroxyl groups that form on the surface of fused silica: (a) at room temperature, hydrated vicinal groups (due to physisorbed water) occur. Heating removes the physisorbed water, and anhydrous vicinal groups remain; (b) vicinal groups, hydrogen bonded; (c) geminal silanol groups; (d) single silanol groups – a hydroxylated surface; (e) siloxane groups – a dehydroxylated surface formed by heating. Adapted from Iler[13] and Senn *et al.*[15]

It has significant potential in other fields which require control over hydroxyl concentration. The process also has advantages over the conventional oven heating method, as process times are significantly shorter (seconds rather than hours), and it allows for selective treatment (patterning) of the surface by either direct write or masking techniques. In addition, only the surface of the glass is irradiated, thus minimizing problems which may arise from changes in the bulk material from heating of the entire glass substrate.

The ToF-SIMS system can be used to make a good relative measurement of the dehydroxylation of the silica samples. Wood et al.[12, 18] had shown that from the mass spectra obtained, it was possible to compare the ratio of the $SiOH^+$ and Si^+ ion peaks to determine if the surface hydroxyl groups were removed after treatment with high temperatures. A decrease in the $SiOH^+/Si^+$ ion ratio meant that surface hydroxyls were being removed. This is also the technique used by Halfpenny, and in this work, to measure the dehydroxylation of the laser irradiated samples.

In Halfpenny's experiments, the UV-CVL was used to irradiate silica samples at fluences in the range 0.1 - 0.5J/cm^2 per pulse, for several hundred pulses. The average laser irradiance was between 0 - 600mW/cm^2, at a repetition rate of 4.25kHz. Beam rastering was used to obtain a large area for analysis, and the equivalent dwell time at each point was 100ms. The study measured the reduction in the $SiOH^+/Si^+$ ion ratio as a function of average laser irradiance (mW/cm^2). It was found that the $SiOH^+/Si^+$ ratio decreased with increasing irradiance, and was zero (total dehydroxylation) and almost zero in several cases. Damage (pitting and cracking) of the surface was observed for average irradiances greater than 250mW/cm^2 during the 100ms exposure. However, the evaluated $SiOH^+/Si^+$ ratio for the ***untreated*** surface of three different samples was not consistent and ranged between 0.04 – 0.25. This was thought to be a result of variations in the physisorbed water on the silica surface.

Halfpenny initially performed low resolution scans of the silica surface, which showed significant peaks of hydrocarbon contamination which may have been masking the $SiOH^+$ and Si^+ ion peaks. The contamination was attributed to exposure of the sample to the atmosphere and storage in non-ideal packaging (polyethylene bags)[9]. High resolution scans allowed the $SiOH^+$ and Si^+ peaks to be resolved from the hydrocarbon peaks, and wiping the surface with methanol before analysis reduced the contamination and further enhanced the resolution but there was still the potential of an unknown impact of hydrocarbon masking on the results[9].

Halfpenny proposed that there were two possible mechanisms which could cause the dehydroxylation of the silica surface. The first proposal was that it was a thermal process, where high temperatures were required at the surface, similar to those needed to achieve dehydroxylation of silica in an oven. It was believed that such high temperatures could be reached, as when high irradiances greater than 250mW/cm^2 were used, the temperature at the surface was high enough to cause physical damage. However, as optical damage needs to be avoided, the required temperatures can only be achieved by cumulative heating using many pulses at a fluence well below the threshold for optical damage. This means that the laser pulse repetition rate (frequency) is an important experimental parameter, and as the UV-CVL experiments were only performed at a repetition rate of 4.25kHz, this variable needs further examination. The dependency on pulse repetition rate combined with fluence also introduces a threshold in the average laser power required at which dehydroxylation will commence. Furthermore, if the dehydroxylation is due to a thermal mechanism, it should have an exponential (1/T(K)) dependency, provided that the surface temperature T(K) increases linearly with laser fluence[10].

Alternatively, if the process is a photochemical effect, then the photon energy of the UV light would actually break the OH bond. This would mean that there would be a limiting laser wavelength above which dehydroxylation would not occur (or would be significantly less efficient if relying on a multi-photon process), and that the amount of dehydroxylation possible would be dependent on the photon flux energy. The investigations detailed in this chapter use a KrF laser (248nm), which has a photon energy of 5.00eV, while the UV-CVL used by Halfpenny had a photon energy of 4.86eV. Both are greater than the bond dissociation energy of the OH bond (E_D = 4.43eV) meaning that bond breaking is possible with either laser.

It was not possible to study the process systematically and to draw well-defined conclusions based on the UV-CVL experiments performed by Halfpenny[9, 10]. The UV-CVL had a maximum pulse energy that was significantly less than optimal for the experiment, and the level of dehydroxylation achieved for a given set of experimental parameters was not reproducible, despite evidence that total dehydroxylation could be achieved. The lack of reproducibility also meant that it was not possible to determine conclusively whether the dehydroxylation was caused by a thermal or photochemical process.

The aim of this work was to determine whether dehydroxylation of silica surfaces was possible using a KrF excimer laser, and if methodologies could be developed from which reproducible and quantifiable results are obtained using ToF-SIMS. This chapter discusses the systematic experiments which have been developed and performed in order to gain a better understanding of the process. The KrF laser allowed for investigation of a greater range of fluences and treatment of larger sample areas without beam rastering. Furthermore, it was possible to vary the repetition rate (albeit at lower frequencies compared to the UV-CVL). The issue of hydrocarbon masking of spectra was also examined in more detail, which has led to the development of new processes for preparing and handling the silica samples for both laser treatment and ToF-SIMS analysis, thus allowing reproducible results to be obtained.

Quantitative measures of the level of dehydroxylation were achieved, and the dependency on laser parameters evaluated. This led to the determination that the mechanism by which laser dehydroxylation occurs is a ***thermal*** process which requires sufficient ***cumulative heating*** of the surface. The work investigating the effect of laser repetition rate showed that there is a laser power threshold where dehydroxylation commences and that the level of dehydroxylation is a function of the laser repetition rate. If the laser pulses are delivered at too long an interpulse interval, less or no dehydroxylation occurs. This shows that a cumulative heating effect is essential. That is, the laser-induced dehydroxylation is a thermal effect. Hence care must be taken to irradiate within certain laser pulse energy/fluence windows to both achieve dehydroxylation and to avoid optical damage at high repetition rates.

Note that although the dehydroxylation increased as a function of laser pulse fluence, it did not follow an exponential dependency as would be expected for a thermal mechanism. This was because insufficient cumulative heating was achieved with the laser parameters used.

Some preliminary dehydroxylation results using the KrF laser were published in [11]. The results presented here include new data as well as expanding on the earlier work. The investigation into dehydroxylation as a function of laser pulse fluence in §4.1 has been expanded to include results obtained using a laser repetition rate of 100Hz rather than just at 20Hz, while in §4.2 the dehydroxylation as a function of total pulse fluence is also investigated at 50Hz and 100Hz. Furthermore, an investigation into the effect of the total number of laser pulses used on dehydroxylation has been done and is detailed in §4.3.

2. Guidelines for Sample Preparation

2.1. *Pre-treatment of samples before laser irradiation and sample storage*

High quality ultraviolet fused silica samples (Corning UV 7980) were used for these experiments. These are highly transmitting (> 95%) at the KrF laser wavelength of 248nm[19, 20]. The dimensions of the samples were $20x10x2mm^3$, so that they could fit in the analysis chamber carousel of the ToF-SIMS system. They were polished to a flatness of $\lambda/2$.

To minimise hydrocarbon contamination from the handling and packaging processes as observed by Halfpenny[9, 10], all samples were ultrasonically cleaned in xylene for ten minutes and rinsed in methanol, before being irradiated. A special storage system was designed and constructed to store these samples and for transporting them between Macquarie University and the University of New South Wales for ToF-SIMS analysis. Exposure to the local atmosphere for a day in this system in addition to the laser irradiation process led to minimal hydrocarbon presence.

Each sample was mounted on its own stainless steel mounting disc, and kept in aluminium trays in aluminium boxes. This was to avoid contamination from plastics[21]. When irradiating the samples, the mounting discs were attached to an aluminium holding mount. All mounts, storage trays, and handling equipment (tweezers, for example) were ultrasonically cleaned in xylene for at least ten minutes, and rinsed with methanol before use, again to minimise hydrocarbon contamination of the samples.

2.2. *The hydrogen peroxide cleaning process*

The hydrogen peroxide cleaning process is the first step in a standard RCA wet cleaning technique for silicon[22, 23], and it is regularly applied to glass[24]. This method was used to treat the UV fused silica slides in order to effectively remove contamination, including surface hydrocarbons. This was a necessary requirement in order to be able to use ToF-SIMS to quantify the hydroxyl concentration of the silica surface.

The relative concentration of surface hydroxyls was monitored by evaluating the ratio of the $SiOH^+$ peak (at 45a.m.u) to the Si^+ peak (at 28a.m.u.) on the obtained spectra (see §3.1 for further details on elemental sensitivities). Non-irradiated slides which were hydrogen peroxide cleaned were found to have a reproducible characteristic $SiOH^+/Si^+$ ratio of 0.19±0.02. However, the presence of hydrocarbons was detected by ToF-SIMS on both irradiated and non-irradiated slides which had been exposed to the environment for more than one day. The hydrocarbon peaks dominated the spectrum obtained for the sample and meant that an accurate analysis of the surface hydroxyl concentration could not be obtained. In order to overcome this, any such sample was hydrogen peroxide cleaned to remove the hydrocarbons. The effect of hydrocarbon masking when evaluating the $SiOH^+/Si^+$ ratio is detailed in §3.2. The hydrogen peroxide process consists of a number of steps to remove surface contamination, including any hydrocarbons. This process was performed at the University of New South Wales, immediately before samples were placed into the ToF-SIMS for analysis, to minimize contamination due to transporting from Macquarie University.

The glass samples are first soaked in ethanol for ten minutes, before drying them with dry nitrogen gas. The slides are then placed in a beaker containing hydrogen peroxide with one-tenth volume of ammonium hydroxide. The beaker was placed in a water bath under a fume hood, and heated to 50°C. At this temperature, the hydrogen peroxide and ammonium hydroxide mixture reacts, forming bubbles. After approximately 20 minutes, the bubbles subside, indicating that the reaction has finished. The samples can then be removed from the heat.

Our samples were then rinsed with Milli-Q® water, before being boiled in Milli-Q® water for a further hour. Milli-Q® water was used because de-ionised water contains organics which affect the surface chemistry, and can make the contact angle change[24]. Milli-Q® water is high-purity water which meets specific water quality requirements for use in laboratories. Millipore's Milli-Q® Ultrapure Water Purification Systems use a three-step purification process and periodical recirculation to meet these standards[25, 26].

The samples were then rinsed again with Milli-Q® water, and stored in a beaker containing Milli-Q® water while waiting to be loaded into the ToF-SIMS for analysis. The samples were dried with dry nitrogen gas immediately before being placed in the loading chamber.

2.3. *Laser irradiation experimental setup*

Figure 2 is a basic depiction of the laser system used to irradiate the fused silica samples. The laser used was a GSI Lumonics Pulsemaster 848 KrF excimer (248nm, 12ns), which has a 'Stabilase' feature that allows the average pulse output to be selected, whilst keeping the pulse to pulse variation to a minimum. An Optec AT4030 attenuator was used to vary the output fluence (energy per unit area). This allowed the laser beam dimensions, and hence the area irradiated by the beam, to remain the same.

Each silica slide was attached to a thin stainless steel mounting disc and during irradiation, the slide/disc were affixed to an aluminium holding mount. This was to avoid contamination from plastics as mentioned in §2.1. The holding mount was attached to a *xyz* translation stage so that different regions of the slide could be irradiated. A spherical lens with a focal length of approximately 10cm was used to focus the laser beam onto the UV fused silica slides.

The slides were irradiated in different ways, by changing the laser repetition rate, the laser fluence, and the total number of pulses used. A discussion of the laser irradiation processes on the dehydroxylation process, and evaluation of the possible mechanisms by which it occurs, is detailed in §4.

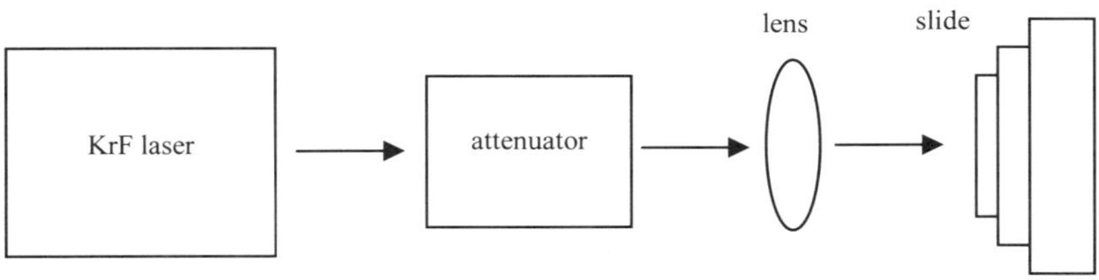

Figure 2: Schematic diagram of the experimental setup for sample irradiation. The laser irradiation is attenuated, before being focused onto the silica slide. Different areas of the slide can be irradiated by moving the translation stages.

3. Time-of-Flight Secondary Ion Mass Spectrometry (ToF-SIMS)

3.1. *An introduction to ToF-SIMS*

The fused silica samples were analysed using the Kratos Time-of-Flight Secondary Ion Mass Spectrometry (ToF-SIMS) system in the Department of Surface Science and Technology at the University of New South Wales. SIMS is a technique that analyses the chemical composition of the top few micrometres of a surface. In these experiments, the different ions present on the surface of the silica samples are determined using a Time-of-Flight mass spectrometer, hence the name 'ToF-SIMS'.

The ToF-SIMS system requires samples to be placed into a main analysis chamber under high vacuum, via a separate loading arm. This is to allow for 'fast' sample exchange without affecting the vacuum of the large main chamber. The sample could be moved inside the chamber which allowed for different regions of the slide to be analysed. However, as the number of slides in a given experimental test could be larger than the limited number of slides that could be loaded into the ToF-SIMS in a single day, the issue of contamination for the remaining slides in a test group had to be considered.

The ToF-SIMS method applied to analyse the silica samples used a focussed beam of Ga^+ primary ions to bombard and erode the surface. The secondary ions formed by the impact are then swept away from the surface by an electric field. The mass and energy of these secondary ions are then analysed by the ToF spectrometer, and are displayed as a function of mass[27]. Note that a spectrum is only obtained when a high yield, or ion count, is detected by the system. A GRAMS/32 software program was used to monitor the ion count, and once optimised, was able to produce a mass spectrum of the surface ions detected.

In order to obtain quantifiable data, an understanding of the sensitivity of element detection is required. A characteristic of SIMS is that the ionization efficiencies for the sputtered secondary ions vary with ionization potential and electron affinity. That is, the different elements are detected with varying sensitivities and will therefore have different respective ion counts[21, 27]. For example, the sodium ion (Na^+, 23a.m.u.) and the potassium ion (K^+, 39a.m.u.) are both very sensitive to SIMS and may appear as very large peaks on the spectrum (high ion counts in the order of thousands over the analysis period). However, the $SiOH^+$ ion (45a.m.u.) has much lower yields (in the order of hundreds during analysis), and despite there actually being a greater quantity of it on the surface, it will appear as a relatively smaller peak.

Furthermore, the total ion count for a sample affects the height and area of each peak on the spectrum, and may vary at different points on the sample, despite being optimised. Therefore, to obtain quantitative data from different spectra, it is necessary to compare ratios of the height and/or area of selected peaks, rather than observing a single peak. It is also necessary to obtain a number of readings/spectra for each test area to determine the average ratio, due to the hydroxyl concentration varying across the surface of the silica, as mentioned in §1.

In these experiments, the area of the $SiOH^+$ (45a.m.u.) and Si^+ (28a.m.u.) ion peaks are monitored and evaluated using the GRAMS/32 software program. The ratio of the peaks $SiOH^+/Si^+$ allows monitoring of the relative concentration of surface hydroxyls, enabling the level of dehydroxylation achieved by laser treatment to be measured as a reduced $SiOH^+/Si^+$ ratio.

3.2. *Analysis of the surface hydroxyl concentration and the effect of hydrocarbon masking*

The surface of silica can still be contaminated by hydrocarbons, despite attempts to reduce the amount of contamination caused by packaging. To investigate the non-irradiated silica surface, several samples were cleaned using the hydrogen peroxide method. Figure 3 is a typical spectrum obtained using ToF-SIMS for such a slide. Repeated testing of such samples were found to have a reproducible $SiOH^+/Si^+$ ratio of 0.19 ± 0.02. This was used as a characteristic reference value for the surface hydroxyl concentration of the UV fused silica used, and may differ for other silicas. A similar spectrum was obtained for samples which were also cleaned using the hydrogen peroxide method, but analysed one day later. This was found to not affect the evaluation of the characteristic reference ratio.

Samples which were exposed to the environment for longer than one day suffered significant hydrocarbon contamination. This resulted in 'false' $SiOH^+/Si^+$ ratios being obtained which were higher than the characteristic reference ratio of 0.19±0.02. This effect was referred to as 'hydrocarbon masking' of the sample, as the true hydroxyl concentration of the surface could not be determined. An example of this effect is depicted in Figure 4, which is a spectrum of a non-irradiated slide that was cleaned with the hydrogen peroxide method but was then exposed to the environment for one week, before analysis with ToF-SIMS. The spectrum shows the significant presence of hydrocarbons due to this exposure, and has a $SiOH^+/Si^+$ ratio of 0.47 (the average ratio for this slide was 0.38±0.09). This ratio is much higher than the characteristic reference ratio of 0.19±0.02, and indicates that the hydrocarbon contamination covers or 'masks' the surface thus preventing the ToF-SIMS from detecting the true nature of the sample. It is important to note that the characteristic reference value could be again obtained after cleaning the samples with the hydrogen peroxide method. That is, the removal of the hydrocarbon contamination via this technique did not affect the 'true' surface hydroxyl concentration.

Samples which had been laser irradiated, and were subsequently found to have a $SiOH^+/Si^+$ ratio lower than the characteristic ratio, were said to be dehydroxylated. The lower the ratio obtained was, the more dehydroxylated a surface was said to be. A surface was said to be completely dehydroxylated if a $SiOH^+/Si^+$ ratio of zero was obtained. Figure 5 is a spectrum for a surface which has been irradiated with 400 pulses at a fluence of 850mJ/cm^2, at a laser repetition rate of 20Hz. This sample was not able to be analysed immediately with ToF-SIMS and required cleaning with the hydrogen peroxide method to remove the hydrocarbon contamination caused by the delay. The average $SiOH^+/Si^+$ ratio for this sample was 0.12±0.04, indicating that the surface was partially dehydroxylated.

Note that hydrocarbon masking also affected irradiated slides which had been exposed to the environment for more than a day after laser treatment. Analysis with ToF-SIMS showed that irradiated samples with hydrocarbon contamination also produced $SiOH^+/Si^+$ ratios which were higher than the characteristic reference ratio for the UV fused silica samples used.

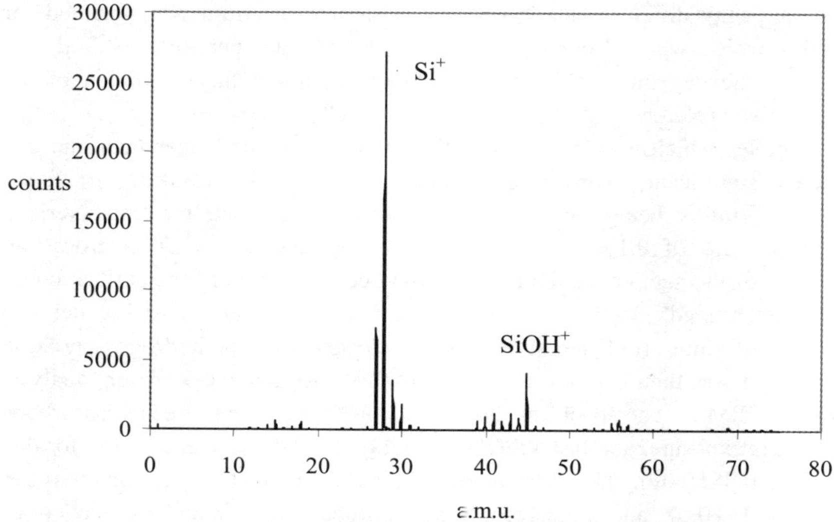

Figure 3: A typical spectrum for a non-irradiated sample which has been cleaned using the hydrogen peroxide process and analysed immediately.

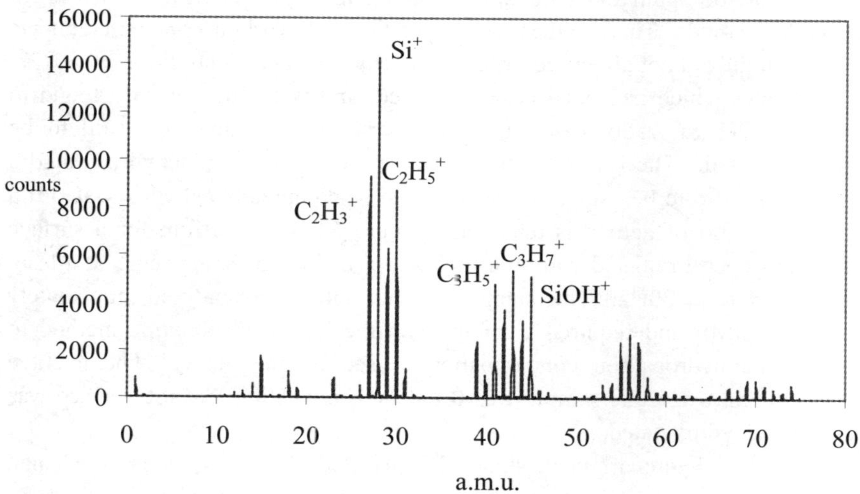

Figure 4: A typical spectrum for a non-irradiated surface, cleaned using the hydrogen peroxide process, and analysed one week after environmental exposure. The hydrocarbon contamination is significant

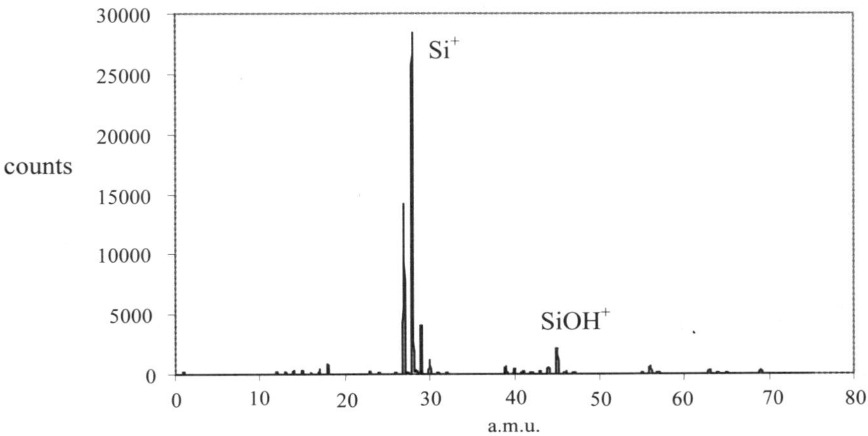

Figure 5: A typical spectrum for an irradiated sample. This sample has been partially dehydroxylated as the average $SiOH^+/Si^+$ ratio was 0.12±0.04 and was lower than the characteristic reference value.

Comparison tests were done where an irradiated sample was analysed by ToF-SIMS a day after irradiation firstly, then analysed again after it had been exposed to the environment for a week, and finally analysed once again immediately after cleaning with the hydrogen peroxide method. The first analysis tests indicated that the sample was dehydroxylated as it had a $SiOH^+/Si^+$ ratio lower than the characteristic reference value. Exposure to the environment caused masking of the surface due to hydrocarbon contamination and $SiOH^+/Si^+$ ratios higher than the characteristic reference value. Final analysis after hydrogen peroxide cleaning produced a similar $SiOH^+/Si^+$ ratio to the initial analysis value. This showed that it was possible to remove the hydrocarbon contamination masking the surface without significantly affecting the hydroxyl concentration measured with ToF-SIMS.

In conclusion, from these tests it was determined that samples which has been subject to environmental exposure for more than one day after preparation/irradiation needed to be cleaned to remove the hydrocarbon contamination before analysis. Reproducibility tests showed that treatment with the hydrogen peroxide method did not affect the hydroxyl concentrations measured by ToF-SIMS and the $SiOH^+/Si^+$ ratio. This was significant due to the limited number of samples that could be loaded into the ToF-SIMS in one day. By understanding the masking effects of hydrocarbons, and taking experimental precautions to minimise them, it was possible to obtain meaningful quantitative analysis of the concentration and removal of hydroxyls on the surface of the UV fused silica slides.

4. Experimental Results

4.1. *Dehydroxylation as a function of laser pulse fluence*

The effect of the laser pulse fluence on hydroxyl removal was investigated. The fluence of a single pulse was calculated by the average energy incident on the glass surface, divided by the surface area irradiated by the pulse. The fluence was varied between 100 – 1200mJ/cm^2. Each test area on the slide was irradiated at a set fluence, with 400 pulses at a laser repetition rate of 20Hz. The experiment was also repeated using a repetition rate of 100Hz.

Figure 6 shows the average $SiOH^+/Si^+$ ratios determined for regions of silica samples which had been irradiated at the different fluences, both at 20Hz and 100Hz. The tests show that the applied irradiation removes surface hydroxyls, as the $SiOH^+/Si^+$ ratios obtained are lower than the characteristic reference value of 0.19±0.02. At both repetition rates, there is a general trend of the $SiOH^+/Si^+$ ratio decreasing with increasing laser pulse fluence. This means that the silica surface has a lower hydroxyl concentration ('more dehydroxylated') when higher fluences are used to irradiate it. Although these results do not show an exponential dependency of the dehydroxylation on the laser pulse fluence, the lower $SiOH^+/Si^+$ ratios obtained at 100Hz, compared to those obtained when using a pulse repetition frequency 20Hz, indicate that cumulative heating is required for hydroxyl removal. This suggests that there is a threshold fluence at which the surface dehydroxylation begins to occur as a result of a laser-induced thermal mechanism where cumulative heating is required, rather than being caused by photochemical means (bond-breaking due to the laser wavelength).

These results indicate that the lowest relative hydroxyl concentrations achieved when using 400 pulses at either 20Hz or 100Hz, occurs at fluences between 800-1000mJ/cm^2. Furthermore, the relative hydroxyl concentration is lower at 100Hz than at 20Hz, for the same total fluence. ***Note that complete dehydroxylation (within the experimental uncertainty) occurs at fluences approximately between 800-1000mJ/cm^2 at 100Hz.*** It is therefore hypothesised that if the same total fluence (a set number of pulses at the same fluence) is applied in a shorter time span, that is, at a higher (faster) repetition rate, then an accumulative heating effect can occur causing an increase in the surface dehydroxylation. Note that care must still be taken to avoid unwanted damage to the surface. Damage is also more likely to occur at the same or lower fluences if a more focussed beam is used.

The proposal that the laser-induced surface dehydroxylation of silica occurs via a thermal mechanism is further investigated in the following sections.

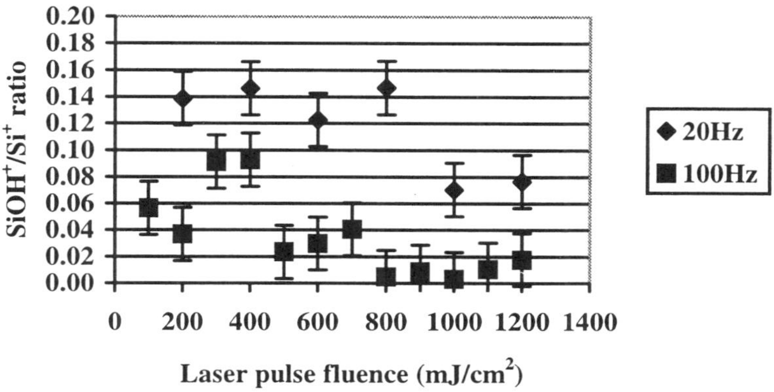

Figure 6: Dehydroxylation as a function of laser pulse fluence.

4.2. *Dehydroxylation as a function of total fluence applied*

The total fluence incident on a given area was calculated by multiplying the pulse fluence with the number of pulses used. This experiment investigated how the total fluence applied to the surface affected the hydroxyl concentration. The total fluence was set to $360J/cm^2$, so that at lower fluences more pulses were used, as compared to fewer pulses at higher fluences. The following pulse fluence/number combinations were used at a repetition rate of 20Hz: $600mJ/cm^2$ x 600 pulses, $700mJ/cm^2$ x 514 pulses, $800mJ/cm^2$ x 450 pulses, $850mJ/cm^2$ x 423 pulses, $1000mJ/cm^2$ x 360 pulses, $1100mJ/cm^2$ x 327 pulses, and $1200mJ/cm^2$ x 300 pulses.

Figure 7 shows the dehydroxylation achieved as a function of the pulse fluence at 20Hz, while keeping the total fluence deposited constant. A plateau region is observed between $700\text{-}1100mJ/cm^2$, where the $SiOH^+/Si^+$ ratio is approximately constant. This indicates that if a given amount of thermal energy is deposited, a corresponding reduction in the hydroxyl concentration is achieved. However, at a pulse fluence of $600mJ/cm^2$, the higher $SiOH^+/Si^+$ ratio means that less dehydroxylation has occurred, despite the same total fluence being applied. The energy is dissipating as it is being applied over too great a period of time (too many pulses). This supports the proposal that the laser-induced dehydroxylation occurs via a thermal mechanism.

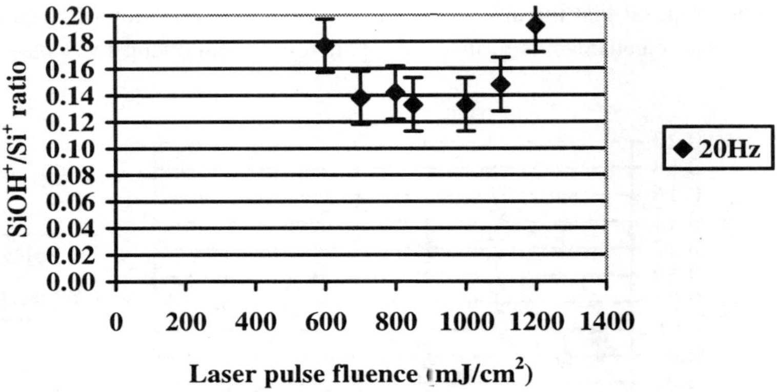

Figure 7: Initial experiments investigating dehydroxylation as a function of total pulse fluence.

At 1100mJ/cm^2 and greater, damage to the surface is visually observed. This exposes regions of the glass which have not been treated by the irradiation, and are therefore not dehydroxylated. The $SiOH^+/Si^+$ ratio obtained at these fluences is the same as for untreated silica. Note that while the fluence required for damage here is 1100mJ/cm^2, it is likely to be lower when higher pulse repetition rates are used.

These results show that there are operating limits to which the laser pulse fluence and number of pulses applied must be set, in order to ensure that the dehydroxylation process is maximised for a given total fluence without causing optical damage to the glass.

This experiment was repeated, again with a total fluence of 360J/cm^2, using the following pulse fluence/number combinations: 400mJ/cm^2 x 900 pulses, 500mJ/cm^2 x 720 pulses, 600mJ/cm^2 x 600 pulses, 700mJ/cm^2 x 514 pulses, 800mJ/cm^2 x 450 pulses, and 900mJ/cm^2 x 400 pulses. It was performed at laser repetition rates of 20Hz, 50Hz, 100Hz, and 200Hz, in order to see how this affected the operating limits and the removal of hydroxyls. It was hypothesised that the higher repetition rates would increase the level of dehydroxylation of the slide, despite the same total fluence being applied in the same way, due to the irradiation being applied in a shorter time.

Figure 8 shows the $SiOH^+/Si^+$ ratios as a function of the pulse fluence when applying the same total fluence at 20Hz and 50Hz. The $SiOH^+/Si^+$ ratios at 50Hz are generally lower than those at 20Hz, for the corresponding fluences. At 20Hz the $SiOH^+/Si^+$ ratio is ~0.04 at 800mJ/cm^2, while at 50Hz the optimal

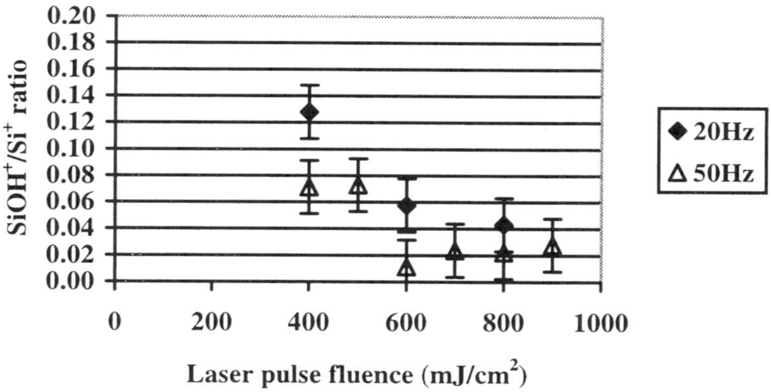

Figure 8: Further experiments investigating dehydroxylation as a function of total pulse fluence.

window for dehydroxylation is between 600-900mJ/cm^2 where a ratio of ~0.02 is achieved. The lowest $SiOH^+/Si^+$ ratio (best dehydroxylation) was achieved at 600mJ/cm^2 at 50Hz.

These results show that by applying the irradiation at a faster repetition rate the optimal window for dehydroxylation occurs at lower pulse fluences, as well as allowing for more hydroxyls to be removed from the silica surface at lower fluences, despite the same total fluence being applied. This supports the idea that the dehydroxylation occurs via a thermal mechanism which requires cumulative heating.

Figure 9 shows the same data as Figure 8, but also includes the results at 100Hz. The error bars for the earlier data have been removed for clarity. The optimal window for dehydroxylation at 100Hz is between 400-600mJ/cm^2, where a ratio of ~0.01 is achieved. The lowest $SiOH^+/Si^+$ ratio obtained occurs at 400mJ/cm^2, and it is possible that complete dehydroxylation may occur at a slightly lower fluence when using 100Hz. This requires a more detailed investigation involving the use of more fluences than were applied in this systematic set of results. Note that the optimal window here occurs at lower fluences compared to the results at 20Hz and 50Hz, and that the $SiOH^+/Si^+$ ratio is lower still. This indicates that the increased repetition rate improves the ability to remove surface hydroxyls. When using a pulse repetition rate of 100Hz, erosion (damage) of the surface begins to occur at pulse fluences above ~700mJ/cm^2.

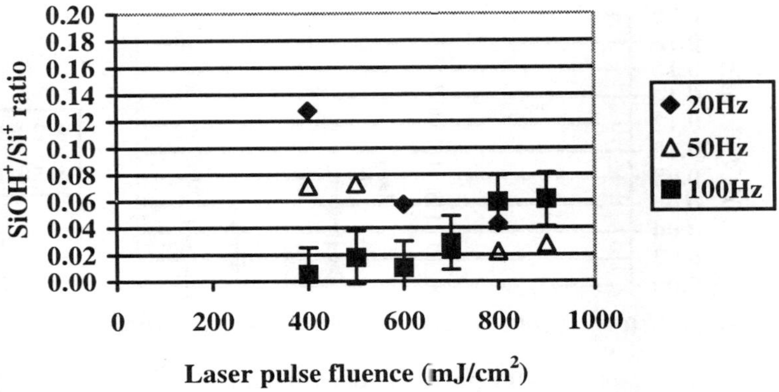

Figure 9: Further experiments investigating dehydroxylation as a function of total pulse fluence.

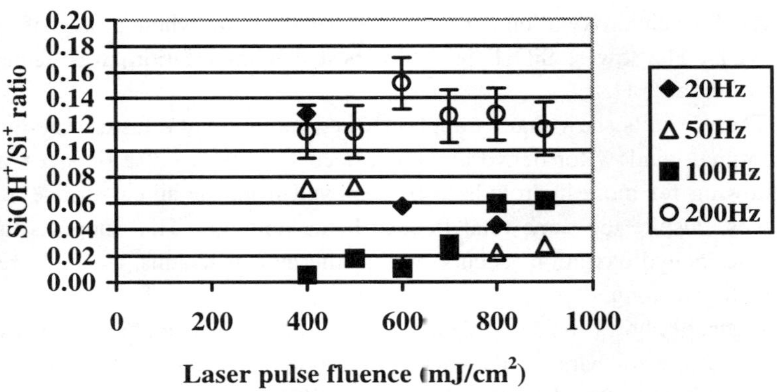

Figure 10: Dehydroxylation as a function of total pulse fluence.

Figure 10 includes the results at 200Hz, and the error bars for the previous data have again been removed for clarity. The $SiOH^+/Si^+$ ratios here are higher than for the similar pulse fluences at 50Hz and 100Hz. Upon visual examination of the surface, damage is observed. The rapid heating induced by the irradiation is causing pitting and ablation of the surface, and results in exposure of the lower regions of the surface which are unaffected by the laser radiation. Thus there is a limit to how much the laser repetition rate can be increased to aid dehydroxylation.

In conclusion, the best laser parameters for applying the total fluence to achieve optimal high levels of surface dehydroxylation in these experiments are to use 900 pulses at a fluence of approximately 400mJ/cm^2 at 100Hz. An increase in the laser repetition rate can aid the dehydroxylation process by heating the surface faster and allows for lower pulse fluences to be used. However, increasing it too much may cause surface damage, which may be evidenced by an increase in the $SiOH^+/Si^+$ ratio at high fluences, and may also be visually observed in the form of pitting and cracking of the surface.

4.3. *Dehydroxylation as a function of laser pulse number*

The effect of varying the number of pulses used to irradiate the surface was investigated. The number of pulses applied ranged from 1000 to 10000 pulses, using a laser repetition rate of 20Hz. The pulse fluence was set to 1200mJ/cm^2 as preliminary testing found that samples irradiated with 300 pulses at this fluence (20Hz repetition rate) gave a $SiOH^+/Si^+$ ratio of 0.05±0.02. The $SiOH^+/Si^+$ ratio results shown in Figure 6 for 20Hz also indicate that the optimal dehydroxylation (within the experimental uncertainty) requires a similarly high fluence. Note that while visual damage was not observed with the human eye, although more detailed examination of the surface with surface analysis tools may have indicated that this was not the case. Furthermore, the effect of laser beam focus was not factored into the experiment, and may have had a greater effect than initially anticipated. The results which were subsequently obtained in §4.2 for 20Hz indicate that using a fluence between 800-1000mJ/cm^2 may have been more appropriate for investigating the dehydroxylation as a function of pulse number.

Figure 11 shows the $SiOH^+/Si^+$ ratio as a function of the number of pulses. The $SiOH^+/Si^+$ ratio over this range appears to be constant, at an approximate value of 0.12±0.02. As this is lower than the characteristic reference value of 0.19±0.02 for the untreated surface, this indicates that the silica samples are partially dehydroxylated.

From this investigation into the effect of the number of laser pulses used, it is not clear if the limited removal of hydroxyls is due to the dehydroxylation being a photochemical process limited by the laser wavelength. It is possible that the fluence used is too high, although damage was not observed with the naked eye. Further testing is required with lower fluences. However, in conjunction with the earlier results, it is believed that the dehydroxylation occurs by a thermal mechanism utilising cumulative heating. In this experiment, the applied radiation is enough to heat the surface and reduce the hydroxyl concentration, but the removal process is limited. This is because the temperature of the surface is not rising sufficiently as the heat is dissipating away between pulses.

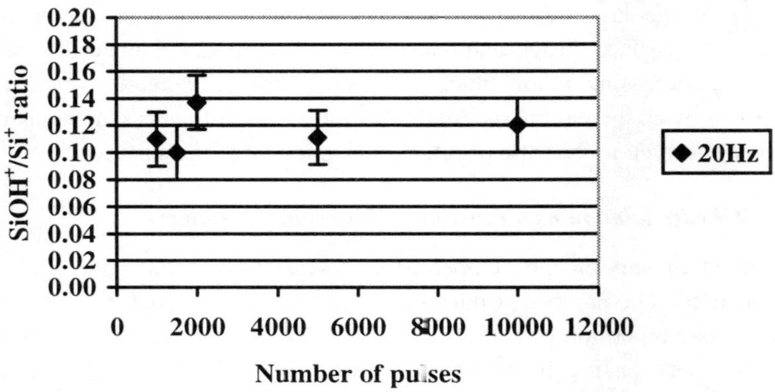

Figure 11: Dehydroxylation as a function of the number of laser pulses used.

4.4. *Dehydroxylation as a function of laser repetition rate*

The previous experiments detailed above indicated that applying the same amount of irradiation in a shorter period of time could improve the dehydroxylation process by causing a greater temperature increase in the substrate, although this effect was limited by damage being induced if the fluence or repetition rate were too high. The relationship between the laser repetition rate and the dehydroxylation process is further investigated in this section.

Each test area on the slides was irradiated with 400 pulses at a pulse fluence of 900mJ/cm^2. The repetition rate was varied between 2Hz to 200Hz, and the $SiOH^+/Si^+$ ratio was determined. Figure 12 shows that this ratio decreases (and therefore the dehydroxylation improves) with increasing laser repetition rate. At low repetition rates (below 10Hz) the energy dissipates too quickly to cause the temperature to rise sufficiently for dehydroxylation. At 150Hz and 200Hz the $SiOH^+/Si^+$ ratio rises again. This is due to the high repetition rates and fluences causing optical damage to the surface. Cracking and pitting of the surface exposed 'new' surfaces on the slide which had not been dehydroxylated by the irradiation.

Plotting these results on a logarithmic scale (see Figure 13) yields the following equation describing the relationship:

$$\log[SiOH^+ / Si^+ ratio] = -0.71 \times \log[rate(Hz)] - 0.48 \quad (1)$$

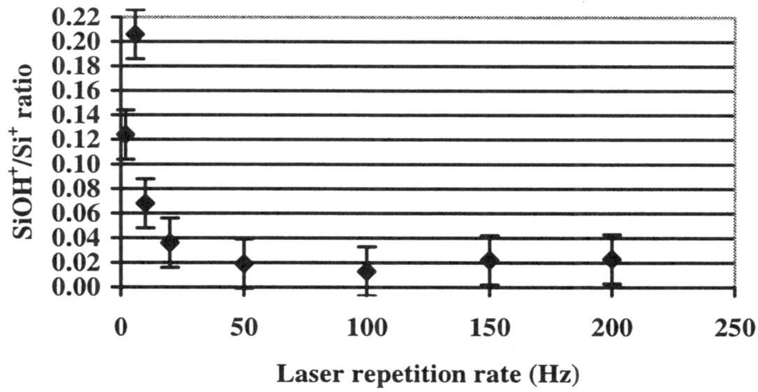

Figure 12: Dehydroxylation as a function of repetition rate.

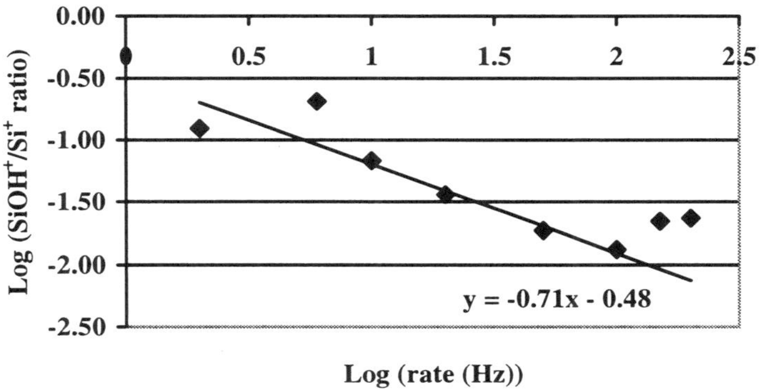

Figure 13: Dehydroxylation as a function of laser repetition rate.

The characteristic reference value for the non-irradiated fused silica samples used in these experiments was determined to be 0.19±0.02 (see §3.2). The logarithm of this value is *log(0.19) = -0.72.* From and and Figure 13, this corresponds to a laser repetition rate of 6±4Hz, which is the threshold laser repetition rate at which dehydroxylation begins to occur, using a pulse fluence of 900mJ/cm^2 and 400 pulses. Note that this threshold value is a characteristic of this experiment, and may vary at different fluences, and with other surfaces and lasers.

The presence of the threshold laser repetition rate also supports the proposal that the dehydroxylation is occurring via a thermal mechanism. Although the same pulse fluence and number of pulses are used, irradiation applied in a shorter time (at higher repetition rates) removes more surface hydroxyls. The highest repetition rate that can be used is limited by optical damage to the surface. Frequencies up to 100Hz are appropriate in this study.

5. Discussion

This work has shown that it is possible to cause dehydroxylation of the surface of high quality UV fused silica with KrF laser irradiation. Quantitative analysis of the surface hydroxyl concentration was possible with ToF-SIMS by comparing the ratio of the $SiOH^+$ and Si^+ ion peaks. Hydrocarbon contamination could affect the analysis process and mask the true hydroxyl concentration of the surface. Such contamination could be removed by treating the samples with a hydrogen peroxide wet clean, without affecting the underlying hydroxyl concentration. This allowed for procedures to be established which allowed for quantifiable and reproducible results to be obtained and for systematic experiments to be carried out.

The non-irradiated samples were hydrogen peroxide cleaned and found to have a $SiOH^+/Si^+$ ratio of 0.19±0.02, which was characteristic of the fused silica glass used. Irradiated samples which produced lower ratios were said to be dehydroxylated. Given certain irradiation parameters, this ratio can be decreased to zero, indicating that the surface has been completely dehydroxylated (for example, see Figure 6 at 1000mJ/cm^2, where complete dehydroxylation within the experimental uncertainty was achieved using 400 pulses at 100Hz). Samples with ratios higher than the characteristic reference value were found to be hydrocarbon contaminated, and required treatment with the hydrogen peroxide process.

Dehydroxylation was improved with higher pulse fluences and with increasing laser repetition rates, as these caused a greater temperature rise in the surface. While the dehydroxylation does not have an exponential dependency on the laser pulse fluence, the relationship between dehydroxylation and the laser repetition rate shows that a cumulative heating effect is required. This means that the removal of surface hydroxyls via laser irradiation is a thermal process involving cumulative heating, rather than a photochemical mechanism. (If the dehydroxylation was a photochemical process, the hydroxyls would be removed via bond breaking due to the photon energy of the UV light. There would be no threshold pulse fluence or repetition rate required for dehydroxylation, but rather a scaling with the number of photons applied).

As the dehydroxylation occurs via a thermal mechanism involving cumulative heating, the laser parameters such as pulse fluence and repetition rate must be considered carefully. These need to be set within certain operating windows, to ensure that the removal process is optimised without damage to the substrate surface. If the irradiation is applied too slowly or if low fluences are used, the temperature of the surface does not rise sufficiently to completely dehydroxylate the surface (although partial dehydroxylation may be possible). However, if the irradiation is applied at too high a rate, and/or at too high a fluence, visible damage to the substrate may occur.

By manipulating the irradiation parameters, the level of hydroxyl removal can be controlled, making laser irradiation a quicker and more versatile alternative to the conventional thermal treatment. The most favourable parameters are those which achieve the most dehydroxylation (lowest $SiOH^+/Si^+$ ratio) at the lowest fluence and/or the lowest pulse number and/or the lowest laser repetition rate. For example, 1000mJ/cm^2 with 400 pulses at 100Hz (see Figure 6), 600mJ/cm^2 with 600 pulses at 50Hz (see Figure 8), and 400mJ/cm^2 with 900 pulses at 100Hz (see Figure 9), all obtain produced highly dehydroxylated surfaces in these studies. Note that these parameters will vary not only between studies using different lasers and surfaces, but also with changing laser beam focus.

Importantly, as only the surface of the glass is irradiated, issues of phase changes in the bulk from the heating of the entire glass can be avoided. The laser can also be used to selectively treat areas of the surface, allowing for the ability to spatially pattern or mark the glass without visible damage.

6. Conclusions

Complete and partial dehydroxylation of UV fused silica surfaces is possible with KrF laser irradiation. Quantifiable analysis of the surface is possible using ToF-SIMS, and hydrocarbon contamination can be removed using a hydrogen peroxide cleaning method without affecting the hydroxyl concentration.

The dehydroxylation occurs via a thermal mechanism, and requires sufficient cumulative heating of the surface. More hydroxyls are removed at higher laser fluences and/or repetition rates, but care must be taken to operate within certain parameters so that optical damage does not occur.

Acknowledgments

This research has been supported by the Australian Research Council, Macquarie University, and The University of New South Wales. A. Fernandes would like to acknowledge the assistance of a Macquarie University Research Award for Areas and Centres of Excellence scholarship, and a Macquarie University Postgraduate Research Fund Grant.

References

1. Y. F. Lu, L. Zhang, W. D. Song, Y. W. Zheng and B. S. Luk'yanchuk, *JETP Lett.* **72**, 457 (2000).
2. H. J. Münzer, M. Mosbacher, M. Bertsch, O. Dubbers, F. Burmeister, A. Pack, R. Wannemacher, B. U. Runge, D. Bäuerle, J. Boneberg and P. Leiderer, *J. Microscopy* **202**, 129 (2001).
3. M. Mosbacher, H. J. Münzer, J. Zimmermann, J. Solis, J. Boneberg and P. Leiderer, *Appl. Phys. A* **72,** 41 (2001).
4. S. M. Huang, M. H. Hong, B. S. Luk'yanchuk, Y. W. Zheng, W. D. Song and Y. F. Lu, *J. Appl. Phys.* **92**, 2495 (2002).
5. Y. F. Lu, L. Zhang, W. D. Song, Y. W. Zheng and B. S. Luk'yanchuk, *Proc. SPIE* **4426**, 143 (2002).
6. H. J. Münzer, M. Mosbacher, M. Bertsch, J. Zimmermann, P. Leiderer and J. Boneberg, *Proc. SPIE* **4426,** 180 (2002).
7. R. Denk, K. Piglmayer and D. Bäuerle, *Appl. Phys. A,* **76**, 1 (2003).
8. F. Lang, M. Mosbacher and P. Leiderer, *Appl. Phys. A* **A77**, 117 (2003).
9. D. R. Halfpenny, *Ultraviolet laser cleaning of glass*, PhD Thesis, Macquarie University, Sydney, (2000).
10. D. R. Halfpenny, D. M. Kane, R. N. Lamb and B. Gong, *Appl. Phys. A* **71**, 147 (2000).
11. A. J. Fernandes, D. M. Kane, B. Gong and R. N. Lamb, *COMMAD 2002 Proceedings,* M. Gal, ed., **IEEE Cat. Num. 02EX601**, 433 (2002).
12. B. J. Wood, *Surface Analytical Studies of Silica and Alumina,* PhD Thesis , The University of New South Wales, Sydney, (1998).
13. R. K. Iler, *The Chemistry of Silica: Solubility, Polymerization, Colloid and Surface Properties, and Biochemistry*, Chapter 6, John Wiley and Sons, New York (1979).
14. J. N. Israelachvili, *Intermolecular and Surface Forces*, 2nd edition, Academic Press, London, (1991).
15. B. C. Senn, P. J. Pigram and J. Liesegang, *Surf. Interface Anal.* **27**, 835 (1999).
16. S. P. Godfrey, J. P. S. Badyal and I. R. Little, *J. Physical Chemistry B* **105**, 2572 (2001).
17. A. S. D'Souza and C. G. Pantano, *J. Am. Ceramic Society* **85**, 1499 (2002).
18. B. J. Wood, R. N. Lamb and C. L. Raston, *Surface Science and Analysis* **23**, 680 (1995).
19. HPFS® Fused Silica Standard Grade, http://www.corning.com/semiconductoroptics/products_services/pdf/H0607_HPFS_Standard_ProductSheet.pdf (Accessed 1st April 2004).
20. Average Transmittance Curves for Corning 7980 http://www.technicalglass.com/tpg03cat.pdf (Accessed 1st April 2004).

21. D. Briggs and M.P. Seah, *Practical Surface Analysis – Vol. 2: Ion and Neutral Spectroscopy*, 2nd ed., John Wiley and Sons, England, (1992).
22. SOP for RCA Clean, http://www.mines.edu/fs_home/cwolden/chen435/clean.htm (Accessed: 31st October 2002).
23. RCA Clean, http://www.ece.gatech.edu/research/labs/vc/processes/rcaClean.html (Accessed: 31st October 2002).
24. Professor Robert Lamb, Surface Science and Technology, University of New South Wales, Sydney, Australia, private communication.
25. Millipore Catalogue – Milli-Q® Ultrapure Water Purification Systems, http://www.millipore.com/catalogue.nsf/docs/C7658 (Accessed: 24/3/2004).
26. Millipore – Technical Publications – Milli-Q Systems, http://www.millipore.com/publications.nsf/docs/pb1104en00 (Accessed: 24/3/2004).
27. R. J. MacDonald and B. V. King, *SIMS (Secondary Ion Mass Spectrometry), (*Chapter 5*, Surface Analysis Methods in Materials Science,* Ed. D. J. O'Connor, B. A. Sexton and R. St C. Smart), Springer-Verlag, Germany, (1992).

Chapter 7

REMOVAL OF SILICA MICROSPHERES FROM GLASS AND SILICA SUBSTRATES BY DRY LASER CLEANING

S PLEASANTS AND D M KANE

Department of Physics, Macquarie University ~ Sydney, NSW 2109, Australia

Laser cleaning of 5.0 μm diameter silica spheres from silica surfaces was investigated. It was found that for three different slide preparation methods that the particles did not adhere strongly to the silica substrate. Experimental laser cleaning particle removal thresholds were 4 orders of magnitude less than that predicted by theoretical modelling based on thermal expansion and Van der Waals adhesion. There are two reasons put forward to explain this discrepancy. The first is that there are other forces acting on the particle besides the Van der Waals force. This is supported both by the present study which found that silica spheres did not adhere strongly to silica slides for three different preparation methods and by previous atomic force microscope (AFM) and surface force apparatus (SFA) studies which showed the existence of a short range repulsive force between silica spheres and silica substrates. The second explanation is that the spheres have a greater effective absorption coefficient than that of bulk silica. This was supported by the fact that at high fluences small hillocks of silica remained after irradiation indicating that temperatures in excess of 1900 K had been achieved. Both factors taken together can account for the discrepancy between theoretical predictions and experimental results in the present study.

1. Introduction

Laser cleaning is one of a range of techniques for removing contamination from surfaces. Dry laser cleaning offers advantages over other laser cleaning and cleaning techniques through being fast and a non-contact cleaning process that does not require the use of solvents. It can also be spatially selective if required. Laser cleaning has been demonstrated to clean both metallic and non-metallic surfaces and to remove particles as small as 65-80 nm in diameter [1]. On glass surfaces removal of particles as small as 0.3 μm has been demonstrated [2-4]. To date a significant amount of work has been done investigating the removal of silica spheres from highly absorbing substrates such as silicon [5-8]. One of the reasons why silica spheres are used is that they are assumed to be transparent at the wavelengths usually employed for laser cleaning. Thus, in the thermal expansion models of laser cleaning it is assumed that the particle expansion is negligible and that all the force required for particle removal originates from the expansion of the substrate. It has been demonstrated both theoretically [9,10] and experimentally [10] that when small, transparent spheres are irradiated they focus the radiation down to a small region near the base of the particle. This is known as near-field focussing and, for particles that have radii small compared to the wavelength of the radiation, it can be calculated using Mie scattering theory. Laser cleaning models which account for this effect have been developed for both highly absorbing substrates [9] and for low absorbing substrates [11]. Compared to earlier models for dry laser cleaning [12,13], the models that have been published recently are quite sophisticated and can account for effects such as near-field focussing, and thermal expansion and elastic deformation of both the particle and the substrate [9,11]. However, it is important to model not just the particle removal mechanism but also the adhesion of the particle to the substrate. This is not a trivial matter as the adhesion of a particle to a substrate is a complex topic and is still an active area of research [14]. Most laser cleaning models assume that the predominant adhesion force is the Van der Waal's force and that the separation distance between the particle and the substrate is usually taken as 0.4 nm. Some work has also suggested that hydrogen bonds also play a significant role in some systems [15-17]. Recent AFM studies of adhesion between inorganic oxide surfaces has shown that, in certain cases, the adhesion force cannot be adequately explained in terms of Van der Waals force alone. One study found that the adhesion energy of iron oxide to silica was more than two orders of magnitude less than that calculated from London-van der Waals interactions [18]. Another study investigated the adhesion of silica spheres to silica substrates and found evidence of a short- range repulsive force which agreed with earlier SFA studies [14]. The importance of pretreating slides prior to conducting laser cleaning experiments was demonstrated by the contrast between two studies conducted using alumina particles on glass and silica substrates. The first work used slides that had been exposed to normal laboratory environment for several weeks [19]. The second study used three different methods to pretreat the 3 slides. In the latter case the laser cleaning threshold fluences for silica were 8 times greater than those for glass, whereas the previous work had found no significant difference between the two substrates [11]. At fluences well above those required for particle removal other effects besides cleaning have been observed. Complete particle removal at low fluences followed by partial ablation of the particle at higher fluences has been reported for polystyrene particles on silicon [20]. Small hillocks about 100 nm in size were also observed after 0.5 μm diameter silica spheres on a silicon substrate were irradiated with 248 nm radiation having a fluence of about 340 mJ/cm^2 [21].

The work presented in this paper looks at the laser cleaning of silica spheres on glass and silica substrates. The purpose of this work is to compare the predictions made by a model published recently [11] which accounts for the near-field effect on low absorbing substrates with experimental results. The effect that the substrate pretreatment has on particle adherence to the substrate is studied. Additionally, effects at higher fluences such as melting and ablation of the particles and optical damage to the substrate caused by cracking are described.

2. Experimental Work

The substrates used in the laser cleaning experiments were silica slides made from Corning 7940 silica, a high purity synthetic amorphous silicon dioxide manufactured by flame hydrolysis, and measured 55.0mmx20.0mm*x*2mm. The 3.5 μm and 5.0 μm particles were silica spheres manufactured by Kromasil [22]. For the silica spheres with nominal diameter of 5.0 μm the average diameter was measured to be 5.89 μm with a standard deviation of 0.84 μm. An image of Kromasil 3.5 μm silica spheres is shown in fig. 1 [22].

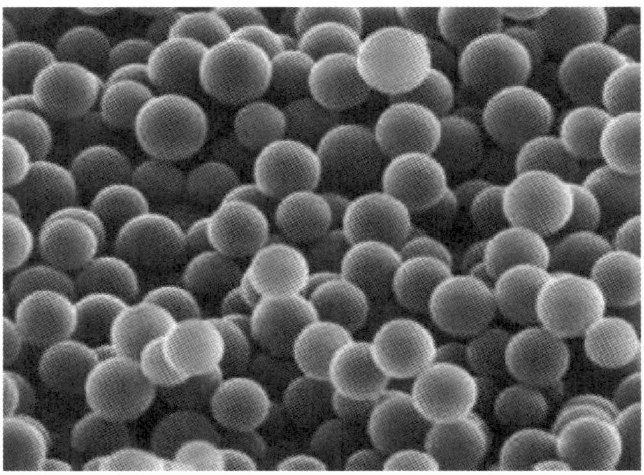

Fig. 1: Manufacturers Image of Kromasil 3.5 μm silica spheres [22].

The 0.5 μm diameter silica spheres were from Duke Scientific [23]. These spheres are dispersed in water (2% by weight) and have a certified mean diameter of 0.49 ± 0.03 μm with a standard deviation of 0.02 μm.

Three different methods of preparing the slides were used. They all entailed initially treating the slide with hydrogen peroxide and ammonium hydroxide. This involved soaking the slides in warm methanol (95%) for 10 minutes. The slides were then dried in air and put in a solution of hydrogen peroxide (27.5%) and one-tenth volume of AR grade ammonium hydroxide solution. This solution was then heated to 45-50 °C for 20 minutes. The slides were dried and rinsed in about 300 ml of milli-Q water. They were then boiled in milli-Q water for one hour. Finally they were rinsed once more in milli-Q water. After cleaning, the slides were stored under milli-Q water to prevent hydrocarbon deposition from occurring [24].

The first preparation method consisted of drying the slides using nitrogen gas and then depositing the silica spheres onto the surface of the substrate. This was done by using a spatula to spoon a layer of particles onto the surface and then tapping the substrate against a hard surface several times to remove any loosely adhered particles.

The second method included the additional step of coating the slides with silane-3-aminopropyltriethoxysilane (3-APS) to make the slides sticky. This was done by dipping the slides in a 2% solution of 3-APS in EM-grade acetone for 1 minute. The slides were transferred into EM-grade acetone and left for 1 minute. They were then submerged in distilled water for 1 minute and rinsed vigoursly by squirting them with distilled water. The slides were then dried and particles deposited on the surface as described above.

The third method involved taking two slides that had been treated with hydrogen peroxide and dried, depositing particles on one of the slides, then pressing the two slides together using a mechanical press for 10 seconds. The purpose of this was to force the particles into close contact with the slides at the same time allowing any differential charging of the particles and surface to dissipate. Prior research on iron oxide particles and silica surfaces found that there was a marked increase in the adhesion force with increasing loading force [25]. Figure 2 shows the experimental set up used to measure the laser cleaning efficiency. A single pulse of 248 nm radiation from a GSI Lumonics PM-848 KrF excimer laser was used. The pulse length was 8.1 ns which was found by fitting the pulse shape with the function given by

$$I(t) = I_0(t/\tau)\exp(-(t/\tau))$$

The fluence was varied using an Optec AT4030 attenuator. The fluences used ranged from 0 to 520 mJ/cm^2 for the laser cleaning experiments, while fluences up to 3600 mJ/cm^2 were used to produce high fluence effects. A spherical lens having a focal length of 96 mm at the laser wavelength was placed after the attenuator. The sample was located 9 mm behind the focal point of the lens. This gave beam dimensions of 7 mm by 3 mm on the sample. The rectangular beam had a Gaussian profile along the short axis and a flat profile along the long axis. The viewing system was set up to image an area of 530 μm x 350 μm - four such regions within the treated area were monitored. These images were taken from an array of 2 by 2 contiguous areas close to the center of the laser beam. The images were obtained using a Hirox Hi-Scope KH-2600 microscope which was coupled to a computer via a frame grabber board for an overall magnification of x250. The sample was mounted in a sample holder that was attached to a Physik Instrumente precision translation stage. This enabled the sample to be translated laterally for viewing and then to be accurately relocated in the same place for a comparison of the same area before and after irradiation. Image Pro Plus v4.1 (Media Cybernetics) image analysis software was used to measure the total particle number for both before and after irradiation images. This was done for all four images and an average cleaning efficiency was calculated from these figures. The Hirox microscope system was also used to acquire images of the silica slides before and after they were irradiated with high fluences (600 mJ/cm^2 to 3600 mJ/cm^2) to observe effects besides particle removal.

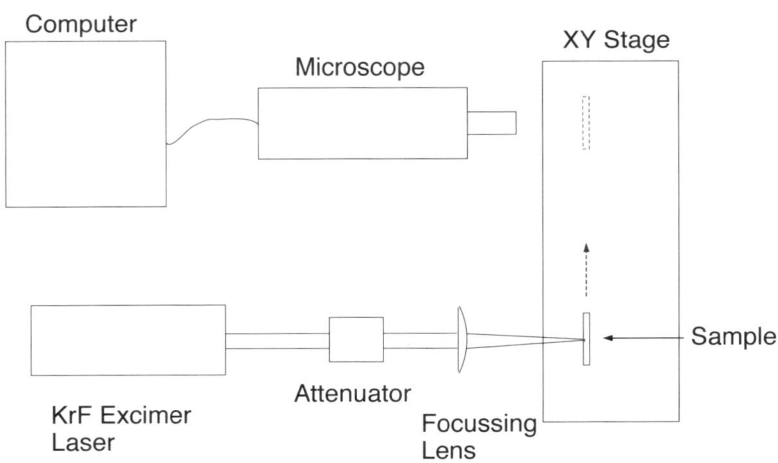

Fig. 2. Schematic diagram showing the experimental arrangement for laser cleaning.

3. "Dry" Laser Cleaning Results and Discussion

It was found that when the 5 μm Kromasil spheres were deposited on silica substrates that had been prepared using the hydrogen peroxide cleaning method that the particles did not adhere to the substrate surface. The particle density was so low that it was not possible to perform meaningful cleaning experiments. The slides that had been coated using 3-APS prior to attaching the particles exhibited greater particle density than the slides which had been prepared with the hydrogen peroxide method, although the density was still relatively low from the point of view of carrying out laser cleaning experiments. The greater adherence is due to the effect of the 3-APS film which makes the slides hydrophilic. In order to obtain meaningful statistics it was necessary to select areas on the slide which had a high particle density. Figure 3 shows images of the same area taken before and after irradiation with a 204 mJ/cm^2 pulse. It demonstrates that damage-free cleaning of silica spheres from silica is possible. Figure 4 shows the laser cleaning efficiency (defined as the percentage of spheres removed) as a function of fluence of one such slide. The threshold fluence is about 25 mJ/cm^2.

Fig. 3. Images of the same area on a silica slide taken before and after irradiation with a 204 mJ/cm^2 pulse. Area imaged ~80 microns x 50 microns.

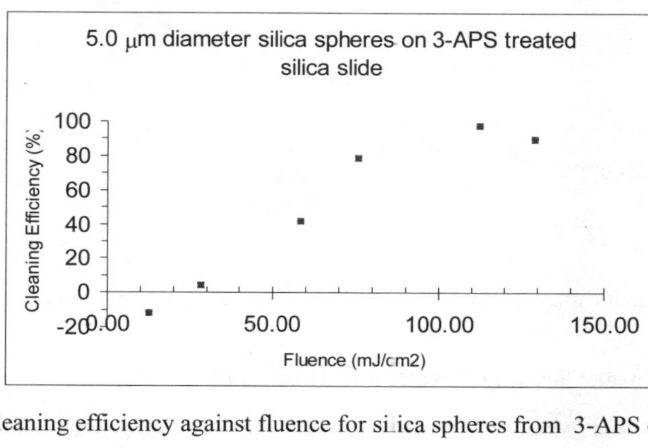

Fig. 4. Cleaning efficiency against fluence for silica spheres from 3-APS coated silica.

In contrast, depositing 3.5 μm Kromasil spheres on silica substrates that had been prepared using the hydrogen peroxide cleaning method resulted in higher densities of adhered spheres. Figure 5 shows images of the same area taken before and after irradiation for four different single laser pulse fluences. These images also demonstrate damage-free cleaning of silica spheres from silica. Figure 6 shows the laser cleaning efficiency (defined as the percentage of spheres removed) as a function of fluence of one such slide. The threshold fluence is about 140 mJ/cm^2.

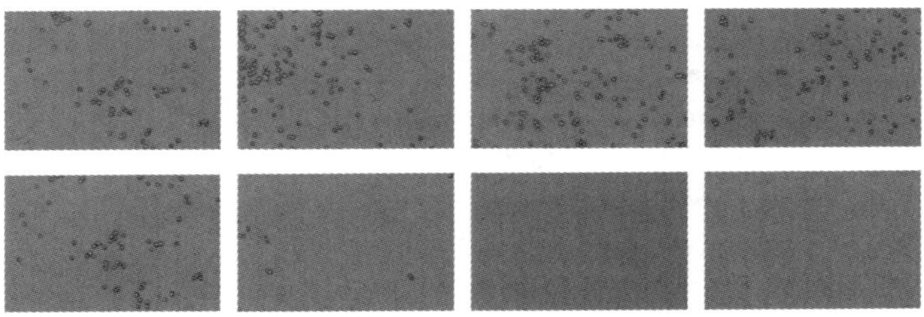

Fig. 5: 3.5 μm diameter spherical silica particles, introduced dry, on silica. Before (top row) and after (bottom row) for pulse fluence (left to right) 91mJ/cm2, 155 mJ/cm2, 198mJ/cm2, and 210 mJ/cm2. Area imaged ~80 microns x 50 microns.

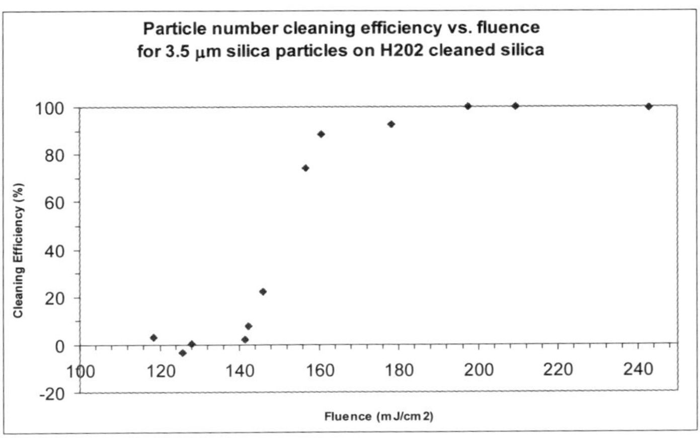

Fig 6: Laser cleaning efficiency vs fluence for 3.5 μm diameter spherical silica particles introduced dry.

Adhesion of the 5 μm diameter silica spheres was increased using the slide "pressing" technique described above. This resulted in higher, adhered particle densities than those which had just been treated with the hydrogen peroxide method. Again it was necessary to choose areas which had high particle densities when measuring the laser cleaning efficiency. Figure 7 shows the laser cleaning efficiency results as a function of laser pulse fluence for two slides which had been prepared using this method. The threshold fluence is about 80 mJ/cm^2. All the measured threshold fluence values are about four orders of magnitude smaller than those predicted by van der Waals adhesion theory. The model described in [11] predicts a threshold fluence of 201 J/cm^2 for silica spheres having a diameter of 5 μm. This is much greater than the damage threshold for bulk silica (about 5 J/cm^2 at 248 nm [26]).

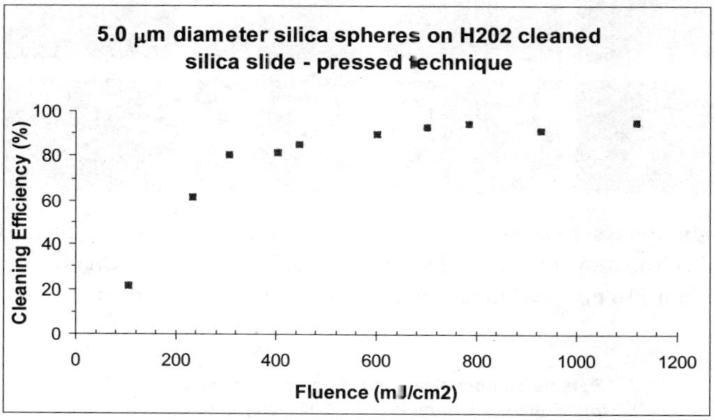

Fig. 7. Cleaning efficiency against fluence for two slides prepared using a pressing technique to achieve silica particle adhesion to the silica surface.

Preparing a suspension in water of the 3.5 μm diameter spherical silica particles, and then dropping and spreading this on the silica slides, followed by evaporation of the suspension fluid also led to increased adhesion of these spheres to the silica surface, relative to the spheres introduced dry, as shown from the laser cleaning results. Fig. 8 shows a before and after image of a laser cleaning of these spheres with a laser pulse of fluence 890 mJ/cm^2. Fig. 9 shows the laser cleaning efficiency, as a percentage of spheres removed, as a function of laser pulse fluence. The laser cleaning threshold fluence has increased to ~500 mJ/cm^2 from 140 mJ/cm^2 for the silica spheres introduced from a water suspension relative to those introduced dry, ie about a three to four fold increase. This indicates that increased adhesion of the silica spheres to the silica surface has been achieved.

The final set of results are for 0.5 μm diameter silica spheres in water suspension from the manufacturer [23]. These spheres have been evaporated onto the hydrogen peroxide cleaned silica slides and laser cleaned with single pulses. A before and after image pair are shown in fig. 10 and the laser cleaning efficiency versus laser pulse fluence is shown in fig. 11. The laser cleaning threshold fluence is about 300-400 mJ/cm^2, similar to that for 3.5 μm diameter silica spheres prepared in suspension (fig. 8). However, there is much greater variation in the cleaning efficiency percentage as compared to the systematic increase with laser pulse fluence obeserved when removing larger silica microspheres.

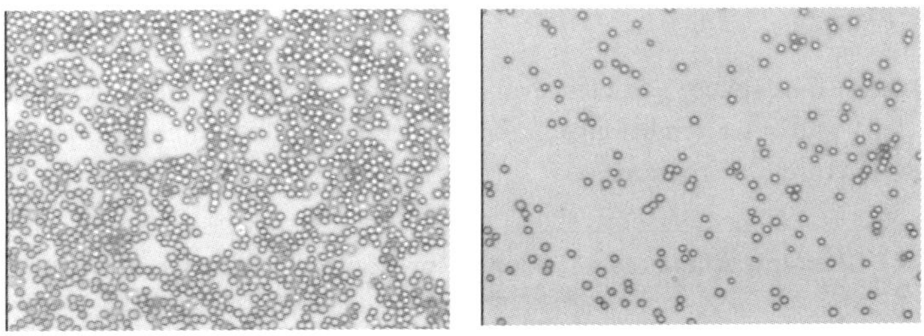

Fig. 8: Laser cleaning efficiency vs fluence for 3.5 μm diameter spherical silica particles evaporated from solution. Left, a before image, and right an after image for a fluence of 890 mJ/cm^2. Area imaged ~80 microns x 50 microns.

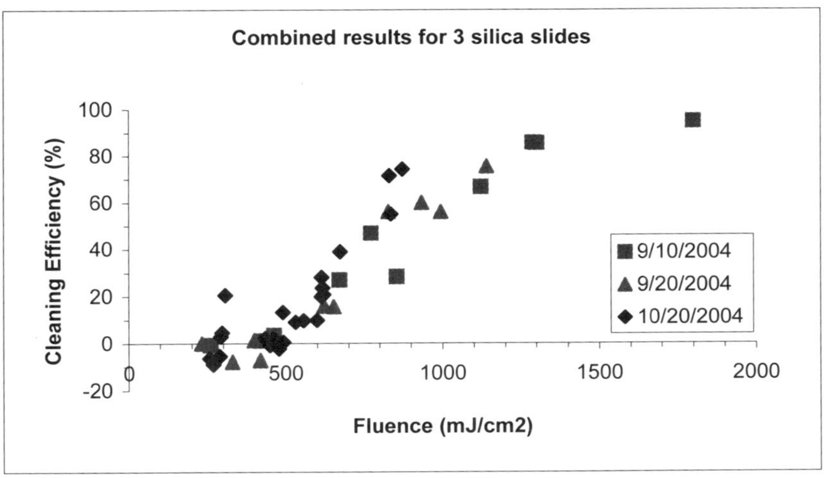

Fig. 9: Laser cleaning efficiency vs fluence for 3.5 μm diameter spherical silica particles evaporated from solution. Results from three different slides laser cleaned on different days.

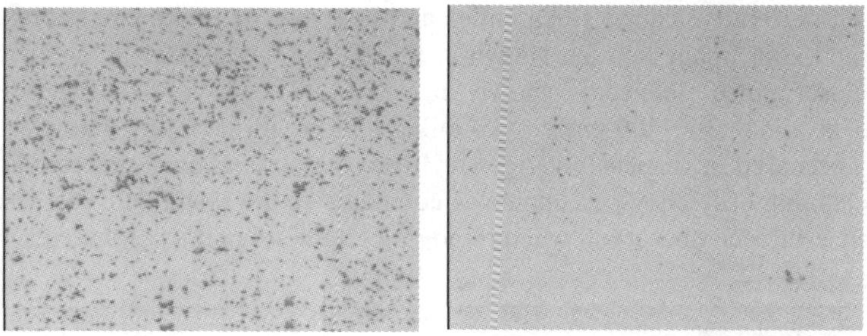

Fig. 10: 0.5μm silica sphere particles, evaporated from suspension, on silica. Before laser cleaning (left) and after (right) image. The laser pulse fluence is 401 mJ/cm^2. Area imaged ~80 microns x 50 microns.

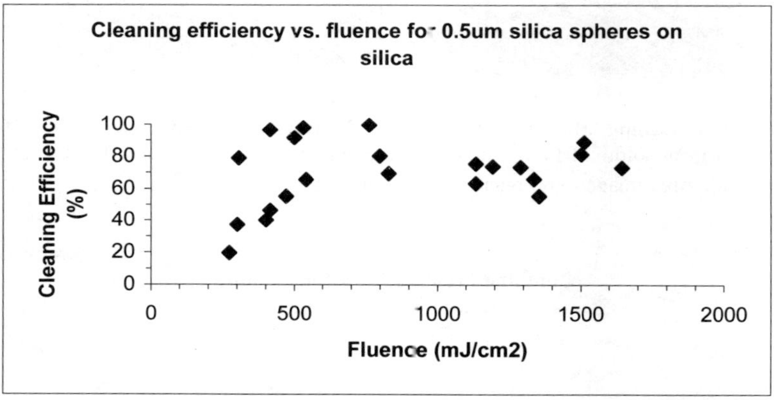

Fig. 11: Laser cleaning efficiency vs fluence for 0.5μm silica sphere particles, evaporated from suspension, on silica.

The discrepancy between theory and experiment for the laser cleaning threshold fluence arises because the adhesion is much lower than the large van der Waal's adhesion used in the models. The low particle densities observed on all pre-treated slides is also evidence of this. In the theoretical model it is assumed that the particle is a distance of 0.4 nm from the substrate surface and at such distances the van der Waal's force dominates. However, other forces which are repulsive in nature appear to be operating, preventing the particles from coming into such close contact with the substrate surface. Possible candidates for such forces include electrostatic forces (silica becomes negatively charged) and solvation forces. Previous AFM studies of silica spheres on silica give support to the existence of a very short range (about 1 nm) repulsive force [14]. Another study found that the adhesion energy of iron oxide to silica was more than two orders of magnitude less than the calculated London-van der Waals interactions [25]. These results highlight the need to consider other forces besides van der Waals force when considering the adhesion of inorganic oxide surfaces.

Even allowing for a reduced adhesion force the threshold fluences are extremely small when it is considered that both the substrate and the particle are nominally transparent at 248 nm. The absorptivity of the substrate was measured directly using the laser radiation and a power meter. It was found to have an absorptivity of 0.0063, which is equivalent to an absorption coefficient of 3.2 m^{-1}. It was not possible to accurately determine the absorptivity of the particles due to their small size. In the theoretical model they were assumed to be completely transparent and are heated only through thermal contact with the substrate. This assumption of completely transparent particles may well be inaccurate and in fact the absorption coefficient of the particles could well be higher than that of the substrate. Similar evidence of high absorption of alumina particles was found in a recent study [27].

4. High Laser Fluence Effects

Further evidence of the higher than expected absorption of the silica spheres was found when the particles were irradiated with even higher fluences, ie beyond the "damage-free", "dry" laser cleaning regime. Images taken before and after irradiation, of 5 μm diameter silica spheres on silica, at 450 mJ/cm^2 are shown in Figure 12. The image taken after irradiation shows that there are residual hillocks of silica remaining. These hillocks, which are approximately 1 μm in diameter, occur at the same spot where a sphere was previously situated. One mechanism for their formation is that the temperature of the particles becomes so high during irradiation that the silica melts and upon cooling it is caked on to the substrate. This would mean that the temperature of the particle must be greater than the softening point of silica which is in the range 1900-2000 K [26] indicating that significant absorption is occurring. An alternative mechanism would be that the bottom of the particle is melted onto the slide and then the remainder of the particle is cracked away. That is, stress cracking after rapid heating may also be occurring. Further research is needed to determine which mechanism is responsible for these features. In particular their topology should provide evidence of how they were formed. This phenomenon of complete particle removal at low flences followed by partial ablation of the particle at higher fluences is very similar to the high fluence effects reported for polystyrene on silicon [20] and silica on silicon [21] systems. At even greater fluences (above about 2500 mJ/cm^2) cracking and removal of shell-shaped volumes of substrate glass was observed. This is seen in Figure 13. In this image a trigonal bi-valve, shell-shaped volume has been carved out from the glass substrate. Comparing this image with the image taken before irradiation it is evident that this removal occurs only at sites where there was previously a particle. At this stage it is not clear what determines the geometry of the "pits", but it is more likely to be a cracking phenomena. Further work is also required to discover the underlying mechanism for this phenomenon.

Fig. 12. Images of the same area before and after irradiation with a 450 mJ/cm^2 pulse. Silica spheres are nominally 5.0 μm in diameter and are on a silica substrate. After image show residual hillocks of silica. Area imaged ~80 microns x 50 microns.

Fig. 13. Images of the same area before and after irradiation with a 3540 mJ/cm^2 pulse. Silica spheres are nominally 5.0 μm in diameter and are on a silica substrate. After image shows shell-shaped "removal following cracking" of the silica surface. Area imaged ~80 microns x 50 microns.

5. Conclusions

The threshold fluence for laser cleaning removal of 5.0 μm diameter silica spheres from silica substrates has been shown to be 4 orders of magnitude less than that predicted by theoretical modeling based on thermal expansion and Van der Waals adhesion. In fact, it was necessary to develop sample preparation techniques to achieve adhesion of the particles to the substrate at all. Two reasons have been proposed to explain this discrepancy. The first reason is that the adhesion force can not be modelled solely in terms of the Van der Waals force. Repulsive electrostatic and/or solvation forces and/or other forces are contributing. This is also supported by prior AFM and SFA studies which indicated the existence of a short range repulsive force between silica spheres and silica substrates. Additionally, it appears that the spheres have a greater absorption coefficient than that of bulk silica. Evidence for this was found when the spheres were irradiated at fluences greater than about 400 mJ/cm^2 with a single laser pulse. This left small hillocks after irradiation indicating that local temperatures of at least 1900 K had been reached. The low adhesion and the high absorption of the silica spheres, when considered together, can account for the discrepancy between theoretical predictions and experimental results in the present study.

Acknowledgements

The authors would like to thank the Australian Research Council (through an ARC Large Grant) and Macquraie University for funding this research. They would also like to thank the Kromasil company for kindly donating the silica spheres used in this study.

References

1. D. M. Kane, A. J. Fernandes, D. R. Halfpenny, in: B. Luk'yanchuk (Ed.), Laser Cleaning, World Scientific, New Jersey, London, 2002, Ch. 4, pp. 181-228), and A J Fernandes and D M Kane, Ch 1, Laser Cleaning II, Ed D M Kane.
2. D. R. Halfpenny, D. M. Kane, J. Appl. Phys. **86** (12) (1999), 6641-6.
3. D. M. Kane, A. J. Fernandes, Proc. SPIE **4426** (2001), 334-339.
4. A. J. Fernandes, D. M. Kane, Proc SPIE **4426** (2001), 290-295.
5. P. Leiderer, J. Boneberg, M. Mosbacher, A. Schilling, O. Yavas, Proc. SPIE **3274** (1998), 68-77.
6. Y. F. Lu, Y. W. Zheng, W. D. Song, J. Appl. Phys. **87** (3) (2000) 1534-1539.
7. Y. F. Lu, Y. W. Zheng, W. D. Song, J. Appl. Phys. **87** (1) (2000), 549-552.
8. G. Schrems, M. P. Delamare, N. Arnold, P. Leiderer, D. BÄauerle, Appl. Phys. A **76** (2003), 847-849.
9. B. S. Luk'yanchuk, N. Arnold, S. M. Huang, Z. B. Wang, M. H. Hong, Appl. Phys. A **77** (2) (2003), 209-215.
10. B. S. Luk'yanchuk, M. Mosbacher, W. Y. Zheng, H. J. Munzer, S. M. Huang, M. Bertsch, W. D. Song, Z. B. Wang, Y. F. Lu, O. Dubbers, J. Boneberg, P. Leiderer, M. H. Hong, C. T. Chong, in: B. S. Luk'yanchuk (Ed.), Laser Cleaning, World Scientific, Singapore, 2002, pp. 103-178.

11. S. Pleasants, B. S. Luk'yanchuk, D. M. Kane, Modelling laser cleaning of low-absorbing substrates: the effect of near field focussing, Appl. Phys. A **79**, #4-5, (2004), 1595-1598.
12. J. D. Kelley, F. E. Hovis, Microelectron Eng **20** (1993), 159-170.
13. Y. F. Lu, W. D. Song, B. W. Ang, M. H. Hong, D. S. H. Chan, T. S. Low, Appl. Phys. A **65** (1) (1997), 9-13.
14. W. R. Bowen, N. Hilal, R. W. Lovitt, C. J.Wright, Colloid Surface A **157** (1999), 117-125.
15. M. Meunier, X.Wu, F. Beaudoin, E. Sacher, M. Simard-Normandin, Proc. SPIE **3618** (1999), 290-301.
16. X. Wu, E. Sacher, M. Meunier, J. Appl. Phys. **86** (3) (1999), 1744-1748.
17. X. Wu, E. Sacher, M. Meunier, J. Appl. Phys. **87** (8) (2000), 3618-3627.
18. S. Veeramasuneni, M. R. Yalamanchili, J. D. Miller, J. Colloid Interf. Sci. **184** (1996), 594 - 600.
19. D. R. Halfpenny, PhD Thesis, Macquarie University, Sydney, Australia (2000).
20. S. I. Kudryashov, S. D. Allen, J. Appl. Phys. **92** (9) (2002) 5159-5162.
21. Y. F. Lu, L. Zhang, W. D. Song, Y. W. Zheng, B. S. Luk'yanchuk, JETP Letters **72** (9) (2000), 457 - 459.
22. http://www.kromasil.com/products/products/kromasil/index.htm, Introduction to Kromasil Packing, accessed November 2004.
23. www.dukescientific.com, Duke Scientific Corporation, accessed Novmber 2004.
24. T. J. Horr, J. Ralston, R. S. Smart, Colloid Surface A **97** (1995), 183-196.
25. G. Toikka, R. A. Hayes, J. Ralston, J. Colloid Interf. Sci. **180** (1996), 329-338.
26. D. N. Nikogosyan, Properties of optical and laser-related materials – a handbook, Wiley, Chichester, 1997.
27. S. Pleasants, N. Arnold, D. M. Kane, Appl. Phys A **79,** #3, 507-514, (2004).

Chapter 8

THE EFFECT OF PULSE SHAPE ON 3D MODELLING OF LASER CLEANING FLUENCES*

S PLEASANTS AND D M KANE

Department of Physics, Macquarie University ~ Sydney, NSW 2109, Australia

BORIS S. LUK'YANCHUK

Laser Microprocessing Laboratory, Data Storage Institute, Agency for Science, Technology and Research, Singapore 117608

Prior research in the field of laser cleaning has suggested that shorter pulses are preferable to achieve low laser cleaning threshold fluences. These predictions were based mainly on exponential pulse shapes. In the present work the three-dimensional model of laser cleaning developed by Boris Luk'yanchuk (B. S. Luk'yanchuk et al., Appl Phys A, 77, 2, 209) which accounts for near-field focussing, has been used to calculate the laser cleaning threshold fluence for three different pulse shapes. These were rectangular, sinusoidal and exponential. For each pulse shape, the threshold fluence was determined as a function of pulse width (1- 200 ns) and height (1-15 GW/cm^2).

It was found that the threshold fluence is strongly dependent on the laser pulse shape, particularly for pulses greater than 100 ns in width. The threshold fluence of the rectangular pulse oscillated with a period equal to that of the period of oscillation of the particle on the substrate. In contrast, for both the exponential and sinusoidal pulses, the threshold fluence increases monotonically with pulse length.

1. Introduction

Laser cleaning is one of a range of techniques for removing contamination from surfaces. Dry laser cleaning offers advantages over other cleaning techniques through being a fast, non-contact cleaning process that does not require the use of solvents. It can also be spatially selective if required. Laser cleaning has been demonstrated to clean both metallic and non-metallic surfaces and to remove

---

* Modified from Proceedings of the 1st Pacific International Conference on Application of Lasers and Optics, *Eds. Milan Brandt and Erol Harvey, (*PICALO 2004 Proceedings on CD, *Laser Institute of America,* 2004). Reproduced with permission of the LIA.

particles as small as 65-80 nm in diameter [1]. On glass surfaces removal of particles as small as 0.3 μm has been demonstrated [2-4].

There have been several reports of models of the dry laser cleaning process [5-11]. The vast majority of them are based on thermal expansion of either the substrate or the particles. Most of the models are one dimensional, that is, they only consider the case in which there is heat flow perpendicular to the substrate's surface, and consequently they tend to overestimate threshold fluences by 1-2 orders of magnitude [12].

Luk'yanchuk, *et al.* have developed a model that accounts for the field enhancement effect, 3D thermal elasticity and the true pulse shape, which predicts threshold fluences of the correct order of magnitude as those observed in experiments [12,13]. A similar model presented by Arnold *et al.* [14-16] also represents a substantial advancement over earlier models that appeared in the literature. It treats the particle and substrate expansion in a unified manner. Both models account for the elasticity of the particle and substrate.

Recently the model developed by Luk'yanchuk, *et al.* [12,13] has been adapted to describe the case for low absorbing substrates [17]. This was the model that was chosen for this investigation. To date, very little work has been done investigating the effect that pulse shape has on the threshold fluence for particle detachment. Most models that appear in the literature assume either a rectangular or an exponential pulse shape. The former is chosen because it simplifies calculations while the latter is chosen because it is a close approximation to most experimental pulse shapes.

Arnold [18] has looked at the effect of using laser pulses having steep fronts and found that they increase the damage free cleaning window for small particles. He also explored the effect of using modulated pulses and compared the results of using a sinusoidally- and a rectangular-modulated pulse train. He found that for small particles, rectangular modulation gave the best results.

This paper investigates the effect of three different pulse shapes (rectangular, exponential and sinusoidal) on laser cleaning thresholds using the model of [17].

2. Description of Mdel Used

The model used in this paper is described more fully elsewhere [17]. It is a three-dimensional model describing laser cleaning of silica spheres from a glass substrate (refer to Table 1 for the physical properties of these materials). 5 μm diameter silica spheres have been modeled for this report. While the silica-glass system chosen for this study yields relatively high threshold fluences, the overall trends observed are expected to be the same for other systems. The model takes into account near-field focussing of the laser radiation by the particles. The intensity distribution under the particle was found using Mie theory together with the geometrical optics approximation. This permits the estimation of the beam width at the substrate surface and the focal distance of the radiation coming from the spherical particle. These parameters are used to find the distribution of intensity within the low-absorbing substrates from the formula for a focused Gaussian beam. This is in contrast with most other models of laser cleaning which assume that all absorption occurs at the surface of the substrate. Detachment of the particle is assumed to occur when the deformation parameter is equal to zero.

Table 1: Material properties for silica and glass, and the laser pulse characteristics.

Silica - particles	
Density, ρ	2185 kg m^{-3}
Specific heat capacity, c_p	772 J kg^{-1}K^{-1}
Thermal conductivity, κ	1.38 W m^{-1} K^{-1}
Volumetric thermal expansion coefficient, β	15.3 x 10^{-7} K^{-1}
Young's modulus,	7.0 x 10^{10} N m^{-2}
Poisson factor, σ	0.17
Thickness, d	2.0 mm
Glass - substrate	
Density, ρ	2437 kg m^{-3}
Specific heat capacity, c_p	646 J kg^{-1}K^{-1}
Thermal conductivity, κ	0.919 W m^{-1} K^{-1}
Volumetric thermal expansion coefficient, β	212 x 10^{-7} K^{-1}
Young's modulus,	7.0 x 10^{10} N m^{-2}
Poisson factor, σ	0.17
Glass substrate – Silica particle	
Hamaker Constant,	6.5 x 10^{-20} J
Laser pulse characteristics	
Wavelength, λ	248 nm

To determine the effect that pulse shape has on threshold fluence, three different pulse shapes were chosen. The first was a rectangular pulse having a height of I_0 and a width of τ. In this case, the intensity as a function of time, $I(t)$, is given by

$$I(t) = I_0 \quad \text{for } 0 \le t \le \tau$$
$$= 0 \quad \text{otherwise.} \tag{1}$$

In this case, the pulse width parameter τ is identical with the full-width, half-maximum (FWHM) value for the pulse width. The second was an exponential pulse described by the function

$$I(t) = I_0 \frac{t}{\tau} \exp\left(-\frac{t}{\tau}\right). \tag{2}$$

Here, the FWHM is approximately 2.45 τ. The third was a sinusoidal pulse given by

$$I(t) = I_0 \sin^2\left[\left(\frac{\pi}{2}\right)\left(\frac{t}{\tau}\right)\right]. \tag{3}$$

For this pulse shape, the FWHM is equal to τ.

For each pulse shape the maximum intensity and the width were varied over the range 0 to 16 GW/cm^2 and 0 to 200 ns respectively in order to find out for what range of pulse parameters detachment of the particle occurred.

3. Results and Discussion

In all three cases there is a local minimum for the maximum pulse intensity, which occurs at relatively low pulse widths (about 33 ns for rectangular pulses, 27 ns for exponential pulses and 28 ns for sinusoidal pulses). The curves for the exponential and sinusoidal pulses are quite similar to each other. The minimum pulse intensity for the exponential pulse is 8 GW/cm^2 while for the sinusoidal case it is 7 GW/cm^2. The curve for the sinusoidal pulse rises much faster for longer pulse widths than that for the exponential case.

The rectangular pulse shape case show a marked difference from the other two cases in that the threshold height oscillates between the values of 6 and 10.5 GW/cm^2. The period of the oscillations is about 65 ns, which corresponds to the period of the particle oscillations. These oscillations are a result of the particle moving in the potential created by the adhesion force [15,16].

Figure 2 shows the same results but this time replotted to show the threshold fluence as a function of pulse width. It shows that the threshold fluence for all three pulse shapes converges to a value of approximately 0.1 J/cm^2 in the limit as the pulse width tends to zero.

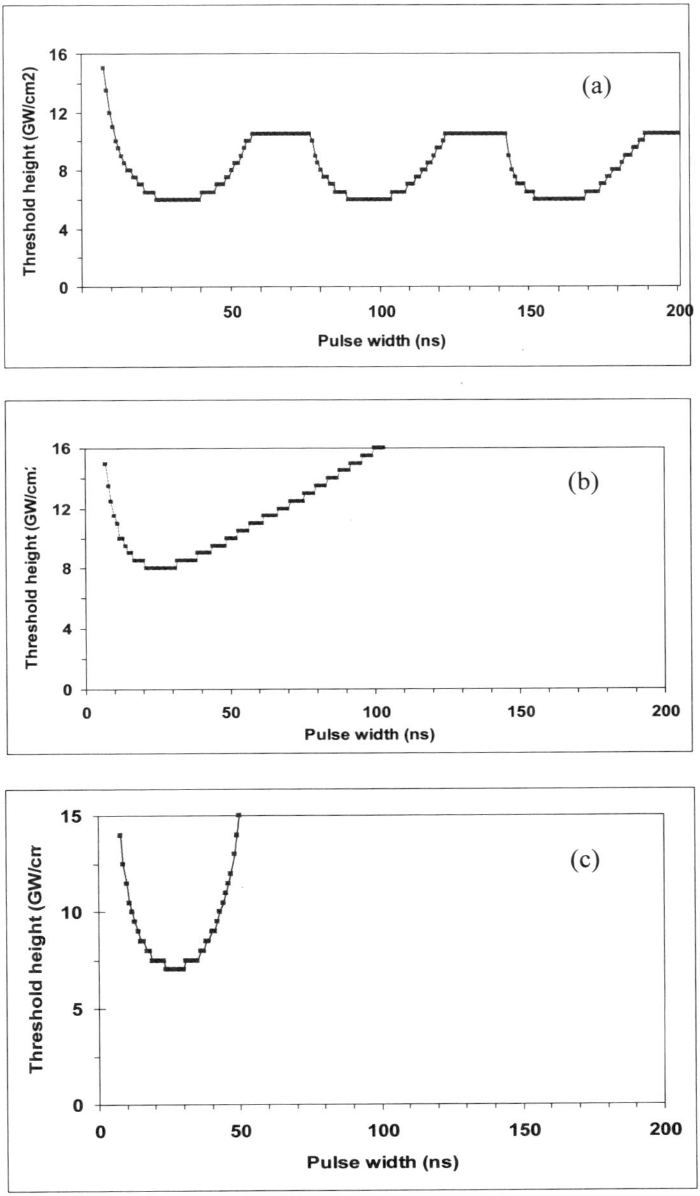

Figure 1: Plot of threshold height vs. pulse width for (a) rectangular pulse shape, (b) exponential pulse shape and (c) sinusoidal pulse shape.

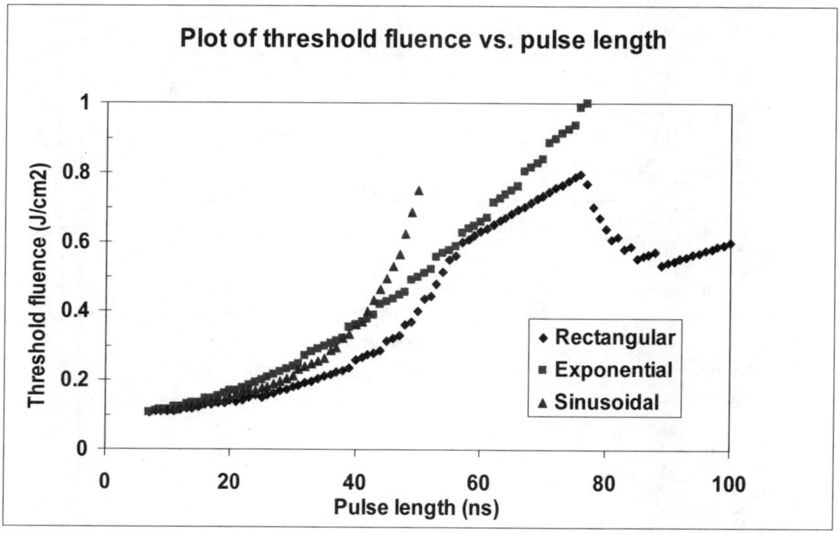

Figure 2: Plot of threshold fluence vs. pulse length for rectangular, exponential and sinusoidal pulses.

As the pulse lengths become longer the threshold fluences of the three pulse shapes behave very differently. The sinusoidal case rises faster than the other two cases. The plot clearly demonstrates the advantage of using exponential pulse shapes at higher pulse lengths greater than about 50 ns compared to sinusoidal pulses.

The rectangular pulse shape shows the same oscillations as observed in the pulse intensity against pulse length plot but this time it is superimposed on a rising background due to the lengthening of the pulse. Thus, for pulses greater than 80 ns, there is great advantage in using a rectangular pulse shape over either an exponential or sinusoidal one.

Clearly, this advantage is linked to the fast rise and fall of the rectangular pulse reinforcing natural oscillations of the system – a driven oscillator with a resonant frequency. In order to investigate this in more detail, plots of the deformation parameter as a function of time were plotted. Figures 3 and 4 show two such plots. Figure 3 is for a pulse length of 96 ns, which occurs in the middle of the dip in the threshold intensity shown in Figure 1(a). It shows that the oscillations are enhanced by a factor of 1.9 when the laser pulse is switched off after 96 ns. In this case particle detachment occurs after the pulse has turned off. A similar plot is shown in Figure 4 for a pulse length of 132 ns, which is in the middle of the peak in the threshold intensity shown in Figure 1(a). In this case the amplitude of the oscillations are damped by a factor of 14 and particle detachment occurs while the laser pulse is on. Thus, the threshold fluence is most dependent on the time at which the laser pulse is turned off relative to the cycle of the particle's forced oscillations. The former case, in which the oscillations are enhanced after the laser pulse is switched off, favours particle removal and results in a considerably lower threshold fluence.

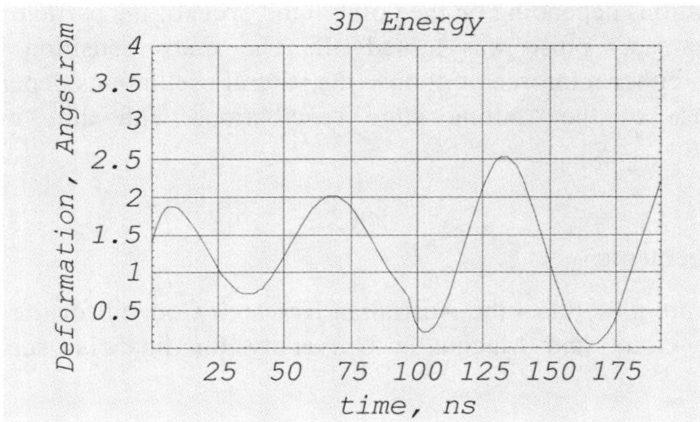

Figure 3: Particle oscillations for τ = 96 ns, rectangular pulse (I_0 = 5.5 GW/cm^2)

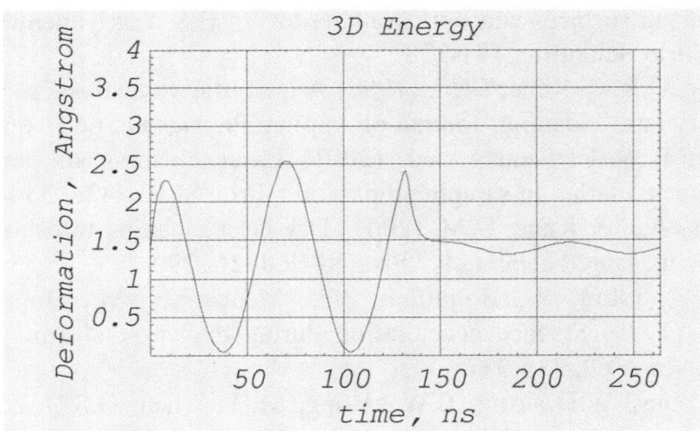

Figure 4: Particle oscillations for τ = 132 ns, rectangular pulse (I_0 = 7.5 GW/cm^2)

4. Conclusions

Using a model developed for low absorbing substrates that accounts for near-field focussing it was found that, for longer pulses, pulse shape has a significant effect on laser cleaning threshold fluences. In the low pulse width limit, the threshold fluence for all three pulse shapes studied converged to a value of 0.1 GW/cm^2. For pulse lengths greater than about 50 ns the exponential pulse shape had significantly lower threshold fluences than the sinusoidal pulse shape. While for pulse widths greater than about 80 ns the rectangular pulse shape resulted in the lowest threshold fluences. In contrast with the other two pulse shapes, the rectangular pulse shape gave threshold fluence which oscillated with pulse length. The cause of this effect is due to enhancement or dampening of the particle vibrations depending on the point in the cycle of the particle's oscillation at which the laser pulse was turned off. The sharp transition in absorbed radiation can either reinforce or oppose the natural oscillations depending on the relative phase of the "falling edge" oscillations with the "leading edge oscillations".

Acknowledgements

The authors are grateful to the Australian Research Council for the award of an ARC Large Grant and Macquarie University for financial support of the research.

References

1. Kane, D.M., Fernandes, A.J., & Halfpenny, D.R (2002) Pulsed laser cleaning of particles from surfaces and optical materials, in B.S. Luk'yanchuk (ed) Laser Cleaning, World Scientific, 191-228.
2. Halfpenny, D.R. & Kane, D.M. (1999) A quantitative analysis of single pulse ultraviolet dry laser cleaning, Journal of Applied Physics **86**, 6641-6646.
3. Kane. D.M. & Fernandes, A.J. (2001) Laser cleaning of particles from surfaces - issues relating to sample preparation, Proc SPIE **4426**, 334-339.
4. Fernandes, A.J. & Kane, D.M. (2001) Dry laser cleaning threshold fluence - how can it be measured accurately? Proc SPIE **4426**, 290-295.
5. Dobler, V., Oltra, R., Boquillon, J.P., Mosbacher, M., Boneberg, J. & Leiderer, P. (1999) Surface acceleration during dry laser cleaning of silicon, Applied Physics A **69**, 335-337.
6. Lu,Y.F., Song, W.D., Ang, B.W., Hong, M.H., Chan, D.S.H. & Low, T.S. (1997) A theoretical model for laser removal of particles from solid surfaces, Applied Physics A **65**, 9-13.
7. Lu, Y.F., Zheng, Y.W. & Song, W.D. (1999) An energy approach to the modelling of particle removal by pulsed laser irradiation, Applied Physics A **68**, 569-572.
8. Lu, Y.F., Song, W.D., Hong, M.H., Teo, B.S., Chong, T.C. & Low, T.S. (1996) Laser removal of particles from magnetic head sliders, Journal of Applied Physics **80**, 499-504.

9. Kolomenskii, A.A., Schussler, H.A., Mikhalevich, V.G. & Maznev, A.A. (1998), Interaction of laser-generated surface acoustic pulses with fine particles: surface cleaning and adhesion studies, Journal of Applied Physics **84**, 2404-2410.
10. Lu, Y.F., Zheng, Y.W. & Song, W.D. (2000), Characterization of ejected particles during laser cleaning, Journal of Applied Physics **87**, 549-552.
11. Vereecke, G., Rohr, E. & Heyns, M.M. (1999) Laser-assisted removal of particles on silicon wafers, Journal of Applied Physics **85**, 3837-3843.
12. Luk'yanchuk, B.S., Zheng, Y.W. & Lu, Y.F. (2002) Basic physical problems related to dry laser cleaning, RIKEN Review **43**, 28-34.
13. Luk'yanchuk, B.S., Arnold, N. Huang, S.M., Wang, Z.B. & Hong, M.H. (2003) Three-dimensional effects in dry laser cleaning, Applied Physics A, **77**, 209-215.
14. Arnold, N., Schrems, G., Muhlberger T., Bertsch, M., Mosbacher, M., Leiderer, P. & Bäuerle, D., (2001) Dynamic Particle Removal by Nanosecond Dry Laser Cleaning: Theory, Proc. SPIE **4426**, 340
15. Arnold, N., (2002) Dry laser cleaning of particles by nanosecond pulses: theory, in B.S. Luk'yanchuk (ed) Laser Cleaning, World Scientific, 51-102.
16. Arnold, N., (2002) Theoretical description of dry laser cleaning, Applied Surface Science **208-209**, 15-22.
17. Pleasants, S., Luk'yanchuk, B.S. & Kane, D.M. (2004) Modelling laser cleaning of transparent substrates - the effect of near-field focusing, Appl. Phys. A **79**, #4-5, 1595- 1598.
18. Arnold, N. (2002) Resonance and steep fronts effects in nanosecond dry laser cleaning, Applied Surface Science **197-198**, 904-910.

Chapter 9

NANOPARTICLES DURING LASER CLEANING OF DECORATION SAMPLES OF SIGISMUND'S CHAPEL

STEPHAN BARCIKOWSKI[†], JÜRGEN WALTER, ANDREAS OSTENDORF
Laser Zentrum Hannover e.V., Hollerithallee 8, 30419 Hanover, Germany

ROMAN OSTROWSKI*, JAN MARCZAK, MAREK STRZELEC
Institute of Optoelectronics, Military University of Technology, 2 Kaliskiego Street, 00-908 Warsaw, Poland

Sigismund Chapel of the Krakow Castle was built in 1525 by Bartholo Berecci. The latest total renovation was executed using laser ablation in order to reconstruct the original appearance of the chapel. Sandstone and gypsum samples from Sigismund Chapel were cleaned using the 'ReNOVALaser 5' system. Air contaminants which emerge during this laser ablation often cause health risks if released at the workplace and a decrease of laser cleaning quality if redeposited at the material surface. At the same time, nanoparticles are generated if short pulses are applied. Consequently, a description of the nanoparticle aerosol generation is given in the presented investigation. Though the emission rates for nanosecond laser ablation are remarkably lower than for conventional macro laser technologies such as cutting or welding, the high respirability of particles can pose health risks. A clear shift of the mean aerodynamic diameter of the aerosols to smaller diameters compared to conventional lasers is observed, so that suitable capture systems near to the processing zone or personal protective equipment such as respiratory masks are required to avoid possible health risks.

1. Introduction

Since laser ablation is a widespread technique for cleaning and it is known from laboratory experiments that it causes fumes, the question arises whether these fumes emerge also under conditions of a cleaning process in practice, and what is the size distribution (and therefore the hazard potential) of particles in the "fumes"?

One of the Wawel Cathedral's numerous chapels – the Sigismund Chapel, built by Bartolomeo Berrecci of Florence in the years 1519 to 1533 – is an

[†]Phone: ++49 511 2788 377, Fax: ++49 511 2788 100, Email: bk@lzh.de
*Email: rostrowski@wat.edu.pl

outstanding achievement of Renaissance architecture and sculpture in Poland and in the whole of northern Europe. Every inch of its stone walls and dome is covered with superbly sculptured, fine floral arabesques, grotesque creatures and mythological scenes. The greenish colour of the walls heightens the deep red of marble statues of St. Peter, St. Paul, St. Venceslas, St. Florian, St. John the Baptist and St. Sigismund, as well as tombstones of the Kings Sigismund I, the Old, and Sigismund Augustus and Queen Anne Jagiellon.

The last partial renovation of the chapel was carried out over fifty years ago. The recent (just finished) renovation was executed using laser ablation, in order to reconstruct the original appearance of the chapel. The application of the laser cleaning technique in renovation processes was connected with the realization of tasks within the frame of the finished EUREKA E!2542 "RENOVA LASER" project.

During laser cleaning, laser generated air contaminants (LGACs) emerge which often cause health risks if released at the workplace and may decrease the laser cleaning quality if redeposited at the material surface [1]. Despite the ablation of the comparable thin outer dirt film (which is partly organic), the main part of the matter ablated during laser cleaning is inorganic [2]. Consequently, the LGAC main components relevant to risk assessment are inorganic particulate matter (PM). Permanent gases such as carbon monoxide, ozone and NO_X also emerge, but in low emission rates which cause concentrations below the workplace threshold limit values, if typical nominal air exchange rates are applied. Concerning the PM, it is known that the mean aerodynamic diameter of the particles becomes smaller if short pulses are applied [3,4], but no information on the influence of the laser parameters and material surface on the particle size distribution during laser cleaning is available up to now. Therefore, a characterization of the nanoparticle aerosol generation is carried out in the presented investigation.

2. Experimental

All measurements were conducted in the experimental setup schematically shown in Figure 1, arranged at Laser Center Hanover (LZH).

The laser system 'ReNOVALaser 5' utilized during the experiments was operated at a constant repetition rate of 10 Hz. The basic technical parameters of the system are listed in Table 1. It is worth emphasizing that the highly multi-mode operation of the laser gives an output beam with a top hat beam profile.

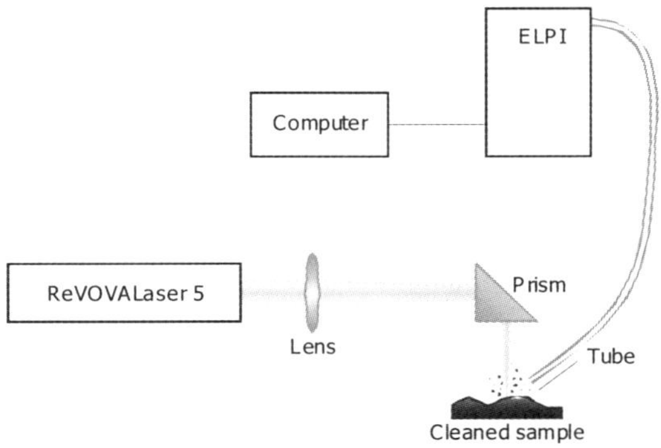

Fig.1. Experimental Setup

The fluence level was adjusted by moving the tested samples and the deflecting prism relatively to the focusing lens. During the experiment, the fluences were varied up to 8.5 J/cm^2. It should be pointed out that such high energy density levels are not applicable in practice in laser cleaning processes. Instead, for the safety of the cleaned object, fluences up to ~1 J/cm^2 are used [5, 6].

Table 1. Basic Performances of ReNOVALaser 5 System

Type	Nd:YAG
wavelength	1064 nm
pulse energy	up to 900 mJ
pulsewidth	15 ns
repetition rate	single shot to 10 Hz
beam width	8 mm

During the laser cleaning process, the particulate emissions were characterized using an online-measurement system. The particulate matter (PM) was captured using a mobile suction nozzle at a volume flow rate of 10 l/min. The aerosols were guided to the online-measuring unit using an antistatic flexible tube. The characterization of the particle size distribution was carried out by an automatic 12-level low-pressure cascade impactor (Dekati Inc.; Type ELPI), which additionally allows post-analysis of the particle morphology by scanning electron microscopy (SEM). The aerodynamic particle size measurement range was 0.03 to 7 μm. Because of the natural sample inhomogeneities, averaging of the single (10 seconds) measurement intervals was necessary. Therefore, a single particle size distribution measurement presented in this investigation is a result of averaging thousands of laser pulses (typically 18,000) up to sufficient saturation of the ELPI stages.

3. Laser Cleaning of Decoration Samples of Sigismund's Chapel

During experiments, samples of sandstone, serving as a basic building material (referred to as 'myslenicki stone'), as well as samples of gypsum, utilized as a filling material in reconstruction processes, were irradiated with high power laser pulses. Examples of such gypsum elements of decoration before and after partial laser cleaning are presented in Figure 2. Figure 3 shows, in turn, the general view of the sandstone sample with and without encrustation.

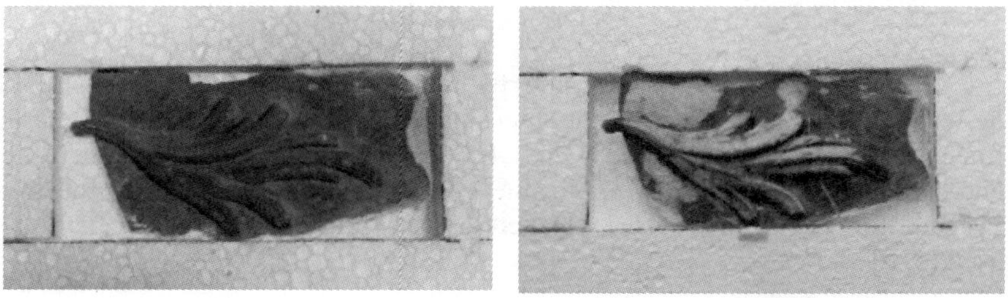

Fig.2. The gypsum samples of Sigismund Chapel decoration before (on the left) and after partial laser cleaning (on the right)

Fig.3. The sandstone sample referred to as 'myslenicki stone' (original stone on the left, encrustation on the right)

4. Characterisation of Nanoparticles During Laser Cleaning

Particulate emissions generated during laser ablation show a median diameter which is very small. It has been shown that particles generated with nanosecond or femtosecond pulses show smaller particle diameters than those generated during long-pulse or cw laser material processing. In the following, the PM captured during laser cleaning is characterized in order to evaluate the dependence of the size distribution on the material surface roughness, laser spot size (fluence) and laser pulse energy.

The decoration samples of Sigismund´s chapel consist of sandstone and gypsum. The sandstone surface was cleaned in areas which were either raw (rough) or polished. Figure 4 shows how the size distribution of the PM during Nd:YAG laser cleaning depends on this surface characteristic. The treatment of the raw rough surface results in a size distribution with an absolute maximum of the relative number concentration at a PM diameter of 100 nm and a secondary maximum at 1,100 nm. In comparison, laser ablation of the polished surface produces a monodisperse size distribution with a maximum at a smaller particle size of 50 nm, as shown in the hatched bars in Fig. 4.

The laser cleaning of sandstone was carried out at a fluence of 8.5 J/cm², whereas the cleaning of the gypsum segment of the object of Sigismund´s Chapel requires less laser fluence in the range of 2.1-3.8 J/cm². The fluence was varied by changing the working distance and therefore the spot size. In Fig. 5, the PM size distribution depending on the laser fluence is shown. In general, at a fluence of 3.8 mJ/cm² (spot size of 4.5 mm), the average particle diameter is smaller compared to the results with a fluence of 2.1 J/cm² (spot size of 6 mm). At the higher fluence, 78% of the PM are nanoparticles (d_P= 30 – 100 nm), whereas at the lower fluence, a lower amount of nanoparticles is generated (65% with d_P= 30 – 100 nm). At the higher fluence, 55% of the overall PM have a diameter of 30 nm (40% at lower fluence), shifting the median particle diameter into the nanometer scale.

Obviously, a higher energy deposition at the surface causes finer particles. To verify this effect, the laser pulse energy was varied at a constant spot size (Fig. 6). It can be clearly seen that a lower pulse energy (510 mJ) causes less nanoparticles (54% with d_P = 30 – 100 nm) and therefore a higher average particle size at the workplace. More than half of the PM have a size of 30 nm or more if 510 mJ is applied, whereas less than a third are in this size range if the pulse power is slightly higher than 600 mJ (which is the upper limit of the laser cleaning working range for this object).

In summary, the particulate matter which emerges during Nd:YAG laser cleaning significantly decreases to the ultrafine particle diameter range (< 100 nm) if a high fluence or a higher laser pulse energy is applied and if the sandstone object surface is polished.

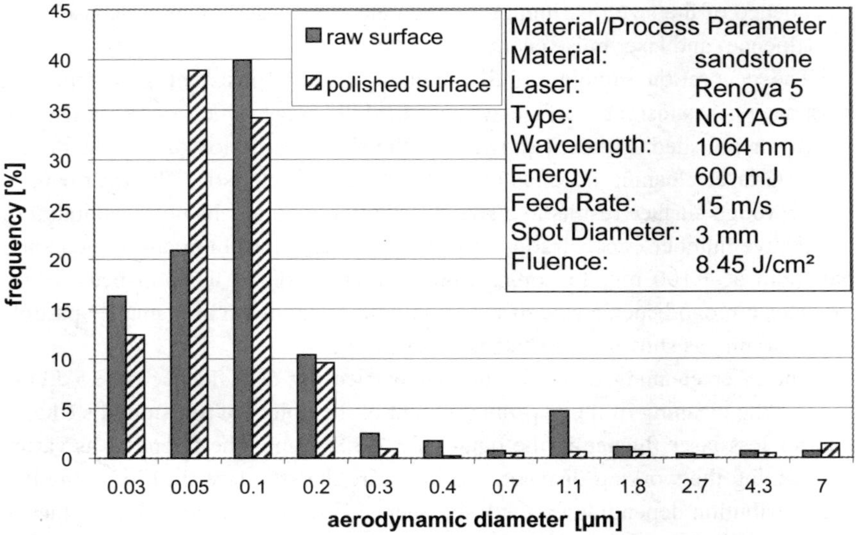

Figure 4. Influence of the object surface on the size distribution of microparticles and nanoparticles during Nd:YAG laser cleaning (fluence: 8.45 J/cm²).

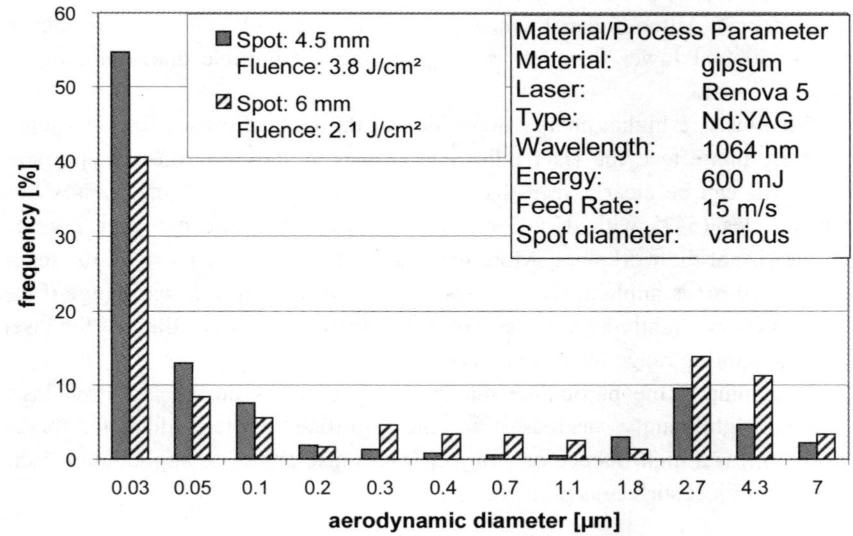

Figure 5. Influence of the laser spot size on the size distribution of microparticles and nanoparticles during Nd:YAG laser cleaning.

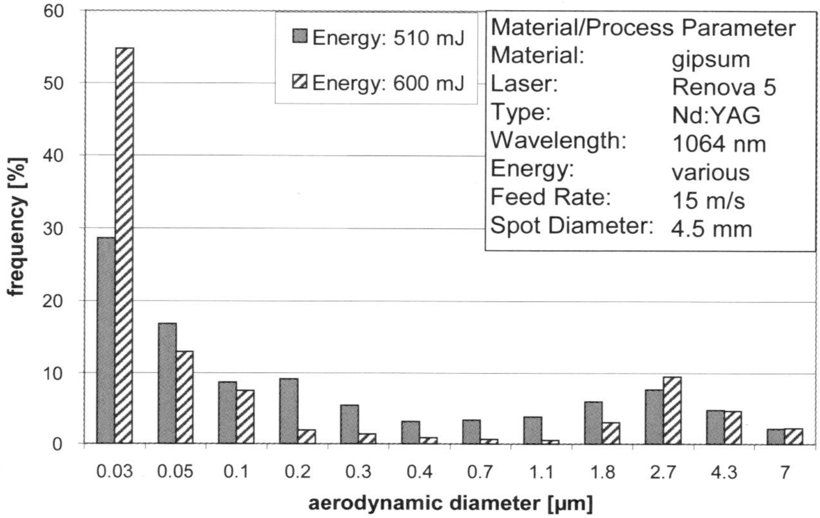

Figure 6. Influence of the pulse energy on the size distribution of microparticles and nanoparticles during Nd:YAG laser cleaning

5. Risk Assessment

Due to the further development of laser cleaning systems towards higher ablation rates, risk assessment and protection against aerosols (nanoparticles) is becoming more and more important [10]. Because of the toxic potential of PM the operator (such as the restorer) has to be efficiently protected from inhaling aerosols. The fraction precipitation rate of the respiratory masks or filters have to be designed to meet these requirements. The fine particles have a negligible settling velocity in air, so that efficient capture near the source of emission is necessary to avoid contamination of the workplace [7, 8]. Even if the emission source (the laser cleaning system) is not subject to an EU-standard or international standard in detail, it is obligatory to apply the 'lowest achievable emission rate' [9, 10].

Depending on the emission mass rate, air cleaning systems have to be used in order to reduce the impact for the environment. As shown above, the LGACs released during Nd:YAG laser ablation are characterised by fine and ultra fine particles and different levels of emission rates. Based upon available information, the maximum achievable control technology (MACT) emission limitation and control technology shall achieve the maximum degree of reduction in emission of hazardous air pollutants which can be achieved by utilising those control technologies. Numerous filter systems have been characterised and qualified for various conventional laser material processing applications [11, 12, 13], but a qualified filter system for nanoparticulate LGACs is still undeveloped.

For chemical processes the plasma filter, surface filter or deep bed filter technology is the MACT for the filtration of nanoparticles. Plasma filtration is capable of reducing odour and nanoparticulate emission at the same time. However, it is not qualified for typical waste gas conditions during laser cleaning (e.g. outdoor or field application) as yet. Deep bed filtration has the capability to separate particles down to < 100 µm, but the filter material has to be disposed after saturation, increasing operational costs. Surface filtration allows the separation of a high particle concentration and automatic cleaning of the filter media, but is often more expensive. The application of these MACT as off-gas technique is recommended to guarantee best employee protection and occupational safety and health.

From the legal point of view, it is important to know that it applies to the owner or operator of the laser cleaning system (source of hazardous air pollutants) to meet the recommended MACT emission limitation even if no emission standard is given by administration [14, 15, 16].

As shown above, process by-products of laser ablation are found to be hazardous due to the small median diameter of the particulate matter. But the overall mass rates are comparably low, so that TLVs for harmful substance in the workplace air will not be exceeded, if an emission capture module is applied. An exhaust volume flow of 50 – 200 m^3/min is appropriate to assure sufficient capture. As a result, safety precautions regarding non-beam hazards will contribute marginally (< 10%) to the operating costs of a laser cleaning system. Even if in future, higher repetition rates and pulse power result in increased emission mass rates, waste gas filtration systems for LGACs are already available as MACT.

6. Conclusion and Outlook

All historical objects need special care and attention during their renovation. The main problem connected with mechanical cleaning methods is the erosion of original substrate in close proximity to the encrustation, sometimes causing irretrievable losses and defects of fine details (relief or artist's tool trace). Laser cleaning offers the ability of avoiding these problems as well as other unique advantages:

- no contact of tool with object enables to renovate fragile artworks,

- selectivity allows to distinguish between encrustation and original material of the object,
- precision connected with the high degree of control of laser beam diameter,
- on-line control, feedback and easy automation through several possible coupled diagnostic optical and laser techniques,
- environment protection through minimisation of removal material, abrasives and chemicals.

Despite these advantages, during Nd:YAG laser cleaning, particulate matter (PM) emerges which has to be captured with regard to occupational safety. At the highest laser fluence applied in the presented laser cleaning investigation, 78% of the PM is in the nanoparticle size range (30 - 100 nm). Lowering the fluence or pulse power results in a lower frequency of nanoparticles in the PM. In addition, a rough surface produces a particle size distribution with a higher average size than during treatment of the polished surface.

It is notable, that during femtosecond laser ablation, the effects are contrary compared to the presented study: Femtosecond laser ablation causes finer particles if lower fluences or lower pulse energies are applied, which may be due to much higher energy density compared to Nd:YAG laser ablation. Another significant difference of both ablation techniques is the emission rate which is one order of magnitude higher in the case of Nd:YAG laser ablation during laser cleaning [17].

In consequence, the assessment of secondary hazards for Nd:YAG laser cleaning applications is important with regard to safety of machinery and workplace. Both the manufacturer of a laser cleaning machine and the operator have to be aware of the hazards. Therefore it is necessary to know that the characteristics of the PM depend on the laser and material parameters.

Since it is up to the owner or user of a laser cleaning system to apply qualified safety measures and since these technical measures have to be designed for handling particles in the micrometer as well as in the nanometer range, the presented study may contribute to the acceptance of Nd:YAG laser cleaning by providing data for the improvement of the laser machine.

Investigations reveal that Nd:YAG laser cleaning poses hazards already known from material processing with conventional lasers, such as hazards from the generation of particulate and gaseous emissions. However, the detailed characteristics of the particles generated during laser cleaning at high fluences show a remarkable shift of the mean aerodynamic diameter of the aerosols to ultra fine diameters (< 100 nm). Since the ultra fine particles are highly respirable, hazards occur especially if materials are processed which release fumes containing carcinogenic or toxic substances. Therefore, suitable capture systems near to the processing zone or respiratory masks are required to avoid possible health risks. Since respiratory masks often constrain renovation work by restriction of the operational head space, suction nozzles will often be the best choice of achievable hazard control techniques.

Since laser users may like to purchase a fully equipped laser cleaning system, it may be advantageous to supply (and at least to recommend) safety equipment such as protective goggles and a fume extraction system. A punctually applied suction would be the adequate fume extraction technique for the emission source, which is a point source as well. In consequence, a typical volume flow of such a system could be lower than 200 m^3/h (or even a quarter of this) if the suction flow tube is guided along with the laser optics.

The investigations show that during Nd:YAG laser cleaning up to 78% of the PM are nanoparticles, so that suitable safety concepts for non-beam hazards must be drawn up on the basis of technical standards for safety of machinery and risk assessment [18, 19]. From the viewpoint of the laser user, a modular integration of such techniques - suction nozzle, fume extraction and filtration - in technical alignment with the laser cleaning system could be quite beneficial.

Acknowledgments

Part of the presented studies were carried out within the framework of the EUREKA RENOVA-LASER project (E!2542).

References

1 H. Haferkamp, M. Goede, M., S. Barcikowski in Lasers in Manufacturing. AT-Fachverlag GmbH, Stuttgart, pp. 716 - 726.H. Haferkamp, M. Goede, M., S. Barcikowski et al.: in Electronics Goes Green 2000+, Berlin, September 11th – 13th, 2000, VDE Verlag Berlin, pp. 613-617
2 M. Strzelec, J. Marczak, A. Koss, R. Szmabelan, Preliminary optoacoustic measurements of air pollution generated during laser cleaning of artworks, Proc. SPIE Vol.5146, pp.210-214, 2003
3 J. Bunte, S. Barcikowski, T. Püster et al. The Ind. Laser User 34, March 2004, pp. 34-35
4 J. Bunte, S. Barcikowski, T. Puester et al. In: Femtosecond Technology for Technical and Medical Applications (Ed.: Dausinger et al.), Springer 2004, pp 309-318

5 R. Ostrowski, J. Marczak, K. Jach, A. Sarzyñski, Selection of Radiation Parameters of Lasers Used for Artworks Conservation, Proc. SPIE Vol.5146, pp.99-107, 2003
6 W. Skrzeczanowski, A. Sarzyński, A. Rycyk, A. Koss, J. Marczak, M. Strzelec, R. Ostrowski, Preliminary LIBS and colorimetry experiments during laser renovation of King Sigismund's Chapel at Wawel Castle, COST G7, Artwork Conservation by Lasers, Xth Management Committee Meeting, IXth Working Groups Meeting and Workshop, 7-9 October 2004, p.30, 2004 – full text will be published in Optica Applicata
7 Haferkamp, H.; von Alvensleben, F.; Seebaum, D.; Goede, M.; Püster, T.: Air contaminants generated during laser processing of organic materials and protective measures. Proceedings: International Laser Safety Conference, 17.-20.03.97 Orlando, USA
8 Haferkamp, H.; Goede, M., Püster, T.; et.al.: The fume hazards in laser cutting and how to deal with it, AILU workshop, "Processing plastics with lasers", 15. Nov. 2000, Warwick, Coventry
9 N.N.: Council Directive 1999/30/EC of 22 April 1999 relating to limit values for sulphur dioxide, nitrogen dioxide and oxides of nitrogen, particulate matter and lead in ambient air. Official Journal L 163, 29/06/1999 P. 0041 - 0060
10 N.N.: Environmental Data Germany 2002. Federal Environmental Agency (Umweltbundesamt). Berlin 2003
11 Haferkamp, H.; Bunte, J.; Barcikowski, S.; Sattari, R.; Hesse, D.; Müller, D.:Geruchsminderung Aerosol enthaltender Abluft - Erfolgreicher Einsatz der Biofiltration für das Laserschneiden von Holzwerkstoffen. Wasser, Luft und Boden Vol. 7/8, 2002. P. 54-58
12 von Alvensleben, F.; Barcikowski, S.; Vitzthum, E.; Goede, M.; Haferkamp, H.: Ablüfte aus der Laserbearbeitung richtig filtrieren. Laser Magazin, June 1998
13 Barcikowski, S.; Goede, M.; Haferkamp, H.: Bioremediation of micron and submicron particular emissions exhausted from laser material processing of polymers. In: Biotechnology 2000, 3.-8. September 2000, Berlin, Germany. Page 354
14 40 CFR Part 63: Part II Environmental Protection Agency. Federal Register: March 23, 2001 (Volume 66, Number 57) Proposed Rules, Page 16317-16360. Revised as of July 1, 2000. From the Federal Register Online via GPO Access: DOCID:fr23mr01-36
15 CAA90: Clean Air Act 1990, Sec. 112 and Title 42 - The public health and welfare. Chapter 85 - Air pollution prevention and control. Subchapter 1 - Programs and activities. Part A - Air quality emission limitations.

Hazardous air pollutants. From the U.S. Code Online via GPO Access :CITE: 42USC7412. (Laws in effect as of January 27, 1998)
16 29CFR1910: Occupational safety and health standards. Title 29 - Labor chapter 17 - Occupational safety and health administration, department of labor (continued) Part 1910. 29 CFR 1910, Page 7-543.
17 T. Burmester, M. Meier, H. Haferkamp, S. Barcikowski, J. Bunte, A. Ostendorf: Femtosecond laser cleaning of metallic cultural heritage and antique artworks. Proceedings of 5th Int. Conf. on Lasers in the Conservation of Artworks - LACONA V, 15.-18.Sept. 2003, Osnabrueck. Springer-Verlag, in press
18 ISO/TR 12100-1:1992, Safety of machinery - Part 1: Basic concepts, general principles of design - basic terminology, methodology
19 DIN EN :1997-01, Safety of machinery – Principles for risk assessment

Chapter 10

FEMTOSECOND LASER CLEANING OF METALLIC ANTIQUE ARTWORKS – ADVANTAGES, LIMITS AND ECONOMIC ASPECTS

STEPHAN BARCIKOWSKI[†], NIKO BÄRSCH, THOMAS BURMESTER, JENS BUNTE, JESSICA ULRICH

Laser Zentrum Hannover e.V., Hollerithallee 8, 30419 Hanover, Germany

ANGELIKA GERVAIS

Norddeutsches Zentrum für Materialkunde von Kulturgut e.V. (North German Centre for the Material Science of Cultural Assets), Scharnhorststr. 1, 30175 Hanover, Germany

MICHAEL MEIER

Niedersächsisches Landesamt für Denkmalpflege (Lower Saxony Department of Preservation of Ancient Monuments), Scharnhorststr. 1, 30175 Hanover, Germany

Corrosion products and other contaminations of metallic artworks can often not be removed sufficiently by conventional conservation techniques. A laser cleaning method for antique metallic artworks using a femtosecond laser technology is presented. Due to the small thermal impact of this technology, compared to nanosecond laser ablation, damage or discolouring of the original surface is avoided. The selective removal of corrosion products from original objects made of copper, bronze alloy and silver, using a Titan-Sapphire femtosecond laser, is shown. Moreover, the influence of the laser fluence and of the number of cleaning cycles on the specific removal efficiency is presented leading to specific fluence thresholds. The maximum surface-specific cleaning rates are evaluated from the viewpoint of restoration in practice, so that an economic balance sheet including operating and labour costs is given for this technology.

1. Introduction

Corrosion products and other contaminations of metallic artworks are often not sufficiently removable with conventional techniques [1, 2]. Nanosecond laser microstructuring of macroscopic workpiece surfaces is being used in several industrial applications as well as in laser cleaning [3, 4]. However, the quality of

[†]Phone: ++49 511 2788 377, Fax: ++49 511 2788 100, Email: bk@lzh.de

resulting structures is limited due to the formation of molten material. It is known that it is possible to avoid melt formation and to minimize the heat-affected zone (HAZ) by shortening the pulse duration into the picosecond and femtosecond regime [5, 6]. Figure 1 shows the regime of pulse duration and pulse power of different laser systems that are usually applied for laser material processing [7].

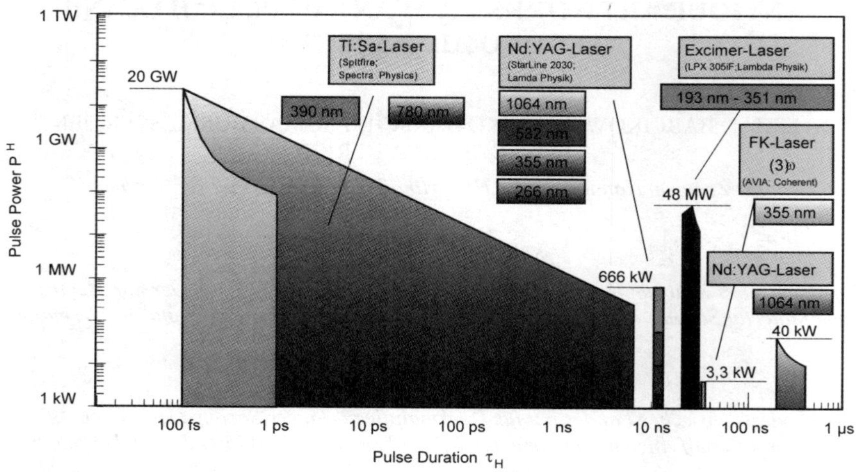

Figure 1. Comparison of pulse duration and pulse power of different laser systems applied for laser ablation, respectively laser cleaning

In the following, a cleaning method for antique metallic artworks using femtosecond (fs) laser technology is investigated, including its advantages, limits, and economic aspects.

2. Experimental Setup

An ultrashort-pulsed laser with a central wavelength of 780 nm (Thales "Bright") was operated at a repetition rate of 1 to 5 kHz and a pulse duration of 150 fs. At 3 kHz, the energy per pulse was 500 µJ. The laser beam was guided using a 2D scanner (Scanlab) to realize different strategies of ablation especially with regard to pulse overlapping. The objects were positioned using a 3D motor drive. They were provided by the Lower Saxony Department of Preservation of Ancient Monuments (NLD) and the Northern German Centre of Material Science of Heritage (ZMK) as well as the restorers Wolfgang Conrad and Habner & Brandner GmbH. Details of the experimental setup have been reported before [8].

3. Femtosecond Laser Cleaning of Bronze and Copper Objects

For the restoration of objects made from copper, bronze alloy, and silver, the applicability of ultrashort laser pulses has been investigated. In the course of the demonstrations, fluence and repetition rate influences on process quality and especially efficiency are shown. Figures 2 and 3 demonstrate laser cleaning results on a corroded bronze object in a pre-treated (Fig. 2) and non-treated (Fig. 3) condition, each comparing the surfaces after different numbers of laser cleaning cycles.

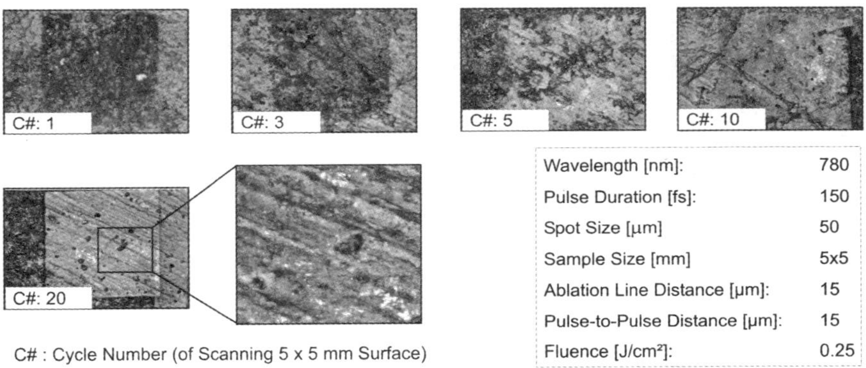

Figure 2. Laser cleaning of a pre-treated bronze object ("Plümecke-Grabplatte"): influence of cycle numbers on surface layer removal

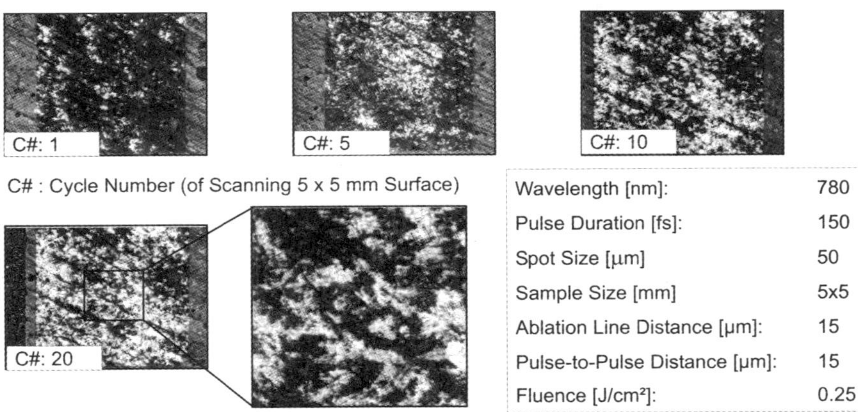

Figure 3. Laser cleaning of a non-treated bronze object ("Plümecke-Grabplatte"): influence of cycle numbers on surface layer removal

By choosing laser fluences slightly above the ablation threshold of the corrosion products, but not exceeding the bulk material thresholds, it is possible to precisely and selectively ablate corrosion layers. Figure 4 shows samples that demonstrate such fs laser removal of single corrosion layers without influencing the colour or structural properties of layers underneath. After the cleaning process, the chemical composition of the surfaces has been analyzed by EDX spectrometry.

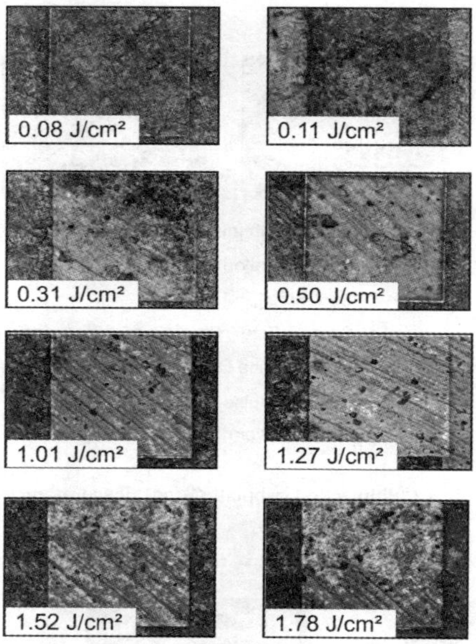

- Application of threshold fluences may lead to selective removal of different corrosion layers

- Ablation or irreversible damage of resp. deeper layer is avoidable

Wavelength [nm]:	780
Pulse Duration [fs]:	150
Spot Size [μm]	50
Sample Size [mm]	5x5
Ablation Line Distance [μm]:	15
Pulse-to-Pulse Distance [μm]:	15
Cycle Number:	10

Figure 4. Samples of femtosecond laser ablation of different corrosion layers using appropriate laser fluences

Using femtosecond laser pulses as specified, varying just numerical apertures or pulse energies, the threshold laser fluences leading to material ablation have been found for different relevant types of corrosion layers for a systematic evaluation. Figure 5 shows that these threshold values constantly increase from organic components and soot to copper sulphates, carbonates, and oxides, to – finally – the original metal matrix. This implies that each of these corrosion products can be removed separately in this order by a respective adaptation of the laser fluence, so that highly selective ablation of corrosion products from the surface of ancient bronze and copper objects is principally possible without damaging the layers underneath.

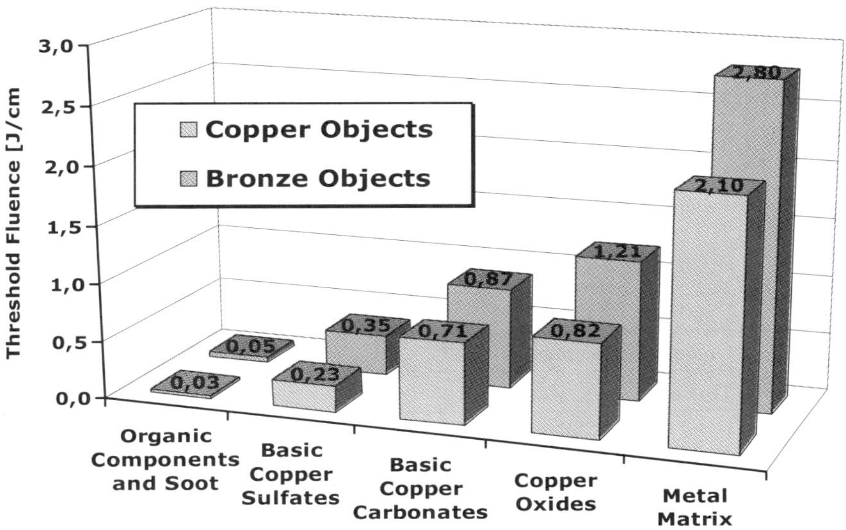

Figure 5. Comparison of femtosecond laser ablation threshold fluences for selective removal of copper and bronze surface layers

4. Economic Aspects

In practice, an application of this cleaning process involves long adjustment times, as the position of the laser focus needs to be adjusted precisely to the surface, taking the major part of the whole processing time for the cleaning of an object. Assuming that adjustment measures took five minutes per 100 mm^2 surface area, the area of an A4 page would take 68 hours if a tin-bronze surface was to be cleaned from organic compounds and soot. Table 1 lists the estimated total processing time for different corrosion layers on tin-bronze and copper objects – taking into account typical thicknesses of the layers as well as specific ablation rates for the laser fluences to be applied. On the basis of seven-hour working days, total cleaning times add up to values between 140 and 250 days per square metre. Apart from organic compounds and soot, corrosion products on copper can generally be ablated more quickly than on bronze.

To evaluate the competitiveness of the cleaning approach with femtosecond laser pulses for metallic antique artworks, a cost analysis has been done. The analysis was made according to VDMA BwB7 and therefore uses the categories and the quantitative assumptions summarized in Table 2, including work conditions, necessary consumables, and machine costs [9]. In an estimation of

Table 1. Required time for selective fs laser removal of different corrosion products on tin-bronze and copper bronze objects.

Corrosion Products and Layers		Theoretical Required Time of Ablation Including 5 Minutes Adjustment Time per cm² Surface Area: Per Area of an A4-Page (630 cm²) [h]	Theoretical Required Time of Ablation Including 5 Minutes Adjustment Time per cm² Surface Area: Per Square Meter [Days à 7 h]
Tin-Bronze Objects	Organic Compounds and Soot	68	155
	Tin Oxide	76	171
	Basic Copper Sulfates (Bronchantie Antlerite)	79	179
	Basic Copper Carbonates (Malachit)	76	171
	Tenorite	110	250
Copper Objects	Organic Compounds and Soot	104	236
	Basic Copper Sulfates (Bronchantie Antlerite)	61	138
	Basic Copper Carbonates (Malachit)	65	148
	Tenorite	62	140
	Cuprite	97	219

the total processing costs per square centimetre, the machine costs – on a basis of 47.13 € per hour – clearly make up almost the whole cost of the process. This is an over-estimation including possible outages and the high establishment costs of the laser system.

To evaluate the economic competitiveness of fs laser systems for this, so far unusual, material processing field, the current strong development of ultrashort-pulsed laser sources must also be considered – for example, femtosecond laser sources with pulse specifications comparable to those used in these investigations, but average powers twice as high, are already available on the market. Also, similar laser sources with longer pulse durations (in the range of 500 fs) that offer significantly higher pulse repetition rates and average powers are currently being developed.

In addition, contrary to most other applications in the field of femtosecond laser material ablation, the maximum pulse energy that can be used for good

Table 2. Cost analysis summary of the femtosecond laser cleaning in terms of surface-specific operating costs including the cost category breakdown.

Pos.	Material: Tin-Bronze, Copper Sulfate	
(1)	Shifts	1-shift operation
	Conditions	
(3)	Assist Gas	200 l/h
(4)	Power Supply	1.5 kW
(5)	Surface Specific Removal Rate	0.29 cm²/min
(6)	Surface Specific Removal Rate	17.14 cm²/h
(7)	Lens Change	1,000 h/Lens
(10)	Area per Lens	17,140 cm²/Lens
	Gas and Power Supply	
(12)	Process Gas	11.7 l/cm²
(13)	Power	87.51 Wh/cm²
	Cost Basis	
(15)	Assist Gas	2.00 €/m³
(16)	Power	0.16 €/kW h
(17)	Lens	600.00 €/Piece
(18)	Tools	–
(19)	Machine Cost	47.13 €/h
(20)	Labour	–
	Cost per cm²	
(23)	Process Gas	0.02 ct/cm²
(24)	Power	0.01 ct/cm²
(25)	Lens	0.03 ct/cm²
(27)	Sum	0.06 ct/cm²
(28)	Machine Cost	270.86 ct/cm²
(31)	Total Sum	270.92 ct/cm²
(32)	**SUM per cm²**	**2.71 €/cm²**

results is not restricted. When doing planar ablation with low laser fluences, the laser focus does not need to be on the material surface. Instead, the irradiated area can be increased by a focus shift, in order to reduce the energy density on the surface. In contrast to cutting applications, this enables the use of short-pulsed systems with highest pulse energies without restrictions regarding the processing quality.

Especially when comparing different laser types, the progress in laser source development is a relevant issue. The efficiency and reliability e.g. of conventional Nd:YAG lasers might be higher at first sight, but the specifications of available sources are not subject to such fast and relevant changes. Also, the presented fs laser results demonstrate higher achievable qualities than with any other laser source, which is not part of the economic calculations.

5. Practicability

When it comes to using femtosecond lasers for artwork cleaning in practice, the mobility of the laser source is a major restriction. Because the lasers are comparatively large laboratory units, objects to be cleaned need to be taken to the respective lab.

With regard to process automation and efficiency, flexible beam guidance and an automatic adjustment of the laser focus are important necessary developments to achieve sufficient flexibility and practicability of the process. The laser focus needs to be positioned with accuracies down to several hundred microns, which requires an automatic distance detection and focus position control for the machining of curved geometries. At the same time, the laser pulses need to be flexibly delivered to the focussing device, which could – ideally – be done by fibre optics in the future (however, fibre transmission of amplified femtosecond laser pulses has not been realized yet) or, as the best current alternative, with use of a hinge construction with mirrors in combination with a partially manual operation.

When using a femtosecond laser cleaning approach – as specified for the illustrated experiments – for the cleaning of a complex bronze statue of a total surface area of 0.5 m^2, the estimation in Table 1 leads to a total processing time of 75 days, assuming a cleaning rate of 187.5 days per square meter. In addition to manpower costs, machine costs in the range of 13,500 Euro are to be expected. Furthermore, complex objects can make necessary an additional manual cleaning process for some parts of the object. Using conventional sculpture cleaning methods for the whole statue, the processing time, depending on the applied methods, can vary between 30 and 100 days.

For the cleaning of a silver coin with a surface of 25 cm^2, the femtosecond laser method requires a processing time of about 50 minutes and technical costs of 70 Euro, while for a thorough cleaning with conventional methods, a time of 4 hours must usually be expected.

As a general conclusion, cleaning of antique artworks with femtosecond lasers is not a mature process in the present state and only partially leads to satisfying results from the view of practical monument preservation. This is due to the current requirement of partial flat planes on the object surface that are needed for constant parameters of laser and optics. When strongly irregular surfaces need to be processed, the results become respectively inhomogeneous, leading to unwanted effects like holes, accidental removal of patina, or an insufficient removal of the corrosion products. However, positive processing results for the cleaning of silver coins have shown that the main challenge with regard to practicability lies in the field of beam guidance, surface detection, and focus adjustment, and not in the laser ablation process itself.

6. Conclusion and Outlook

For restoration purposes, a continuation of investigations on fs laser cleaning is desirable, since the results at least on planar surfaces are convincing. The precise cleaning of large and complex curved shapes is a matter of opto-mechanical, metrological, and control issues that could be optimized in further development efforts.

Current amplified fs lasers are immobile due to their size and complexity. In future development towards mobile systems fs-specific laser safety aspects – in particular protection against beam hazards as well as aerosols (nano-particles) – have to be considered [10, 11].

Laser operation makes up almost the whole of the process costs. However, this means that when reducing these machine costs by a factor of 2 or 3 within the next years in the course of a development of more powerful short-pulsed lasers, the total costs for restoration works will decrease respectively. Besides compactness and output power, development of ultrashort-pulsed lasers must also focus on pulse duration stability and robustness, if they shall compete with conventional lasers in practical use in this context.

Acknowledgments

Part of the presented work was supported by Deutsche Bundesstiftung Umwelt (DBU). The authors would like to thank DBU for its support and also the restorer Wolfgang Conrad and Haber & Brandner GmbH for providing the original copper and bronze objects.

References

1. B. Stöckle, A. Krätschmer: Die atmosphärische Korrosion von Kupfer und Bronze. In: *Metallrestaurierung*, Arbeitshefte des Bayerischen Landesamtes für Denkmalpflege, Bd. 94, München 1998.
2. M. Mach: Bildatlas typischer Oberflächenphänomene von frei bewitterten Bronzen. In: *Bronze- und Galvanoplastik*, Hrsg. Landesamt für Denkmalpflege Sachsen, Arbeitsheft 5, pp. 152-155, 2001.

3. K. G. Watkins: A review of materials interaction during laser cleaning in art restoration. In: *Lasers in the Conservation of Artworks I*, Heraklion Crete Greece, Verlag Mayer & Comp. Wien, pp. 7-15, 1997.
4. J. Jandeleit et al.: Fundamental Investigations of Micromachining by nano- and Picosecond Laser Radiation. In: *Applied Surface Science* 127-129 (1998), pp. 885-891.
5. B. N. Chichkov et al.: Femtosecond, picosecond and nanosecond laser ablation of solids. In: *Appl. Phys. A* 63, pp. 109-115, 1996.
6. F. Dausinger, F. Lichtner, H. Lubatschowski: *Femtosecond Technology for Technical and Medical Applications*. In: Springer Books "Topics in Applied Physics", Vol. 96 / 2004, pp. 105ff.
7. C. Kulik: Short and ultrashort laser pulses: an upcoming tool for processing optical and semiconductor materials. In: *Photonics West 2004, LASE, Photon Processing in Microelectronics and Photonics III*, SPIE Vol. 5339, San José, 2004, pp. 35-48.
8. T. Burmester, M. Meier, H. Haferkamp, S. Barcikowski, J. Bunte, A. Ostendorf: Femtosecond laser cleaning of metallic cultural heritage and antique artworks. In: *5th Int. Conf. on Lasers in the Conservation of Artworks - LACONA V*, 2003, Osnabrueck. Springer-Verlag, in press.
9. J. Bunte, S. Barcikowski, T. Burmester, A. Gervais, M. Meier, E. Stadelbauer, W. Conrad, J. Ulrich: *Reinigung und Konservierung von umweltgeschädigten Kulturgütern aus Metall mittels Femtosekunden-Lasertechnik*. Abschlussbericht, DBU Az: 18 772: 06/02 - 02/04. 55 pages, Mai 2004, http://www.tib.uni-hannover.de/.
10. J. Bunte, S. Barcikowski, T. Püster et al.: *Femtosecond Technology for Technical and Medical Applications*. In: Springer Books "Topics in Applied Physics", Vol. 96 / 2004, pp. 309-318.
11. S. Barcikowski, J. Walter, A. Ostendorf, R. Ostrowski, J. Marczak, M. Strzelek: Nanoparticles during laser cleaning of decoration samples of Sigismund's Chapel. In: *4th International Workshop on Laser Cleaning*, Macquarie University, Sydney, 2004 (in press).

Chapter 11

ULTRAFAST LASER CLEANING OF MUSEUM ARTIFACTS

A V RODE, N R MADSEN, E G GAMALY, B LUTHER-DAVIES

Centre for Ultra-high Bandwidth Devices for Optical Systems, Laser Physics Centre, Research School of Physical Sciences and Engineering, the Australian National University, Canberra, ACT 0200, Australia

K G H BALDWIN

Atomic and Molecular Physics Laboratories, Research School of Physical Sciences and Engineering the Australian National University, Canberra, ACT 0200, Australia

D HALLAM

National Museum of Australia, Acton Peninsula, Canberra, ACT 2600, Australia

A WAIN

Australian War Memorial, ANZAC Parade, Campbell ACT, 2612, Australia

J HUGHES

The University of Canberra, Bruce, ACT 2617, Australia

A new method of laser ablation, having high pulse-rate of the order of MHz, with short ps and sub-ps pulses, has been recently developed at the ANU. As a preliminary test for the short-pulse ablation applied to art conservation for selective removal of unwanted surface layers, we used the ps laser to clean brass samples coated with contaminant layers such as rust, paint and wax. We demonstrated the laser cleaning system with 2nd (λ_2=532nm) and 4th (λ_4=266nm) harmonics of a Nd:YVO_4 ultra-fast laser, combined with a constant-velocity scanning system. Great flexibility in cleaning the surfaces selectively was demonstrated by removal of the wax coating from the black-painted surface without altering the black paint surface sub-structure. Major conservation and restoration challenges were considered where ultrafast laser cleaning might play a vital role.

1. Laser Cleaning of Artworks

Lasers are non-contact ablating tools, which provide direct electromagnetic energy transfer into the material intended to be removed. Cleaning of artworks and antiques by lasers offers the advantage of selective removal of undesired surface layers of sub-micron thickness, without affecting the underlying (often valuable) structure [1-3]. The laser cleaning concept is based on the removal of surface layers with well-defined thickness by scanning a laser beam over the surface under fully controlled conditions. Removal of these surface layers must be carried out in such a way that the integrity of the bulk material of the original art object is guaranteed.

Basic guidelines for choosing the proper laser parameters for safe laser ablation of unwanted surface layers usually consider:

(i) the laser wavelength to ensure low optical penetration depth and thus high energy absorption in the thin surface layer being removed; and

(ii) the laser energy density, or fluence, which determines the ablation rate per pulse and is thus responsible for the overall precision of the laser cleaning process.

Regarding the pulse duration, nanosecond excimer lasers are usually considered for laser cleaning as they are commercially available, have an adequate energy per pulse, and are easily applicable to larger-scale applications. With nanosecond lasers, the removal of the surface layer proceeds in a purely thermal regime of ablation, and the bulk material under the surface layer is exposed to thermal waves and shock waves which originate from the laser heated and stressed surface layer. As a result, thermal, thermo-chemical, and mechanical effects may alter the underlayer in an undesirable manner.

In recent years, however, short ps and sub-ps pulses have been shown to ablate material without any collateral damage to the bulk [4-6]. It was shown that ultrashort laser pulses produce a better etching and printing quality, thus improving the already established method of excimer laser microetching [7]. The development of high-average-power, high-repetition-rate, sub-picosecond pulse lasers has refocused interest in laser dentistry due to the highly efficient tissue ablation and minimal collateral damage offered by using such lasers in the appropriate parameter regime. This short pulse interaction takes place when the laser pulse duration is shorter than the electron-lattice temperature equilibration time, (determined rate of electron-phonon collisions and of electron heat conduction), and generally requires the pulse duration to be less than a picosecond. The precise removal of surface layers without any collateral damage to the bulk of the material has been already demonstrated as a potential painless laser dentistry technique [8,9,13].

A new method of high pulse-rate, short-pulse (ps and sub-ps) laser ablation for deposition of high optical quality films has been recently developed at the ANU [5,10-12]. In optimum conditions – pulse duration, wavelength, and laser fluence which are linked to the parameters of the material to be ablated- the ultra-fast laser ablation results in nearly 100% conversion of absorbed energy into the ablated vapour, and complete elimination of shock waves and thermal waves.

In this paper we present a preliminary test for the short-pulse ablation applied to art conservation for selective removal of unwanted surface layers. We used the picosecond laser to clean brass samples coated with contaminant layers such as rust, paint and wax. We demonstrated the laser cleaning system with 2nd (λ_2=532nm) and 4th (λ_4=266nm) harmonics of a Nd:YVO_4 ultra-fast laser, combined with a constant-velocity scanning system. Great flexibility in cleaning the surfaces selectively was demonstrated by removal of wax coating from black-painted surfaces without altering the surface sub-structure.

2. Experiments with PS-Pulse Laser Cleaning

2.1. *Laser cleaning of metal samples*

A number of experiments were conducted in order to demonstrate the effectiveness of ultra-fast laser ablation in the cleaning of metals with various coatings. An in-house designed and built mode-locked Nd:YVO laser (fundamental λ=1.06μm) was converted to fourth harmonic (λ_4= 266nm) for cleaning experiments. The laser was prepared such that pulses were generated at a rate of 1.5 MHz with a pulse duration of 13 ps. The laser was delivered to the target metal via a telescopic lens arrangement and focussed to a spot size of about 20 microns. The laser was then scanned in a Lissajous pattern with amplitude of 6mm, via the use of x and y scanning mirrors oscillating at about 200 Hz. A second, doughnut shaped spiral pattern (radius 10 mm) was also used since it presented the ability of constant velocity scanning. In this case the linear velocity of the beam was about 5 ms^{-1}. Most of the experiments were conducted with laser pulse trains having a ratio 1:13, where the target sees the laser for the shorter time period. This was done to avoid thermal problems associated with operating the fourth harmonic generation crystal in continuous pulse train mode.

The first material presented is anodised aluminium (Fig.1a). It is clear that after about 25 seconds the laser has had sufficient time to completely cover the scanning area, and hence has completely removed the anodised layer leaving a clean layer of regular aluminium exposed. The same scanning conditions were then used to clean a piece of aged brass (Fig. 1b). The deep extended scratch seen in both cleaned and unclean surfaces demonstrate the important point to note from this experiment, which is a crucial requirement of laser cleaning: the maintenance of surface structure, which is clearly met. In addition, a two-minute exposure was also performed (top left pattern) to demonstrate the ability of the technique to remove macroscopic accumulation of dirt.

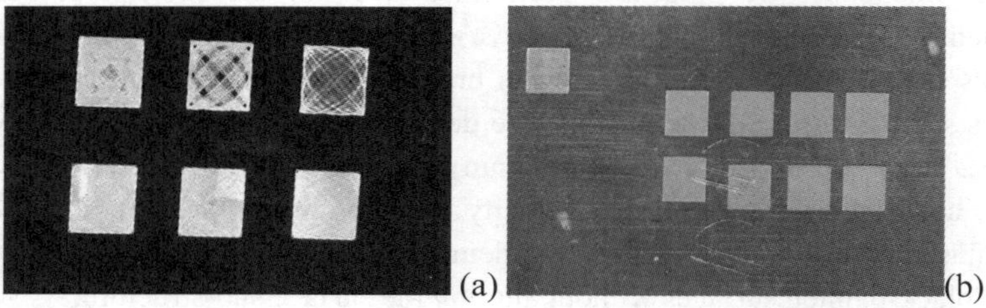

Figure 1. Anodised aluminium (a) and aged brass (b) samples with 1 cm^2 areas exposed to a focused laser beam (for cleaning) for time intervals from 5 to 30 s. For anodized aluminium, the exposure time increases from top right corner anticlockwise. The time required for area coverage is about 25 seconds. Note the image of aged brass on the right shows clearly the maintenance of surface structure (ie no alteration of "scratches" in the surface) using the ultrafast laser cleaning technique.

Next, experiments were performed using a deeply eroded, acid-treated brass sample coated with a much thicker and durable layer of contamination containing a complex mixture of chemicals (Fig. 2a). These results further reinforce the ability of the ultra-fast technique in maintenance of surface structure. It is also worth noting in this experiment that as a consequence of the nature of the coating, it takes many complete scanning area coverage cycles to remove the layer of contamination. At the same time, the use of the Lissajous pattern begins to dig craters into the corners and sides of the scanning area – this is due to the inertia of the galvo-mirrors in the turning points. To assure constant scanning speed of the laser spot over the sample surface the scanning pattern was changed to a constant velocity spiral for the remaining experiments.

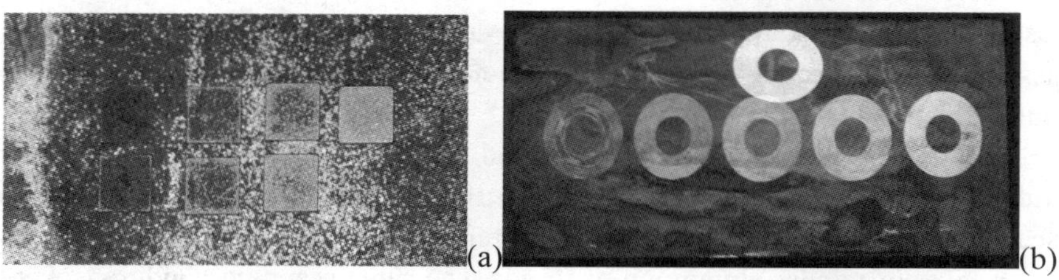

Figure 2. Laser cleaned rusted and chemically damaged brass (a) and oxide-coated brass (b). Cleaning times for the 1 cm^2 squares in the left image are of the order of minutes due to the extremely thick contamination layer. Cleaned areas of the oxide coated brass increase in duration from 10 to 320 seconds.

The first material presented which has been cleaned using the constant velocity spiral is another brass sample this time with an oxide coating – Fig.2b. In the image are seen exposures made for increasing durations from 10 to 320 seconds. The cleaning time for the particular spiral size here is three minutes and it is clear that the surface structure is maintained leaving a clean finish on the original brass surface.

The next sample seen in Fig.3 provided a more complex set of parameters to investigate. This is a brass sample first coated with paint and then coated with a thin layer of wax. Initially (position "a") an 80 second exposure is made (same laser conditions as above) and clearly both the wax and paint layers are removed. However, it is desirable that only one particular layer is removed from the target. Hence an attempt was made to remove only the wax by lowering the fluence to below that required for ablation of the paint.

This was achieved by moving the target to a position 10 mm away from focus, hence increasing spot size. Clearly in position b) the fluence is sufficient to remove some paint, however it becomes more obvious in position c) at 20mm from focus that only the layer of wax is removed. This is confirmed by inspection with a microscope. Having performed these experiments for 80 seconds, for comparison an experiment was conducted for about 5 mins (position "d") 30mm behind the focus. Again, as is the case for position c), we see only removal of the layer of wax.

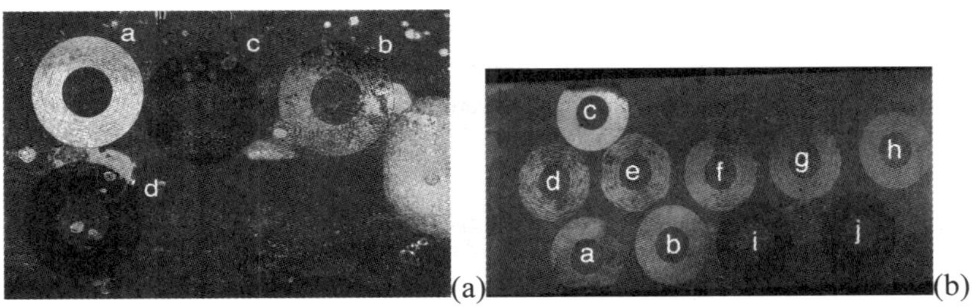

Fig.3. Laser cleaned brass coated with paint and wax (a) and brass coated only with wax (b). Both sets of experiments demonstrate the control over single layer removal achievable by appropriate choice of incident laser fluence.

The final sample investigated was another brass sample coated in a thin layer of wax. Positions d-h in Fig. 3a were experiments conducted with the laser operating with the target at the focus. Exposure times were increased from 10 to 160 seconds again showing the time required for full coverage of the scanning area. To demonstrate the way in which the laser fills the scanning area and to show the relatively low cleaning threshold for wax, the laser was switched to continuous mode operation. Positions a), b) and c) are exposures using the continuous mode of the laser with the target positioned at the focus for 20, 40 and 80 seconds respectively. Clearly, since the target always "sees" the laser, the cleaning time will be less than for the chopped (1:13) pulse train. Here the cleaning time is about one minute as opposed to three minutes for the pulse train. The use of a continuous pulse sequence is somewhat counteracted by the lower single pulse fluence used for the experiment. Hence the cleaning time is not decreased by as much as thirteen times, which one might expect.

In addition, we reduced the laser intensity on the sample surface by moving the sample 20 mm away from the focus (i,j in Fig.3b). As a result, the laser intensity was sufficient to remove the wax, but not high enough to clean the brass surface.

2.2. *Summary of the experiments*

We have demonstrated the effectiveness of the technique of ultra-fast laser ablation for cleaning of metals both in terms of its control over cleaning of particular surfaces and maintenance of surface structure. The control over the removal process was demonstrated by changing the laser intensity on the target surface, and the exposure time. Further, more precise control can be provided through the variation of the laser pulse duration, laser repetition rate, and laser wavelength. All the laser parameters should be tuned precisely to ablate the particular contamination with certain chemical, optical, and thermal properties. In addition, these contaminating layers should be removed from the surface leaving the bulk material untouched. Much work is still required to ensure that the technique is applicable to cleaning of physical metallic artworks

3. Application of Laser Cleaning Technique to the Museum Artefacts

3.1. *Traditional art collectables*

Museum conservators have used nanosecond laser technology for more than 25 years, particularly for removing coloured stains from surfaces (eg stone, paper, paintings) that could be damaged using conventional cleaning techniques involving liquids or physical methods. Particular advantages are foreseen for the new picosecond lased technique in treating artifacts which are composites of several different materials, having differing treatment requirements, such as ethnographic sculptures comprising metals with organic materials such as ivory, wood, feathers, leather, etc. Often, such artifacts cannot be dismantled for treatment without significant physical risk or loss of cultural significance.

A review was conducted with ANU researchers and museum conservators from the National Gallery of Australlia, the National Museum of Australia, and the Australian War Memorial, examining typical intractable treatment problems where it was considered that the layer-by-layer picosecond laser treatment should offer some particular advantages, such as:

- Controlled surface ablation of very small amounts of material and from locations that are difficult to access such as crevices and pits
- Selective removal of damaging stains or accretions from a differently coloured substrate
- Removal of damaging material or undesirable overpaint where liquids (aqueous, solvent) would damage the other materials.
- Extremely fragile materials which cannot easily be washed or separated (eg fraying silk adhered with nylon and synthetic adhesives, as in the Ballets Russes problem, and removal of lacquers from aboriginal bark paintings, etc).

Discussion of problems associated with paper, paintings and textiles suggested that treatment of metal corrosion was the most likely to produce significant, immediate results, with particular interest in removal from surfaces where the corrosion product can easily be distinguished from the parent metal. Samples of corroded metals (including mild steel, stainless steel, bronze, brass, lead and zinc), some significantly affected by pitting or flaking surfaces, were subjected to laser cleaning to identify the ease of removing surface corrosion products without damage or change to the underlying metal. Of particular interest is the ability to completely remove salts that can lead to cyclic corrosion, such as chlorides cause in the notorious problems of bronze disease.

The sub-picosecond pulses and the ability to ablate surfaces without heat is potentially a considerable and unique advantage as conservators particularly want to avoid heat that could change metallographic structure or that could release damaging gases or cause heat stresses which could occur for example with leaded bronzes, for example restoration of Asian Buddhas.

3.2. *Conservation techniques for modern collections*

The Australian institutions hold a wide range of materials, including paper and photographic archival records, artworks, textiles and objects. Most collection objects date from the 19th and 20th centuries and therefore represent an extensive range of relatively modern and often experimental materials and industrial techniques.

Conservation techniques for modern collections are currently in their infancy. Many of the objects in these collections are only just starting to show deterioration or are newly becoming a focus for display. Existing conservation techniques are often either inadequate for the treatment of these materials, or their effects have not been studied, so that damage may be done through inappropriate use of existing techniques.

Development of existing lasers for conservation has been focused on the cleaning of older and art objects, which form a much higher proportion of northern hemisphere collections. These objects present different challenges to the modern and substantially technological collections found in Australia, and thus, the lasers required for their treatment accordingly have different properties.

3.3. *New challenges in conservation and restoration of modern collections at the Australian War Memorial*

Here we would like to illustrate new challenges in restoration of relatively recent collectables at the Australian War Memorial as an example of where laser cleaning technique could play a vital role. Below we present the most challenging restoration problems where traditional restoration technique can not be applied for various reasons.

3.3.1. *Removal of overlying paint, wax or varnish from original paint*

Many of the War Memorial large technology objects have been overpainted after their wartime service. To restore them to their appearance during the most significant phase of their history we need to remove the overpaint to allow the original paint layers to be seen. Attempts to do this mechanically or with solvents have proved unsuccessful as they cause extensive damage to the underlying original paint layers.

3.3.2. *Removal of paint, oil or dirt from plastics such as Perspex*

Many plastic components of our large technology items have been contaminated by paint overspray, dripping oil and other substances during decades of poor storage and restoration. As well as being inappropriate to the wartime service of these objects, these contaminants are often damaging to the original materials. Removal of these contaminants, either mechanically or with solvents, damages the material surface either by scratching or by allowing solvents to penetrate the physical matrix of the material which can lead to swelling and self-propagating, irreversible crazing of the material.

Fig. 4. Model AC 1 Sentinel Tank. This model shows copper corrosion spots erupting through chrome plating. Cleaning this mechanically would cause scratching and be extremely time consuming; cleaning it chemically (citric acid) is liable to leave acid in the pores and crevices, which will cause further damage.

3.3.3. *Removal of paint, polish residues and dirt from organic materials with complex, easily damaged surfaces*

Poor museum numbering and restoration techniques in the past have often resulted in organic items such as wood, leather and textiles being disfigured with paint. Examples of this are a buff leather belt, originally whitened with kaolin clay and later painted with white paint to simulate the original clay filled surface, and textiles which have had metal polish applied to the surface in an attempt to remove tarnish from the metal threads. Cleaning these surfaces mechanically causes extensive disruption, while using solvents flushes dirt and polish residues further into the material and may cause dyes to run.

Fig.5. Patent leather shoulder belt (1880s). The metals threads are tarnished and, although hard to see in the photograph, heavily soaked in metal polish. It is virtually impossible to remove either of these because any sort of handling damages the metal threads and solvent chemicals will drive the metal polish residues further into the object and damage the patent varnish.

3.3.4. *Cleaning of photographic emulsions*

Cleaning of negatives required to remove surface silver mirroring, fungal attack residues, fibres, starch/ dextrin adhesives and sticky tape residues. Removal of these contaminants either mechanically or with solvents can damage the emulsion, particularly where the emulsion is fragile and delaminating.

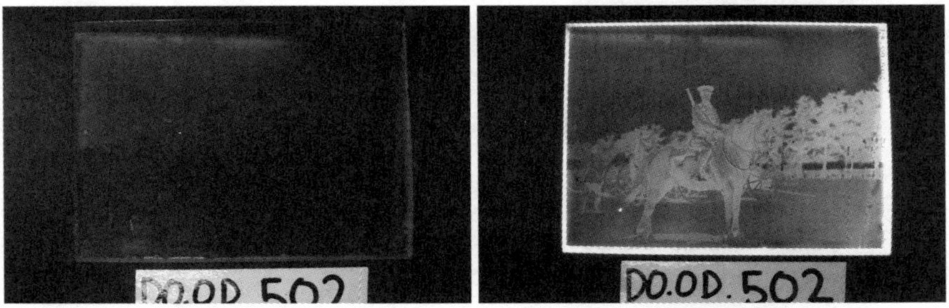

Fig. 6. A First World War glass plate negative before and after chemical treatment. Wet treatment is not possible as gelatin is soluble while dry treatment involves rubbing excess fixer off the surface. This later technique is time consuming, marks the surface and has still left a substantial amount of silvering.

3.3.5. *Removal of corrosion*

Removal of corrosion from delicate metal surfaces such as highly polished sword blades, blued metal, painted metal (where it is desirable to remove the corrosion but retain the paint) and finely machined technological components. Removal of corrosion mechanically or chemically from these surfaces is extraordinarily time consuming due to the care required, and is still liable to disfigure the surface with subtle differences in the colour and texture of the cleaned areas.

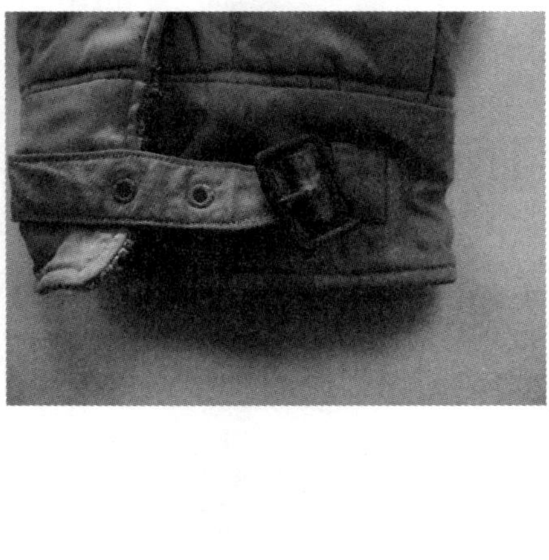

Fig. 7. RFG Australia flying suit, 1914-1918. Sir Gordon Taylor. Corroded metal fittings. Any attempt to clean the metal distributes corrosion particles over textile even further or introduces undesirable chemicals to the textile.

3.3.6. *Removal of radioactive material from contaminated objects*

The Memorial's collection contains a very significant number of items – mostly optics and aircraft instruments – which have areas of radioactive paint. There is currently neither technology nor legal way exist to remove this radioactive material. In addition, contamination in the form of radioactive paint flakes and dust particles, can be found in areas where radioactive material has been stored or handled. This contamination can only be removed by swabbing, generating further low level radioactive wasted in the form of contaminated gloves, swabs and plastic bags. If radioactive material could be ablated using lasers, and reliably captured in a shielded chamber, radioactive objects and areas could be effectively decontaminated and the radioactivity contained in a small shielded package. This would be a major improvement in reducing occupational health and safety risks, would substantially reduce the staff time to monitor, track, package and report on radioactive items, and would reduce the cost of licenses for storing radioactive material through reducing the number of sites at which radioactive items are held. The technology would also be applicable to industrial and research facilities, which use radioactive materials.

Fig. 8. Aircraft Altimeter for Beaufort aircraft. This dial has radioactive paint on the surface. There is currently no legal way to remove this paint (and therefore decontaminate the object) or to dispose of this object, as no facility for long term storage of radioactivity has yet been built. The object must be removed from the aircraft, packed in a shielded container and risk continued exposure each time the object needs to be accessed or prevent the object being accessed. Also significant annual license fees must be paid to keep it.

4. Conclusion

We have demonstrated the potential of MHz-range repetition rate laser with 10-ps pulses for laser cleaning of surface layers on metal substrates with accurate depth precision. The control over the removal process was demonstrated by changing the laser intensity on the target surface, and the laser exposure time. The flexibility of short-pulse laser ablation for precise removal of the unwanted surface layer was demonstrated by removal of optically transparent wax coating from a black-painted metal surface without altering the paint.

As a result of detailed consultations with conservation specialists from the National Museum of Australia, the National Australian Gallery, and the War Memorial, the most challenging problems were identified where traditional techniques are not applicable, and where the ultrafast laser cleaning may have a vital role as a potential solution to the problem.

References

1. C. Fotakis, V. Zafiropulos, D. Anglos, S. Georgiou, P. V. Maravelaki, A. Fostiridou, M. Doulgeridis, *Lasers In Art Conservation*, in: *The interface between science and conservation*, Ed. S Bradley, p.83 (The Trustees of the British Museum, 1997).
2. S. Georgiou, V. Zafiropulos, D. Anglos, C. Balas, V. Tornari, C. Fotakis, *Appl. Surf. Sci.* 127-129, 738 (1998).
3. V. Zafiropulos, *Laser Ablation in cleaning of artworks*, in: *Laser Cleaning*, Ed. B. Luk'yanchuk, p. 343 (World Scientific, New Jersey, 2002).
4. C. Momma, S. Nolte, B. N. Chichkov, F. V. Alvensleben, A. Tunnermann, *Appl. Surf. Sci.* **109/110**, 15-19 (1997).
5. E. G. Gamaly, A. V. Rode, V. T. Tikhonchuk, and B. Luther-Davies, *Physics of Plasmas*, **9**, 949-957 (2002).
6. J. Bonse, S.Baudach, J. Kruger, W. Kautek, M. Lenzner, *Appl. Phys. A*, **74**, 19-25 (2002).
7. S. Mailis, I. Zergioti, G. Koundourakis, A. Ikiades, A. Patentalaki, P. Papakonstantinou, N. A. Vainos, C. Fotakis, *Appl. Optics* **38**, 2301 (1999).
8. M. D. Feit, A. M. Rubenchik, B. M. Kim, L. B. Da Silva, M. D. Perry. *Appl Surf Sci*;**127-129**, 869-874 (1998).
9. A. V. Rode, E. G. Gamaly, B. Luther-Davies, B. T. Taylor, J. Dawes, A. Chan, R. M. Lowe, P. Hannaford, *J. Appl. Phys.*, **92**, 2153-2158 (2002).
10. B. Luther-Davies, V. Z. Kolev, M. J. Lederer, N. R. Madsen, A. V. Rode, J. Giesekus, K.-M. Du, M. Duering, *Appl. Phys. A* **79**, 1051 (2004).
11. E. G. Gamaly, A. V. Rode, O. Uteza, V. Kolev, B. Luther-Davies, T. Bauer, J. Koch, F. Korte, and B. N. Chichkov, *J. Appl. Phys.*, **95**, 2250 (2004).
12. E. G. Gamaly, A. V. Rode, V. T. Tikhonchuk, and B. Luther-Davies, *Appl. Surf. Sci.* **197-198**, 699 (2002).

Chapter 12

LASER CLEANING OF ENTRANCE WINDOW DURING ULTRA-FAST PULSED LASER DEPOSITION

N R MADSEN, A V RODE, D FREEMAN, V Z KOLEV, B LUTHER-DAVIES

CUDOS, Laser Physics Centre, The Australian National University, Canberra, ACT 0200, Australia

A new mode of ultra-fast pulsed laser deposition has been recently proposed as a solution for deposition of highest quality optical films, such as those required for fabrication of optical waveguides. This mode of laser ablation employs short ps and sub-ps pulses with rather low energy, of the order of μJ, delivered at MHz repetition rates, as opposed to conventional low-repetition-rate ns pulse deposition. The low energy pulses have to be tightly focused to a focal spot of the order of tens of microns to achieve the ablation conditions on the target surface. This tight focusing together with short time between the pulses, results in much broader expansion of the laser plume when compared to a narrow-shaped plume in conventional ns-pulse laser deposition. An advantage of broad plume expansion is a more homogeneous film coating, however this comes at the cost of subsequent deposition of the ablated material on the laser beam entry window. To overcome this problem, a laser cleaning technique was developed for in-situ deposition cleaning, and applied to deposition of chalcogenide glasses comprised of Ge-As-Se and As-S used for creation of nonlinear optical films for photonics applications.

1. Introduction

The necessity for the development of a film deposition technique, which can create atomically smooth films free of particulate contamination, has driven the implementation of a new mode of ultra-fast pulsed laser deposition [1,2]. This mode employs short ps and sub-ps pulses with rather low energy, of the order of μJ, delivered at repetition rates of the order of MHz. This is compared to conventional laser ablation methods, which use pulses in the ns range delivered at repetition rates ranging from 10's of Hertz up to kHz. This transition to an ultra-fast regime of laser ablation changes the laser-matter interaction mode from thermal evaporation to non-equilibrium ablation, which in turn improves the efficiency of the laser-target coupling and gives greater control over the plume conditions and the end product of an ablation process [2-5].

It is this higher degree of control over plume conditions and the laser-matter interaction that leads to better atomisation in the laser plume and enables the technique of ultra-fast pulsed laser ablation to create films with a significantly reduced amount of particle contamination than conventional techniques. By reducing the energy per pulse, the number of atoms evaporated per pulse is also reduced. Put simply, in the ultra-fast mode it is less likely for droplets to form in the laser-ablated plume, and also less likely for pieces of the target material to be ejected and deposited upon the substrate. Ultimately with the appropriate laser parameters, one can set up conditions where the deposited film is completely free of particulates. It is this benefit which makes the ultra-fast technique extremely useful for deposition of optical films and in particular those made from the nonlinear optical chalcogenide glasses.

Chalcogenide glasses are transparent from the visible up to infrared. These glasses have a high refractive index (2.4–3.0), which makes it possible to fabricate diffractive as well as waveguide structures within the deposited film. In particular, deposition of arsenic trisulphide (As_2S_3) optical films with minimal particle contamination has been performed using the ultra-fast technique, with the produced high quality films used for fabrication of very low loss channel waveguides [6,7].

In addition to the production of high quality chalcogenide optical films, the ultra-fast laser ablation method has been applied to the production of atomically smooth diamond-like carbon films [2]. However in spite of the obvious advantages in producing these high quality optical films over conventional laser deposition methods, a new problem has arisen which is directly related to the nature of the ultra-fast deposition method. This problem stems from the fact that in order to create ablation conditions with such small pulse energies, it is necessary that tight focusing be achieved. This means that the low-energy pulses need to be focused down to spot sizes as small as 20μm. Moreover, short time intervals of the order of tens to hundreds of nanoseconds between the pulses may lead to collision of the atoms ablated by different laser shots in the plume. As a result, the ablated plume has a broad, hemisphere-like angle expansion, in contrast to the forwardly directed plume seen in conventional laser ablation. This leads to deposition of the ablated material on the laser beam entrance window in the vacuum chamber, blocking the target from the laser radiation.

This problem is quite significant for laser deposition experiments. The deposits on the entrance window alter the laser parameters on the target. The fact that laser intensity is not constant upon the target surface is obviously not a desirable situation, especially when a stringent requirement of the film deposition process is the creation of high quality films free of particle contamination.

The non constant incident intensity which results through deposition upon the laser entry window is clearly evident in the early stages of film deposition, when the deposited thickness upon the entry window increases through the range in which interference fringes are visible (see Figure 1). The laser intensity and hence the deposition rate rise and fall periodically until the deposited thickness reaches a point where the fringes become less separated, and thus the maxima and minima less distinctive. This phenomenon is observed during deposition of As_2S_3, and is less pronounced during deposition of another chalcogenide glass $Ge_{33}As_{12}Se_{55}$, known as AMTIR-1. It must be mentioned that the problem of laser cleaning is particularly important to the deposition of AMTIR-1 films since this material has a longer wavelength band edge and thus absorbs the 532 nm laser radiation used for film deposition to a higher degree than As_2S_3. This means that the loss of laser power in the entry window is more pronounced for AMTIR-1 and consequently deposition of films of micron thickness take much longer than those of As_2S_3.

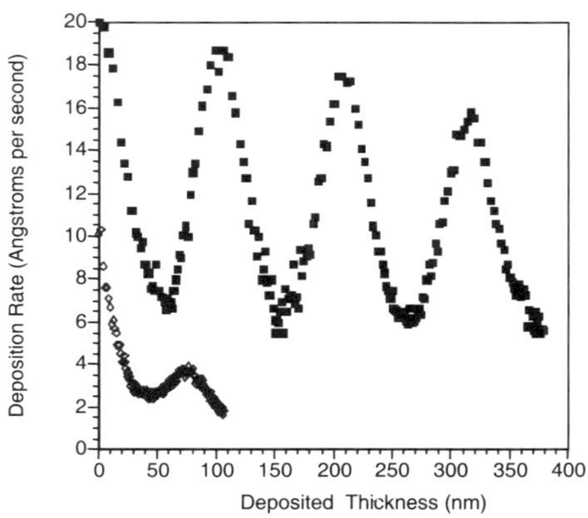

Figure 1. Change in deposition rate through the coating of laser entry window. Deposition rate for As_sS_3 (squares) and AMTIR-1 (open diamonds). Data for both materials is taken for about six minutes. Note the thickness of material deposited for AMTIR-1 is much less than that for As_sS_3 deposited for the same time period.

To overcome this problem an *in-situ* laser cleaning technique was designed for entrance window cleaning during the deposition process. The major advantage of the laser cleaning technique is the possibility of a direct energy transfer into the volume of the material to be removed (contamination on the surface) without altering the material being cleaned. The scheme for laser cleaning is presented in Figure 2. We covered the laser entrance window with a rotating fused silica plate and a fixed protective shielding, both placed between the entrance window and the laser created-plume. The fused silica plate is constantly rotated, so the ablating laser beam will always pass a clean area, while a second laser is used to remove deposited material from the opposite side of the protective plate. The removed debris is re-deposited on the rear side of the protecting shield to avoid contamination of the chamber.

The main requirement for such a technique is to find optimal parameters for the cleaning laser beam focusing and scanning conditions which must satisfy a number of criteria such as: (i) the pulse intensity on the plate surface must be sufficient to evaporate the deposits but to be below the damaging threshold for the fused silica surface; (ii) the dwell time and the scanning velocity should be within the optimal range to provide continuous coverage of the window surface with high scan surface uniformity; and (iii) the evaporated debris from the plate should not spoil the quality of the deposited films in the chamber. In this paper, experiments conducted in order to determine the laser-cleaning threshold for AMTIR-1 are presented and discussed, and the laser cleaning of the entrance window has been demonstrated.

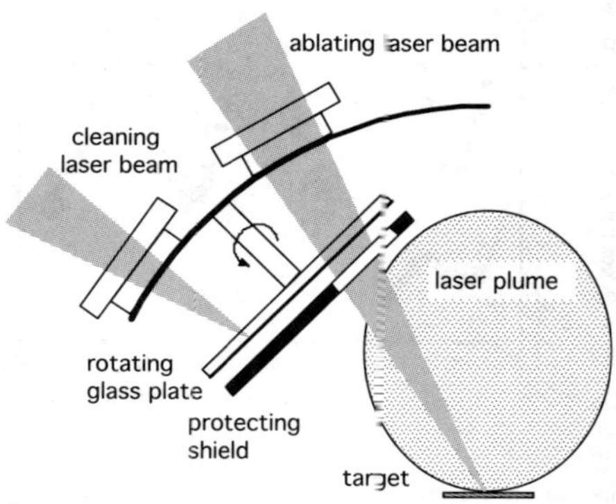

Figure 2. Geometry for protecting the laser entrance window and for laser cleaning of chalcogenide films from the protecting plate performed during deposition.

2. Experimental

2.1. *Experimental setup*

Laser cleaning experiments were performed upon pre-deposited samples of AMTIR-1 on fused silica substrates. Sample thickness varied in the range 400 - 600 nm, found via transmission fringe analysis [8,9] and this was adequate to mimic the cleaning conditions during a deposition. Cleaning was performed with the use of a Coherent "Antares" laser operating at 76 MHz, 60 ps pulse duration with smallest spot size of about 28 μm generating a maximum fluence of 7.5 mJ/cm^2. A range of fluences was achieved by changing the location of the coated sample with respect to the laser beam focal spot. The film coating is placed on the laser beam exit side of the target substrate, as this is the arrangement when performing in-situ cleaning. Experiments were performed under vacuum of $\sim 10^{-6}$ Torr.

It has to be mentioned that the scheme is similar to the well-known geometry of Laser Induced Forward Transfer (LIFT) [10]. The major difference is that the laser energy is deposited in a thin layer of deposited film, heating and ablating the whole film thickness. On the other hand, in LIFT experiments, films are much thicker and the laser energy is spent in heating the layer near the film-substrate interface, and hence breaking the bonding between the film and the substrate.

2.2. *Laser scanning system*

It is necessary that a high degree of control over the scanning of the focused laser beam be achieved. There are a number of general requirements for scanning systems in the high-repetition-rate laser ablation technique:

- it is desirable that each laser shot be located at a fresh surface spot;
- laser shots should move with a constant speed over the target surface;
- the scanning pattern should cover the surface uniformly; and
- the spot size should be the same across the entire scanning pattern.

The first requirement leads to a relatively high scanning speed $v_{scan} = d_{foc} \times R_{rep}$, where d_{foc} is the focal spot diameter, and R_{rep} is the laser repetition rate. For example, 1.5 MHz repetition rate laser ablation with a focal spot of 20 μm in diameter would require the scanning speed of 30 m/s. The second requirement places a constraint on the inertia of the scanning mirrors. To reduce the influence of the mirrors inertia, the scanning pattern is required to be free of sharp 'turning' points, such as in a Lissajous pattern. The third condition should be taken into account when the scanning pattern is computed, so that each point of the surface is exposed to an equal number of laser pulses. The last condition imposes the requirements of using a special flat field telecentric lens in which the focal plane of a deflected laser beam is a flat surface. This last condition can be overcome with conventional long-focus lenses by placing the scanning mirrors after the focusing lens so that the difference in optical path of the deflected beam is within the lens caustic, however this is at the expense of the tight focusing required for low-energy pulses.

A laser scanning system was based on Cambridge Technology galvanometer optical scanners equipped with mirrors with dielectric coating to maximize reflection of the high average power laser beam. A 'doughnut' pattern consisting of the sum of two circular trajectories, one small and fast plus one larger and slower, was chosen as one pattern for use in the laser cleaning experiments to adequately satisfy the requirements of scanning. It requires only analogue electronics, and for its relative simplicity it achieves good uniformity over the scanned area as is reported in Section 3. By scanning one of the circles at a much higher frequency than the other we can maintain a constant velocity to within about 5%. A computer-controlled positioning system has been designed and built in house, utilising a pair of 16-bit digital-to-analogue converters, such that one can accurately generate arbitrary scanning patterns with uniform area coverage and constant linear speed.

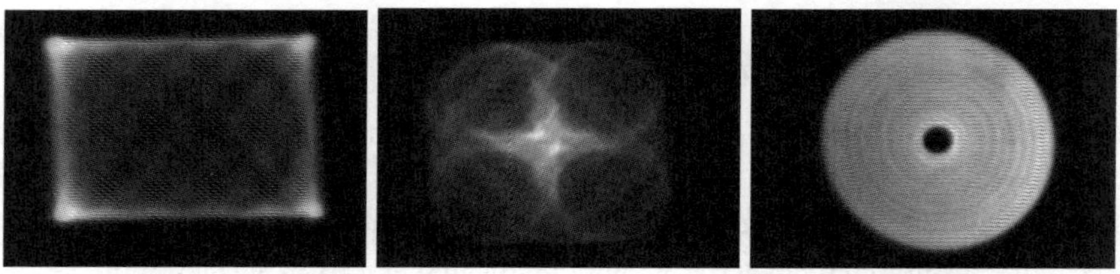

Figure 3. Sample scanning patterns used for laser ablation experiments. Standard Lissajous (left), Lissajous plus circle (middle) and constant linear speed spiral (right).

Figure 3 presents time-integrated images of various scanning patterns, illustrating the shape and the coverage pattern of the laser beam upon the target surface. Note that the third 'spirals' pattern has the highest area uniformity. This pattern is a sequence of spirals with constant distance between turns and constant linear speed. The spirals alternate between increasing and decreasing radial direction, with a phase step between pairs to cumulatively fill the area between the turns over a large number of spirals. Since the angular frequency is greatest at the centre of the pattern, the diameter of the central hole must be chosen to respect the limitations of the scanners. This pattern has been the most successful to date in our laser deposition experiments of the chalcogenide glasses.

2.3. *Irradiation thresholds*

The laser threshold conditions for cleaning and for damage of the particular substrate must be determined for successful cleaning. During a deposition experiment, one can then ensure effective cleaning by keeping the fluence in a range above the cleaning threshold but below the substrate damage threshold. We determined laser cleaning threshold conditions by conducting cleaning experiments from fluences as low as 0.04 mJ/cm^2 up to 7.5 mJ/cm^2 using the "Antares" laser. At these fluences, damage of the fused silica substrate cannot be observed. In order to find the damage threshold, higher fluences were obtained through the use of a laser system which was designed and built in-house for laser ablation experiments, in particular. This laser is capable of fluences up to 3 J/cm^2 at a repetition rate of 1.5 MHz. The pulse duration is 12 ps and the wavelength is 532 nm [11].

2.4. *Cleaning rates*

The time for full area cleaning using a two-circle doughnut pattern can be estimated assuming the linear velocity of the laser spot to be equal to that of the faster circle, which is oscillating at 50 Hz (significantly faster than the second circle). Hence we can obtain the expression for the required minimum time for cleaning,

$$T_{clean} \cong \frac{8a}{\pi d f} \tag{1}$$

where a is the radius of the fast circle, d is the diameter of the laser spot on the target and f is the frequency of the fast circle. For experimental parameters $a \sim 3$ mm, $d \sim 30\mu$m and $f \sim 50$Hz we get a minimum cleaning time of 5 seconds. In reality there is a degree of overlap in the scanning pattern so in order to completely cover a cleaning area we perform experiments for 30 seconds. We show in the following transmission measurements that this time is adequate for complete cleaning of the samples.

3. Results and Discussion

Images of samples of AMTIR-1 laser cleaned in vacuum using the "Antares" laser with wavelength 532 nm are presented in Figure 4. The top sample starts with an incident fluence of 3 mJ/cm^2 and the fluence is decreased from left to right across each sample until a situation is reached in the bottom image where there is no laser cleaning of the sample. This minimum fluence is approximately 0.04 mJ/cm^2. The ablation pattern of the doughnut, which is comprised of two added circles, is clearly evident at lower fluences. In addition the ablation pattern also shows a high degree of uniformity and complete area cleaning at the higher fluences.

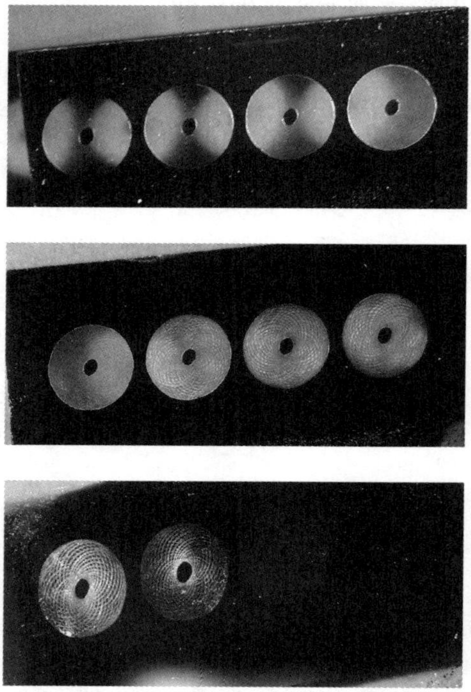

Figure 4. Images of AMTIR-1 laser cleaned areas. Fluence decreases from 3 mJ/cm^2 in the top left cleaned area to 0.04 mJ/cm^2 in the last visible cleaned area in the bottom right image. Note that the cone-like dark areas in the top image are shadows and not unclean parts of the substrate.

Figure 5 plots the degree of cleanliness of the cleaned AMTIR-1 areas as a function of incident fluence. The degree of cleanliness is scaled from zero to one where one is fully cleaned and zero is an unchanged target. This information is obtained by normalizing the measured transmission of 532nm radiation through the area exposed to the laser. For higher fluences it can be seen that during the 30 second cleaning time, the AMTIR-1 film has been completely removed from the substrate. As the fluence is decreased, by moving the focus of the laser beam away from the film surface, a sharp decrease in cleanliness is observed. Investigation of the data indicates the threshold fluence for cleaning to be in the range 3.0 ± 0.5 mJ/cm^2. This film removal threshold fluence found for the 60ps, 76MHz laser, was found to be >100 times lower than that expected for a single pulse. A model of irradiation by high repetition rate pulse trains [5] demonstrates that the energy accumulates in the target surface from several hundred successive pulses lowering the ablation threshold. Hence the lowering of threshold is due to the fact that the material has very low thermal diffusivity (~2×10^{-3} cm^2/s), indicating that the energy from many pulses can accumulate in the surface before cooling via thermal conduction can become significant.

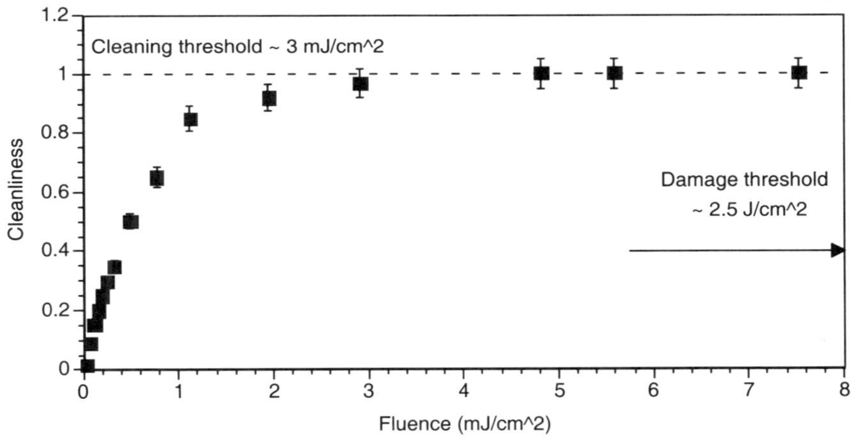

Figure 5. Degree of cleanliness of AMTIR-1 coated slide as a function of incident fluence.

Another possible reason for the reduced cleaning threshold is the particular confinement geometry used in these experiments, which is similar to that of LIFT. In the LIFT geometry, energy is predominantly absorbed at the film/substrate interface, and material removal occurs due to breaking of film/substrate bonds. However, laser cleaning of AMTIR-1 presented here represents a situation where laser energy is absorbed in the entire contaminant layer so this explanation cannot completely account for the results.

Film-coated fused silica substrate surfaces were found to have a damage threshold of approximately 2.5 J/cm^2. Measurements were made using our in-house, high power laser system, with pulse duration of 12 ps [2,8]. Even though this laser system has a shorter pulse duration than that of the laser used for cleaning (60 ps), the measurement does give a good indication for the order of magnitude damage threshold for the longer pulse laser. Clearly this threshold for damage is significantly higher than that for cleaning (~ 1000 times) giving a very broad range of fluences over which it is possible to clean effectively without optical damage of the underlying silica. It is worth noting that in a single pulse mode of ablation one expects an ablation threshold of 4 J/cm^2 [3] for fused silica at 12-ps pulse duration. The fact that our damage threshold is lower than that of the predicted ablation threshold suggests that there is a cumulative ablation effect [2] seen due to the use of the high, 1.5 MHz, repetition rate laser.

4. Conclusion

We developed a laser window cleaning system for the ultrafast pulsed laser deposition technique. Various laser-scanning patterns were considered in search for optimal regime of laser window cleaning. It was found that the collapsing-expanding spiral pattern with constant linear scanning velocity is the best to satisfy strict requirements for an efficient and homogeneous cleaning process.

The system was applied for laser deposition of AMTIR-1 chalcogenide glass films. The laser fluence threshold for total cleaning of film from the surface was found to be 3.0 ± 0.5 mJ/cm^2, while the film-coated fused silica surface damage threshold is at the level of ~2.5 J/cm^2 (intensity 2×10^{11} W/cm^2).

The experimental results and analysis presented demonstrate that the energy threshold for removal of chalcogenide glass from the fused silica substrate using high repetition rate 76 MHz lasers is much lower than that required to ablate the bulk material with a single pulse. It is suggested that this is due to accumulation of energy from successive, high repetition rate pulses, heating many times the same spot on the target surface. It is also suggested that the confinement geometry used in the experiment, where the contaminating film is laser-heated from the substrate-film interface side, also contributes to lowering of threshold.

The developed technique will be applied for ultra-fast laser deposition of various nonlinear optical and other high-quality films.

Acknowledgments

The support of the Australian Research Council through its CUDOS (the Centre for Ultrahigh-bandwidth Devices for Optical Systems) Centre of Excellence and Federation Fellowship programs is gratefully acknowledged.

References

1. E. G. Gamaly, A. V. Rode, B. Luther-Davies, *Appl. Phys.* **85**, 4213 (1999).
2. A. V. Rode, B. Luther-Davies, E. G. Gamaly, *Appl. Phys.* **85**, 4222 (1999).
3. B. C. Stuart, M. D. Feit, S. Herman, A. M. Rubenchik, B. W. Shore, and M. D. Perry, *J. Opt. Soc. Am. B* **13**, 459 ~1996.
4. E. G. Gamaly, A. V. Rode, B. Luther-Davies, V. T. Tikhonchuk, *Physics of Plasmas* **9**, 949 (2002).
5. B. Luther-Davies, A. V. Rode, N. Madsen, E. G. Gamaly, to be published in *Opt. Eng.*, 2004.
6. A. Zakery, Y. Ruan, A. V. Rode, M. Samoc, and B. Luther-Davies, *J. Opt. Soc. Am. B*, **20**, 1844-1852 (2003).
7. Y. Ruan, W. Li, R. Jarvis, N. Madsen, A. Rode, B. Luther-Davies, *Optics Express*, **12**, 5140-5145 (2004).
8. R. Swanepoel, *J. Opt. Soc. Am. A* **2**, 1339–1343 (1985).
9. C. Corrales, J. B. Ramirez-Malo, J. Fernandez-Pena, P. Villares, R. Swanepoel, E. Marquez, *Appl. Optics* **34**, 7907 (1995).
10. J. Bohandy, B. F. Kim, F. J. Adrian, *Appl. Phys.* **60**, 1538 (1986).
11. B. Luther-Davies, V. Z. Kolev, M. J. Lederer, N. R. Madsen, A. V. Rode, J. Giesekus, K. M. Du, M. Duering, *Appl. Phys. A* **79**, 1051 (2004).

Chapter 13

SURFACE CLEANING OF OPTICAL MATERIALS USING NOVEL VUV SOURCES

D M KANE, D HIRSCHAUSEN, B K WARD, R P MILDREN AND R J CARMAN

Department of Physics, Macquarie University ~ Sydney, NSW 2109, Australia

Short pulsed, high peak power VUV output has been obtained from pulsed voltage excitation of high pressure dielectric barrier discharges (DBD). Output with these characteristics has demonstrated advantage over longer pulse DBD VUV sources for several materials processing and surface modification applications. Results using a Xe_2^* DBD source at 172 nm for removing optical mountants from optical surfaces, for removing hydrocarbon contamination from optical and polymer surfaces, and dehydroxylation of fused silica are presented. Guidelines

1. Introduction

At Macquarie University we have pioneered the development of a new mode of operation of dielectric barrier discharge (DBD) lamps which uses pulsed voltage (fast rise-time) excitation of high pressure gas discharges to generate short-pulsed, high-peak-power ultra-violet/vacuum ultraviolet (UV/VUV) output with high pulse-to-pulse reproducibility. This platform technology has been implemented using a Xe_2^* excimer DBD lamp (172 nm) so far [1-6]. Comparative surface treatment studies have been conducted using a commercial pulsed Xe_2^* DBD lamp at equivalent VUV irradiances. For a typical VUV irradiance of 10mW.cm^{-2}, our lamp generates VUV pulses with shorter duration (390ns vs 1720ns FWHM, respectively), an order of magnitude faster risetime (100ns vs 1000ns) and higher (3x) peak power (0.5W.cm^{-2} vs 0.16W.cm^{-2}), than observed using the commercial lamp (figure 1) [7]. Such a short pulsed, high peak power output is expected to have significant advantage compared to the same energy delivered in multiple smaller pulses as is more normally achieved from DBD lamp sources, most typically from AC excited DBDs [8]. Even shorter duration pulses have been generated at higher Xe gas pressures in the discharge [4,6].

There has been some research on the use of DBD excimer lamps for photochemical ablation and etching of polymers [9-11] and for cleaning of hydrocarbon contaminants from surfaces [12]. Xe_2^* DBD lamps are commercially available for the latter application [7, 13]. These incoherent sources represent are an effective and relatively inexpensive choice compared to lasers, for example. However, very little is known about the materials, chemical and physical properties of substrates and contaminants at VUV wavelengths. We have completed studies of removing normal hydrocarbon contamination that condenses on optical surfaces from the air and studies of paraffin wax and thermopolymer mountant removal, using these short pulse (SP) Xe_2^* DBD sources. These have been contrasted with studies using a commercial Xe_2^* DBD lamp [7]. We have observed significant advantage of using the short pulse VUV output in comparative studies but, also, there are new experimental issues to address.

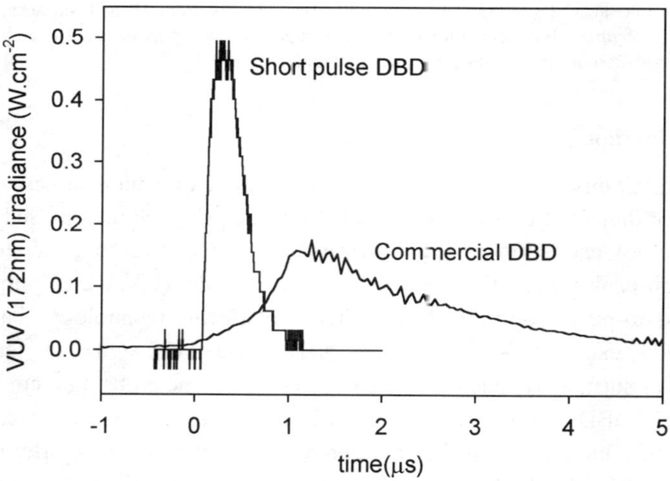

Figure 1: VUV pulse shapes from the Xe_2^* DBD lamps at a typical irradiance of 10mW.cm^{-2}. (Note the overall pulse energy of our lamp (0.21μJ.cm^{-2}) is lower than from the commercial lamp (0.35μJ.cm^{-2}) due to the different pulse repetition frequencies (50kHz and 30kHz, respectively)).

Ongoing studies utilising pulsed voltage excitation of DBD lamps with other excimer gas mixtures also offers the exciting prospect of generating high-peak power output covering a broad range of UV and VUV wavelengths between 88-350nm, as shown in figure 2.

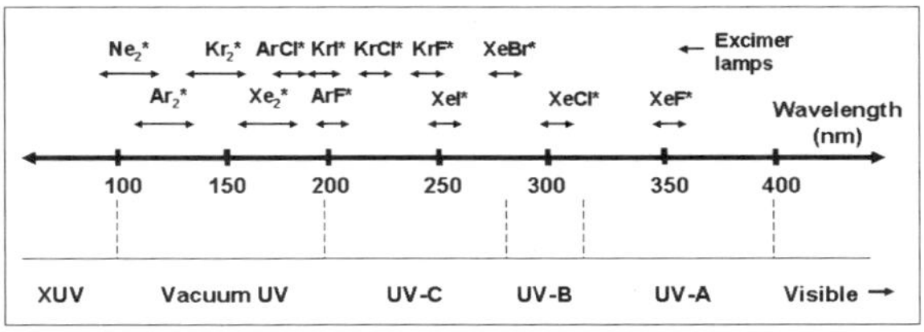

Figure 2: Principal emission wavelengths for the rare-gas and rare-gas/halide excimers ($B_{1/2} \Rightarrow X_{1/2}$ transitions).

2. Wax and Thermopolymer Removal – Experiment

In previous research the removal of paraffin wax in the form of small particles from optical surfaces by laser cleaning using a KrF excimer laser (248 nm) has been successfully demonstrated [14]. Waxes and thermopolymers are used in the optics and photonics industries as mountants and are currently removed by wet chemistry techniques. This occurs off-line and there is a strong drive to develop non-contact cleaning techniques that may be integrated into production lines. Thus, the material system of wax or thermopolymer on glass and optical surfaces is an excellent test system for the VUV/UV short-pulse DBD sources. In order to generate reproducible results it is necessary to adopt standard procedures for sample preparation and treatment.

2.1 *Sample preparation – glass and silica substrates*

The substrates were prepared by ultrasonic cleaning in a 10% isopropyl alcohol solution for 8 mins. The slides were rinsed with excess isopropyl alcohol and either left to dry or dried with a lint-free tissue. The glass slides were standard microscope slides. Fused silica samples of the same dimensions as standard microscope slides, polished to a flatness of $\lambda/20$ (at 550 nm), were the silica substrates.

2.1.1 *Paraffin wax and thermopolymer CB 555 particles on paraffin and silica substrates*

Paraffin wax and Crystalbond™ 555 (see Table 1 for characteristics) are low melting point mountants that are readily prepared in particle form on the glass and silica substrates by atomisation.

This method of sample preparation has been described in [14] where results of laser cleaning of paraffin particles were presented. This method produced a layer of evenly distributed dome-shaped particles upon the substrate surface. The particles ranged in size from approximately 1 – 20 μm in diameter.

Name	Crystalbond™ 555	Crystalbond™ 509
Composition	Nonylphenoxypoly ethanol	Ethylene glycol phthalic anhydride polymer
Flow Point	54°C	77°C
Viscosity	500 cps	6000 cps
Colour	White	Clear
Soluble In	Hot Water	Acetone or MEK

Table 1: Physical characteristics of two thermopolymer adhesives Crystalbond™ 555 and Crystalbond™ 509.

2.1.2 *Thermopolymer CB 509 on glass and silica substrates*

The viscosity and melting point of this thermopolymer adhesive were too high to allow preparation of particles on the substrates using the atomisation apparatus. Film samples were prepared using two different methods that resulted in thicker or thinner films. Thick films were prepared by placing a clean slide onto an electric hotplate. The slide was gently heated and its temperature was monitored with an electronic thermometer. When the slide had reached the appropriate temperature (approximately 10°C below the flow point) a small amount of thermopolymer was melted onto the surface to form a uniform film. The slide was then removed from the hotplate and left to cool. When the sample had returned to room temperature it was labelled and placed within a sealed container to minimise the amount of airborne contamination settling upon the surface prior to testing. Samples were always tested on the same day that they were prepared.

Thin films were prepared in a similar manner except the thick film was thinned by being partially dissolved in an appropriate solvent. Thin films were typically 1 to a few microns thick as determined by microscopy and surface profilometry.

2.2 *VUV DBD removal of wax and thermopolymer – Method*

The SP DBD lamp consisted of two concentric quartz tubes (VUV grade) with an outer diameter of 25mm, an active discharge length of 25mm as defined by semi-transparent metal mesh electrodes, and a 3mm discharge gap. The Xenon gas pressure was 500mb. The lamp was excited by short rise-time (40ns) high voltage (6-10kV) uni-polar pulses at pulse repetitions rates between 10-50kHz.

The measured output irradiance at 172nm was in the range 5-15mW.cm^{-2}. The VUV DBD source (SP or commercial) was mounted in an air tight enclosure made of Perspex. A flow of high purity (99.99%) nitrogen was maintained in the enclosure to ensure the minimum amount of the VUV radiation was absorbed. The maximum fluence achieved with the commercial VUV DBD source was 40mW/cm^2 at the lamp surface. The sample substrate was placed about 28 mm away from the lamp surface to ensure any vapour and particulates leaving the sample did not deposit on the lamp enclosure. At this distance the fluence of the commercial source was 10 mW.cm^{-2}. The fluence of the SP DBD source was reduced to this level by adjusting the flow rate of pure nitrogen. This also has the effect of preferentially absorbing the shorter wavelength end of the emission so there is a small difference in the spectral emission characteristic when both sources are at the same fluence. Any effect of this difference would be expected to be in the favour of the commercial DBD source. The sample was removed from the housing and mounted on a precision translation stage unit which allowed accurate registration of the area imaged at the regular time intervals through the treatment of the sample. A total exposure time of 2-3 hours was typically used.

2.3 *Results and discussion of wax and thermopolymer removal from glass and silica*

Results of the treatment of CB 509 thin film samples are shown in fig. 3. Images of the same area were recorded before any exposure and after 1, 2 and 3 hours of exposure. The before images and those after three hours of exposure are shown, when both sources were set to give 10 mW/cm^2 average power at the substrate surface. After 3 hours the thin film had been completely removed using the SP DBD source while using the commercial DBD source led to a reduced area of coverage and thinning of the film but total removal was not observed, even for longer treatment times. On thick film samples there was evidence of significant removal of material using the SP DBD source but complete removal was not achieved. Very little removal of the CB 509 was achieved on thick film samples using the commercial DBD source. Note the dark spots in the bottom right figure are associated with the optics of the microscope being used to image the substrate and film and are not dirt remaining on the substrate. Thus, it appears thet after a certain exposure time the primary result is consistent with curing of the wax or thermoploymer onto the surface rather than its removal. This could be predicted as one of the likely outcomes of the treatment. What is significant is the fact that the short pulse VUV output does remove the very thin films.

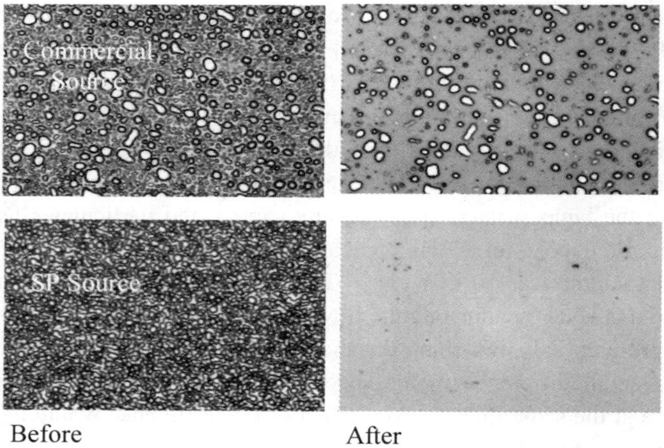

Figure 3: Photopolymer mountant CB 509 before and after 3 hours treatment with 10 mW/cm^2 VUV. The sample in the top row has been irradiated with the commercial source, that in the bottom row with the SP source. The area imaged is 122 μm x 81 μm viewed at 2500 x magnification.

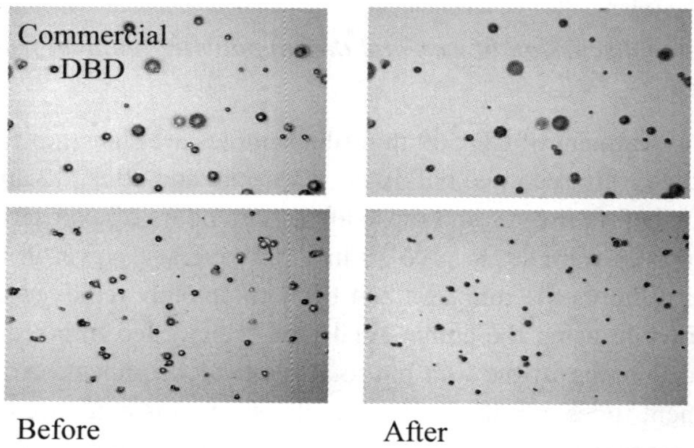

Figure 4: Images of paraffin particle samples before and after 3 hours irradiation with 10 mW/cm^2. The sample in the top row has been irradiated with the commercial source, that in the bottom row with the SP source. The area imaged is 122 μm x 81 μm viewed at 2500 x magnification.

The removal of wax or CB 555 particles give similar results. These show significantly higher removal of the materials using the SP DBD source than achieved using the commercial source, at the same average power. This is shown for paraffin wax particles in fig. 4. The results also show the "self limiting" behaviour similar to that observed for the thin film samples Here only the smaller particles (< 3 microns) were completely removed, while others were reduced in size when using the SP DBD. Using the commercial DBD no particles were completely removed. The more detailed analysis of the particle size distribution before and after treatment is shown in fig. 5 for the commercial DBD source and in fig. 6 for the SP DBD source. This has been carried out using the image processing tools of the Image-Pro Plus with Materials-Pro software.

Results in fig. 5 for the commercial DBD source show that three of the 26 particles in the area imaged actually increased in size due to melting and spreading. The area covered by particles was reduced to 92% of the original area covered. Restricting the particles to those 10 microns in radius or less (as per the sample used for the SP DBD test and shown in the inset in fig. 5) in the analysis gave a result of an area reduction to 93.4% ie the removal was not preferentially from smaller particles.

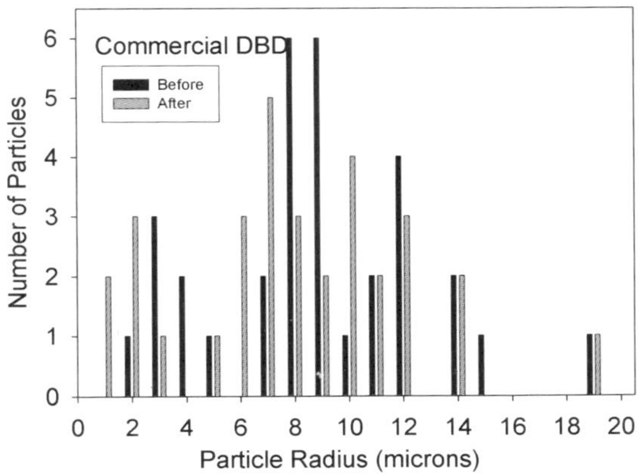

Figure 5: Number of particles in the distribution before and after 3 hours of exposure with the commercial DBD source with an average power of 10 mW/cm^2. The inset shows the distribution for particles with a radius 10 microns or less before treatment for more direct comparison with fig. 6.

Using the SP DBD source 19 of the original 45 particles were completely removed as shown in fig. 6. All 16 particles with radii less that 3 microns were removed. Three particles with radii of 4 and 5 microns were removed but another 8 particles with these sizes were reduced in size rather than being removed. The area of the particles was reduced to 51% of the original coverage, significantly greater than that achieved with the SP DBD source.

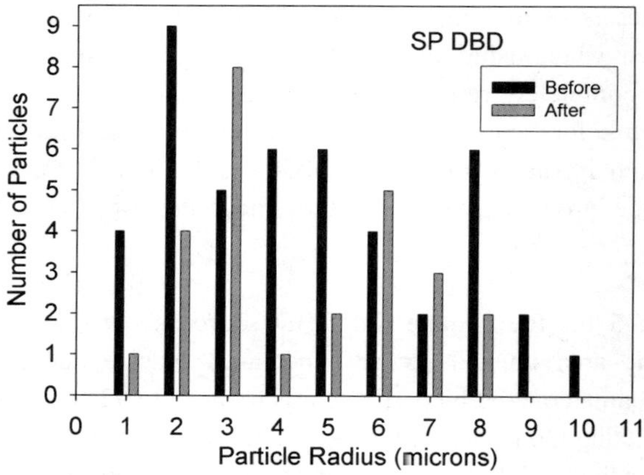

Figure 6: Number of particles in the distribution before and after 3 hours of exposure with the SP DBD source with an average power of 10 mW/cm^2.

3. Hydrocarbon/Moisture Removal from Glass and Polymer Surfaces

3.1 *Experiment*

The use of the VUV DBD sources for removing the normal moisture and hydrocarbon contamination that occurs due to exposure to air in the normal laboratory environment was investigated for both glass & silica substrates, and polymer substrates. Polymers are known to have a porous surface that presents a large surface area, and has semi-enclosed volumes that can retain such contamination once it condenses from the air. Polymers have also been extensively studied for UV/VUV lamp assisted photochemical ablation and dry etching [9-11] where PTFE etch rates of ~1 micron per minute (at 40 mW.cm^{-2}, 172 nm) have been reported by treatment in a low pressure of O_2 (1mb air) in an irradiation reaction chamber [11]. Three polymers were included in the study here – PET, polyethylene terephthalate or polyester; PETG polyethylene terephthalate glycol or glycolised polyester; and PC, polycarbonate. The polymers were in thin sheet form. Samples with an area of 20 mm by 40 mm were prepared. The mass of these samples was ~0.14g for PET, 0.63g for PETG and 0.91g for PC reflecting the relative thickness of the various films. Glass and silica slides of standard microscope slide dimensions were used. In order to prepare the substrates by a common method they were ultrasonically cleaned in a 10% isopropyl alcohol solution for 8 minutes and then rinsed in excess water and dried with a lint free cloth. The removal of the moisture and hydrocarbon contamination caused by VUV exposure was monitored by contact angle measurements for all samples and additionally by changes in mass for the polymer samples where the mass removed was measurable. It was established that this mass removal was not associated with etching or ablation of the polymer. If any such ablation occurred it was at the nanometre scales measured in [9] in a low pressure rare gas background in the reaction chamber. Note, no reaction chamber is being used in these experiments. The samples are exposed in flowing pure nitrogen at atmospheric pressure.

Substrates with hydrocarbon contamination are usually hydrophobic giving large contact angles between the substrate and a 10 μL droplet of distilled water introduced onto the surface using a micropipette. Decontamination leads to a more hydrophilic surface characterised by a decreasing contact angle between the droplet and the substrate. The contact angle was measured from side-on digital images recorded using an optical microscope system. The sample was removed from the exposure housing for the contact angle measurement. A similar geometry for sample housing and placement, during exposure, as that described in section 2.2 was used. An average power of 10 mW/cm^2 at the sample surface was used, and measured with a calibrated silicon diode detector.

3.2 *Hydrocarbon/ moisture removal – Results and discussion*

The results of contact angle measurements for treatment using the commercial and SP DBD sources on polymer surfaces and the SP DBD source on glass and silica are shown in fig. 7. Similar reductions in contact angle were achieved with both sources. Thus, both types of VUV output were effective for removal of hydrocarbon and moisture from the substrate surfaces. The mass removal effected from the polymer surfaces was between 60-160 $\mu g/cm^2$ for the three polymer surfaces. This mass removal was reversed by exposure to the normal laboratory environment for up to three days after the initial treatment indicating there was no ablative removal of any polymer material but just the moisture and hydrocarbon contamination. The changes in contact angle were also reversed after longtime exposure to the normal laboratory environment subsequent to the treatment. One significant difference in the results achieved with the two DBD sources was the degree of coloration of the polymers caused by the exposure to the VUV. All the polymers were nominally clear before exposure, and all showed yellowing due to the VUV exposure. However, the yellowing was significantly greater when the polymers were exposed using the commercial DBD source than when using the SP DBD source. Thus, the SP DBD source has potential to be developed as a source that can remove hydrocarbon and moisture decontamination from polymers with reduced, or possibly no coloration.

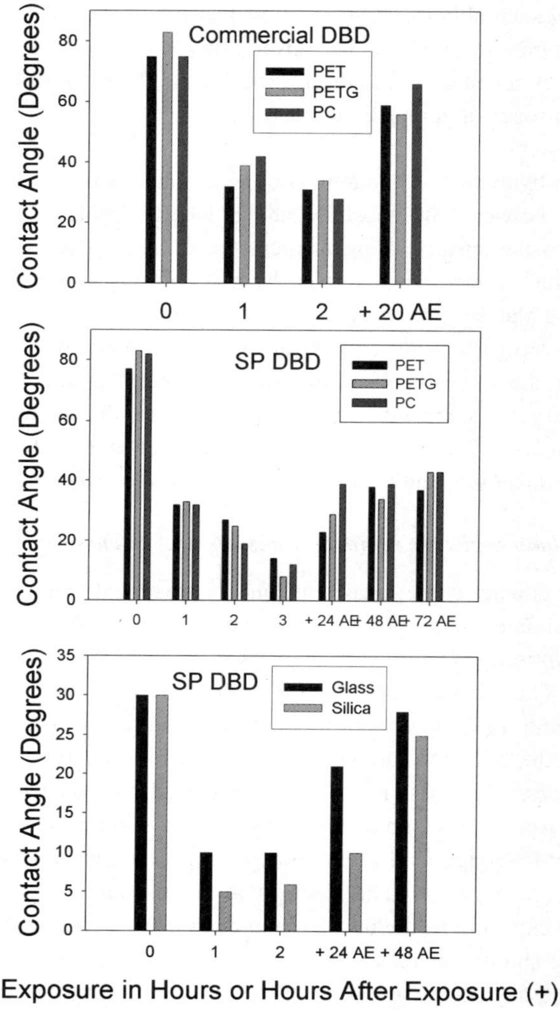

Figure 7: Contact angle measurements as a function of VUV exposure time in hours followed by measurements made at long time intervals after exposure (AE) when the samples are left to recontaminate in air.

In the case of the fused silica samples dehydroxylation of the surface was also observed with extended exposure using the SP DBD but a similar result was not observed with the commercial DBD. This effect has been demonstrated and characterised by UV pulsed laser treatment [15,16]. From the contact angle measurements its occurrence is indicated by a sequence which starts with a contaminated hydrophobic surface. This becomes a hydrophilic surface when the physiosorbed moisture and hydrocarbons are removed, thus exposing a larger density of reactive OH groups on the silica surface. When these OH groups are subsequently removed the surface becomes hydrophobic again. The dehydroxylation has also been shown from time of flight secondary ion mass spectrometry (TOF SIMS) which, when using appropriate experimental methodology [16], can be used to quantify the relative values of the $SiOH^+/Si^+$ ratio at the fused silica surface. An interesting side effect of this dehydroxylation is that when the surface becomes recontaminated with a film of moisture and hydrocarbons by exposure to air in the laboratory environment the contamination initially forms in small beads as shown in fig. 8, rather than a film in the first day or so. Thereafter a film eventually forms.

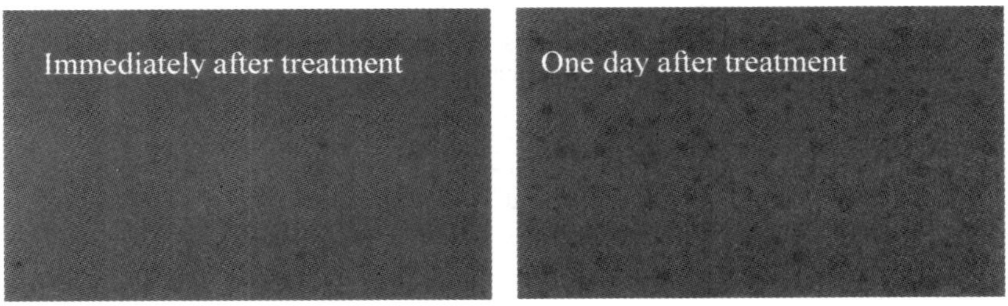

Figure 8: Images of VUV treated fused silica surface immediately after and exposure which achieves a dehydroxylated surface, and ~24 hours later when the sample has started recontaminating, in beaded form, due to moisture and hydrocarbons in the air. The area imaged is 122 μm x 81 μm viewed at 2500 x magnification.

The comparative results of the studies reported here-in are summarised in Table 2 for the SP DBD and the commercial DBD. Overall they show that the SP DBD source is better than the commercial VUV DBD in most of the applications trialled. Both sources are excellent for removing moisture and hydrocarbon contamination due to exposure in air. It is also expected that even better results will be achieved with the SP DBD operated at higher pressure and therefore shorter pulse width. The observation that when using the commercial, longer pulse VUV DBD the exposure leads to a "curing" of the wax or thermopolymer materials rather than removal and that this "curing" effect is observed to onset at a later time when using the SP DBD, after an initial period of exposure that leads to removal of the material, suggests this self-limiting effect in the removal process may occur even later or not at all when using even shorter pulse durations such as can be achieved with higher gas pressure in the DBD lamp.

	SP VUV DBD	Commercial VUV DBD
Removal of few monolayers of normal environmental contamination from glasses	Yes	Yes
Removal of few monolayers of normal environmental contamination from polymers	Yes, with little colouration	Yes, with significant colouration
Removal of thin film of particles of wax or thermopolymer from glasses	Yes, but processing times can be long	No
Removal Wax/ CB555 Particles (atomized) from glasses	Complete removal of particle < 3 microns in diameter Partial removal of others	Very slight reduction in size
Removal of thick films of wax/thermopolymer from glasses	Yes, sometimes depending on the wax/thermopolymer	No
Dehydroxylation of Fused Silica	Yes	No

Table 2: Summary of materials processing and surface cleaning treatments

4. Conclusions

The SP DBD source is shown to give significantly better results than a commercial VUV DBD source of the same average power in removing wax and thermopolymer mountants used in optical fabrication. The SP DBD is able to remove thin films and small particles and has the prospect of achieving even better results with shorter pulse durations, which are achievable. When processing in a flow of pure nitrogen at atmospheric pressure there is no evidence of dry etching or ablation of the glasses and polymer materials studied.

This reinforces previous results [10,11] which show this etching process is photochemical in nature and requires O_2 to proceed efficiently. Thus, using the methodology described here-in it is possible to clean the moisture and hydrocarbon contamination from these surfaces without etching. This aspect will be followed up in more detail in future work.

Both types of VUV DBD are very effective at removing hydrocarbon and moisture contaminants from substrates. In the case of polished optical surfaces no significant difference in the cleaning results achieved with either source have been observed. When removing hydrocarbon and moisture from polymer surfaces the SP VUV DBD achieved this with less coloration of the nominally clear samples, which may be advantageous in some applications. Overall the SP DBD source appears promising for cleaning applications, even at the very modest average power levels and pulse durations that have so far been investigated. Further improvement are to be expected with higher average fluence and shorter pulse durations.

Acknowledgments

This research was supported by a Macquarie University Research Innovation Fund grant.

References

1. R.P. Mildren and R.J.Carman, "Enhanced performance of a dielectric barrier discharge lamp using short-pulsed excitation", J.Phys.D:Applied Physics, **34**, L1-L6, (2001).
2. R.P.Mildren, R.J.Carman, I.S. Falconer, "Visible and VUV images of dielectric barrier discharges in Xe", J.Phys.D: Applied Physics, **34**, 3378-3382, (2001).
3. R.P.Mildren, R.J.Carman, I.S. Falconer, "Visible & VUV emission from a Xenon Dielectric Barrier Discharge using pulsed and sinusoidal voltage excitation waveforms", IEEE Trans. Plasma Science, Special Issue on images in Plasma Science, **30**, 192-193, (2002) (see also **30**, 154-155, (2002)).
4. R.P. Mildren, B.K. Ward, R.J. Carman, "High peak power VUV flashlamps based on short-pulse excited Xe dielectric barrier discharges", Proc. XIV International Conference on Gas Discharges and their Applications, Liverpool, England, 1-6 September, 2002, pp140-143.

5. R.J.Carman and R.P. Mildren, "Computer modelling of a short-pulse excited dielectric barrier discharge Xenon excimer lamp (λ~172nm)", J.Phys.D:Applied Physics, **36**, 19-33, (2003).
6. D.M. Kane, R.P. Mildren and R.J. Carman, US patent 6541924 (2003)
7. http://www.osram.com/products/photo_optic/discharge/xeradex.html (Osram Xeradex Xenon Excimer Lamp, accessed 30/6/03). Xeradex is a pulsed excited DBD lamp with good pulse-to-pulse reproducibility of the VUV output (viz. pulse energy and peak power). It has been used as a benchmark.
8. U. Kogelschatz, B. Eliasson and W. Egli, J. de Physique IV, 7, Colloque C4, 7, pp47-66 (1997).
9. A Yokotani, N Takezoe, K Kurosawa, W Sasaki, H Matsuno and T Igarashi, "A new photo-material processing using incoherent vacuum ultraviolet radiation", Rev. Laser Engineering, p148-152, (1998).
10. J.-Y. Zhang and H Esrom, "Large area photochemical dry etching of polyimide with excimer UV lamps", Appl. Surf. Sci. **69**, 299-304, (1993).
11. Polymer surfaces and interfaces: characterisation, modification and application, Eds. K. L. Mittal and K.-W. Lee, VPS, Utrecht, 1997.
12. Z Falkenstein, "Surface cleaning mechanisms utilizing VUV radiation in oxygen containing gaseous environments", SPIE – Int. Society of Optical Engineering, Proceedings of SPIE **4440**, 246 – 255, (2001).
13. http://www.ushio.co.jp/english/product/index.html (Excimer VUV/O_3 Cleaning System, Ushio Light City, Ushio Inc., accessed 30/6/03).
14. D Hirschausen and D M Kane, Laser removal of paraffin wax from glass surfaces, J. Appl. Phys., **92**, 4201-4208, (2002).
15. D. R. Halfpenny, D.M. Kane, R N Lamb and B Gong, "Surface modification of silica with ultraviolet laser radiation", Appl. Phys. A. **71**, 147-151, (2000).
16. A. J. Fernandes, D. M. Kane, B Gong and R. N. Lamb, "Dehydroxylation of UV fused silica via UV laser irradiation", pp. 433-436, 2002 Conference on Optoelectronics and Microelectronics Materials and Devices Proceedings, Ed. M. Gal, IEEE, Piscataway, 2002 (IEEE Cat. # 02EX601).

Chapter 14

MICRO AND NANO-MACHINING WITH ULTRASHORT LASER PULSES: FROM BASIC SCIENCE TO THE REAL WORLD

PETER BALLING

Department of Physics and Astronomy, University of Aarhus, Ny Munkegade, DK-8000 Aarhus C, Denmark

The interaction of short laser pulses with matter is an interesting study in its own right. Such investigations are preferably performed under well-defined, reproducible conditions. It is by no means straightforward to transfer the knowledge from basic investigations to the "real world" of laser machining. In this paper, a brief overview of the challenges that meet real-world applications will be given, and different solutions to improving important parameters like reproducibility and throughput will be presented.

1. Introduction

Ultrashort laser pulses were originally developed to study basic science at the frontier of temporal resolution of experimental physics. It was, however, soon realized that their specific interaction with materials provided unique machining possibilities for certain applications: the pulse duration is so short that heat propagation can essentially be neglected for the duration of the pulse, and the intensities that are readily available from commercial laser systems correspond to an electric field, which essentially ionizes all materials [1]. This means that laser machining does not rely on traditional (i.e. linear) absorption. Even materials that are transparent to the laser wavelength can be ablated – from the surface or inside their bulk.

The much-reduced heat transport during the laser pulse means that the heat is kept where it is needed: at the location where material removal is desired. The result is that ultrashort pulses are able to machine with a much higher precision than longer laser pulses and that the unwanted thermal effects on the material surrounding the machining region is strongly reduced.

The interaction of short laser pulses with matter is an interesting study in itself. These investigations are preferably performed under well-defined, reproducible conditions as offered, e.g., by single-shot experiments on thin-film or single-crystal samples, possibly under vacuum to eliminate any effects of the ambient atmosphere. A great variety of investigations have been performed in this context, see, e.g. [2].

The challenge for those operating in the field bridging the gap between fundamental studies and applications is to transfer the knowledge from such basic investigations to the "real world" of laser machining. In this endeavor, one is often met with "real" (non-uniform) materials, a large number of laser shots are often needed to obtain the desired geometry, and vacuum is typically not compatible with a production line where throughput is an important parameter.

This paper offers a brief overview of the challenges that meet real-world applications, and different solutions to inherent problems will be proposed [3-5].

2. Basic Properties of Ultrashort-Pulse Laser Machining

2.1. *Ultra-short pulses yield high precision*

Laser cutting is initiated by absorption of light energy by the electrons of the sample. In traditional laser cutting, the laser pulse is so long that during the irradiation the electrons thermalize with the lattice and the heat has time to diffuse over distances that are significantly longer than the length scale over which the light is absorbed. Consequently the precision is limited to the so-called diffusion length $l \approx \sqrt{\kappa\tau}$, which depends on the thermal diffusivity κ of the material and the laser pulse duration τ [6]. It is always several tens of micrometers or longer for nanosecond laser pulses ablating metals. In addition, the heat that propagates away from the machining region heats the surrounding areas to very high temperatures. This possibly molten and re-solidified area, the so-called heat-affected zone, typically exhibits a much-reduced strength compared to the native material.

During an ultrashort laser pulse, the heat-propagation distance is generally smaller than or of the same order of magnitude as the optical absorption length, and the heat-affected zone is absent or at least much smaller than for long pulses. This allows for very precise machining with maximum strength of laser-formed structures.

A very attractive property of ultrashort-pulse machining is its threshold character: Only when a certain threshold fluence is surpassed, the ablation process starts (see also Sec. 2.7 below). This can be used to generate structures that are smaller than the diffraction limit of a given laser focus [7]. The fluence is adjusted so that only the central part of the (typically Gaussian) intensity distribution is above the threshold for ablation. Since the diffraction limit for visible or near-infrared lasers is on the micrometer level, this technique is often dubbed nanomachining. Nanomachining can of course also be carried out with ultrashort ultraviolet laser pulses – even without applying the sub-diffraction-limited technique.

2.2. *Ultrashort pulses are versatile*

Traditionally laser machining can only be done with lasers that are absorbed by the specific sample. This intuitive observation is for example the reason why ultraviolet (typically excimer) lasers are applied for laser cutting of insulators, since only these lasers have photon energies high enough to bridge the band gap of such samples.

This intuition does not hold for ultrashort pulses: In the focus of an ultrashort laser pulse, the intensity is extremely high. For a typical pulse energy of 1 mJ, pulse duration of 100 fs and laser spot size of diameter 10 μm, the intensity is $\sim 10^{16}$ W/cm^2. This intensity corresponds to an electric field vector of magnitude determined by the relation $I = \frac{1}{2} c \varepsilon_o E^2$, which shows that the electric field in the laser focus is $\sim 3 \cdot 10^9$ V/cm, which is comparable to the electric field binding the electrons of atoms to their nucleus. This simple estimate shows that the electric field in the laser focus is strong enough to ionize any material!

For ultrashort-pulse lasers, ablation is thus an intensity effect and not a heat (average power) effect. One spectacular application of this fact is in ultrashort-pulse machining inside the bulk of transparent materials: During propagation through the material, it is truly transparent to the laser light, but at the point of focus, the high intensity (or electric field) leads to a breakdown and allows machining.

2.3. *Point 1: Short pulses mean large bandwidth*

A basic property of an ultrashort laser pulse is that it has a finite bandwidth (see, e.g. Ref. 8). This property, which is of course well known from the scientific point of view, is important for any application. Generally, the finite bandwidth is not a problem, but it does require attention.

A pulse, which may be ultrashort coming out of the laser, will change its properties during propagation: The passage of any dispersive material (i.e. any optical material in the beam path) will introduce different delay for the different frequency components in the pulse ("chirp") and thus lead to a gradual increase of the pulse duration. This effect is called group-velocity dispersion. For most commercial laser systems, with pulse durations in the 100-fs range, any group-velocity dispersion introduced during beam transport can easily be neutralized by the application of a small initial chirp on the laser pulses coming out of the laser. However, the passage of substantial amounts of optical material (several centimeters) will inevitably lead to the accumulation of higher-order phase shifts, which cannot easily be removed by pre-chirping the pulse. Similarly, the use of a setup with large variations in the amount of optical material in the beam path (from, e.g., a laser scanning system) is difficult to implement.

2.4. *Point 2: Short pulses mean large intensities and possibly non-linear propagation effects*

The strength of the ultrashort pulses, the ability to achieve extremely high intensities, is also a subject of concern: Under normal machining conditions, the intensity of the laser beam in front of the sample is typically high enough to introduce non-linear effects in the ambient air. In the simplest description, the non-linearity is described in terms of an intensity-proportional (non-linear) refractive index, n_2, which adds to the normal refractive index, n_0, so that $n = n_0 + n_2 I$. A refractive index of this form can be shown to lead to various non-linear phenomena as, e.g., self-focusing and self-phase modulation [8].

In order to quantify the effect of these non-linear processes, it is useful to introduce the so-called B-integral [8],

$$B = \frac{2\pi}{\lambda} \int_0^L n_2 I(z) dz, \tag{1}$$

where λ is the laser wavelength and the integration proceeds along the beam propagation direction. The B-integral is a measure for the amount of phaseshift, which gets "picked up" due to the non-linear propagation. As long as this phaseshift is substantially smaller than 1, the non-linear effects are small.

The integral in Eq. 1 can readily be calculated for propagation through the focus of a Gaussian laser beam on the beam axis, since the limits of integration can be taken to infinity without introducing any significant error. The resulting expression

$$B = n_2 \frac{\pi^2}{\lambda^2} \frac{E_p}{\tau} \tag{2}$$

depends only on the wavelength and the pulse energy and duration, E_p and τ. It becomes independent of the focal length of the lens, since the higher intensity in a short focus is balanced by the longer propagation of the long focal length. In laser machining, if the sample is positioned in the focus, the expression of Eq. 2

should of course be reduced by a factor of two, while for any imaging geometries, where the sample is after the focus, it gives a good approximation to the non-linear effects.

For typical machining conditions ($\lambda = 800$ nm, $E_p = 0.1$ mJ, $\tau = 100$ fs), the above integral becomes $B = 1.5$ for atmospheric air at ambient pressures [9]. This is enough to introduce significant distortions of the laser beam, spatially and temporally/spectrally. This phenomenon seems to be responsible for the very different conclusions of early feasibility studies of femtosecond laser machining.

An obvious solution to the problem is to work under vacuum: Since the *B*-integral is proportional to the pressure, it is clear that working under even a modest vacuum can remove the problem. Vacuum is, however, generally not compatible with the demands of a production line, in particular the requirements of high throughput. There is, however, another straightforward solution, which is to use a cutting gas with a low non-linear refractive index [9]. If, for instance, a helium atmosphere encloses the laser-machining zone (and thus the beam focus), the above expression for the *B*-integral is reduced to $B = -1.5 \cdot 10^{-3}$ which makes it virtually insignificant. The use of an inert gas also reduces any possible chemical reactions in the laser ablation zone during machining.

2.5. *Point 3: Diffractive optics stretches the pulse*

Diffractive optics is a quite strong tool for laser machining with longer laser pulses: It offers the possibility to modify the intensity distribution more or less at will, e.g. to make beam homogenizers before mask projection or to make special laser intensity distributions on a sample.

Applying the same principles for ultrashort pulses must be done with great care. One must always consider that any diffractive optics stretches the pulse duration. If we take again an 800 nm laser pulse, the stretching introduced *per order* of the diffractive optics is 2.7 femtoseconds. For example, after passing, e.g., a 1-cm-diameter beam through a standard Fresnel lens with 200 lines per inch, the pulse duration has in a rough estimate changed from 100 fs to ~240 fs.

2.6. *Point 4: Throughput is fundamentally limited*

Ultrashort-pulse lasers are not likely to become the workhorses of large-scale industrial cutting and drilling. The reason is simply that at present, the maximum average power is several orders of magnitude lower than that available from, e.g., industrial carbon-dioxide lasers.

For illustration, consider that the full average power of 1 W from a commercial femtosecond laser is used with a – somewhat unrealistic – perfect efficiency to remove material from a metallic sample. Then the amount of metal that can be removed per second can be estimated by taking the energy needed for removal per unit volume as the total enthalpy of evaporation, which is normally the dominant contribution. For copper this is 42.2 kJ/cm^3, which shows that the absolute maximum removal rate is $2.4 \cdot 10^{-5}$ cm^3/s or a dice of side lengths 0.3 mm per second. These are numbers that put natural constraints on the applications of ultrashort-pulse lasers, and show that potential uses of the lasers will be found within micro- or nanomachining.

2.7. *Point 5: Thermal effects during drilling of metals*

The previous section showed that femtosecond lasers are limited in material removal rate by the average power. In addition, it turns out that for metal processing, it is not possible to preserve the celebrated high precision of laser machining at very high pulse energies.

Several experimental investigations have shown (see, e.g., Refs. 10 and 11) that the ablation rate (expressed as the ablation depth d per laser pulse) at low laser fluence follows an approximate logarithmic dependence:

$$d = d_0 \ln\left(\frac{F}{F_{th}}\right), \tag{3}$$

where d_0 is a characteristic absorption length (the optical penetration depth in the simplest approximation) and

$$F_{th} = H_{vap} \cdot d_0 /(1 - R) \tag{4}$$

is a material-dependent threshold fluence determined by the reflectivity R, the absorption length and the enthalpy of evaporation (per volume) H_{vap}. This law can be found from simplified theoretical treatments of the ablation process that neglect heat propagation and thus assume the heat deposition to follow the initial exponential distribution of the absorbed laser light [6].

This approximation is, however, only valid in the limit of quite small laser fluences. As soon as the fluence is increased to ~1 J/cm^2, the ablation depth is becoming strongly influenced by the heat transport of the conduction electrons of the metal [10, 11]. A certain fraction of the heat is then lost by propagation to the surrounding material. This reduces the efficiency (i.e. the removal per unit of laser energy) and leads to the reappearance of a heat-affected zone.

Another way to increase the throughput is by increasing the repetition rate of the laser. In the low-fluence limit, where Eq. 3 is assumed valid, one can optimize the ablation rate per second versus repetition rate ν under the condition of fixed average power P_{av} and fixed spot size A by maximizing the expression

$$d \cdot \nu = \nu \cdot d_0 \ln\left(\frac{P_{av}}{\nu \cdot A \cdot F_{th}}\right). \tag{5}$$

The optimum is found at $\nu' = P_{av} / (A \cdot F_{th} \cdot e)$, i.e. when the repetition rate is adjusted so that the fluence is exactly e times the threshold fluence. At this fluence, the ablation rate per second becomes $(d \cdot \nu)_{\max} = P_{av} \cdot d_0 / (A \cdot F_{th} \cdot e)$ or by inserting F_{th} from Eq. 4, $(d \cdot \nu)_{\max} = P_{av} \cdot (1 - R) / (A \cdot H_{vap} \cdot e)$. It can be seen that the optimum volume ablation efficiency is only a small fraction, $(1 - R) / e$, of the absolute maximum discussed in Sec. 2.6 above. A critical parameter here is the reflectivity, which may differ from the Fresnel reflectivity of a pristine surface, see Sec. 3.3.

In general, performing machining at relatively low fluence and relatively high repetition rate is a promising way to combine high throughput and high precision. This is especially important for the machining of metals with their high heat conductivity. It is a challenge for laser manufacturers to deliver high-repetition-rate high-average-power ultrashort-pulse lasers, but the technological development is still going on, and in particular the development in fiber-based laser systems looks very promising. Note, however, that the assumption of a linear increase in ablation per second with the repetition rate will eventually cease to be valid at high repetition rate, for instance because of residual heat from previous laser pulses or particle-ignited plasma breakdown, see Sec. 3.2 below.

3. Optimizing Ultrashort-Pulse Laser Machining

3.1. *Challenge 1: Reproducibility*

Basic investigations of ultrashort-pulse ablation have shown the process to be extremely reproducible. These studies were, however, typically performed on very homogeneous thin-film or single-crystal samples. In "real" applications of femtosecond machining, this is rarely the case. Since, however, reproducibility is a key parameter for high-accuracy micromachining, it is important to devise methods for process control.

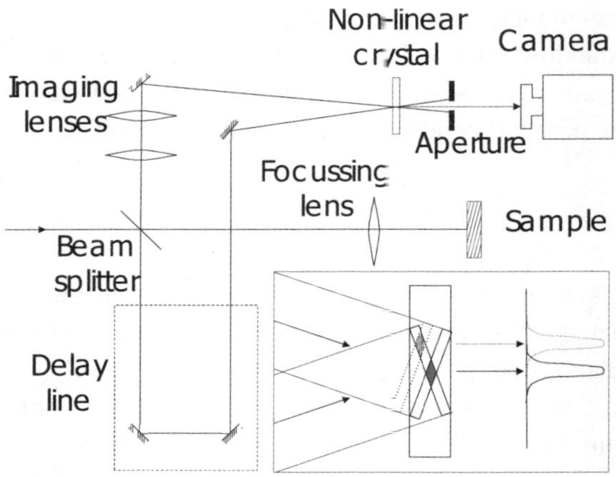

Fig. 1: A schematic drawing of the setup for in-situ depth gauging of laser ablated areas [3]. A fraction of the laser light is split off by a beamsplitter to measure the flight time of backscattered light from the sample. This time determines the depth with micrometer accuracy.

Our group recently demonstrated a new approach to on-line process control [3]: By combining the temporal structure of the backscattered light from the machining region with an imaging system, it is possible to measure the depth-profile of the laser ablated structure on the fly. The ablation and geometrical measurement is thus done with one and the same pulse. As shown in Fig. 1 above, the temporal information is obtained by an optical cross-correlation technique: The output of the femtosecond laser is split in two parts by a beam splitter. One part is focused onto the sample and performs ablation while the second part enters a delay line and is used for gating. Part of the backscattered light is collected by the lens in front of the sample and after being picked off by the beam-splitter sent through an imaging system designed to image the ablation region of the sample onto the non-linear crystal that forms the optical gate. Inside the non-linear crystal, the backscattered, re-imaged light is crossed by the gate beam. As shown in the insert of Fig. 1, part of the image is up-converted to the second harmonic, and the resulting pattern reflects in one direction the time (or depth) coordinate of the backscattered light and in the perpendicular direction an image across the laser-ablated region. Consequently, the pattern represents a depth profile across the ablation region.

Such on-line depth profiles are useful in micromachining and laser surgery since they can be used as a feedback signal to control the ablation process. The depth-resolution depends on the laser pulse duration and the material subjected to ablation. For example for a 100-femtosecond pulse duration, an accuracy of 1 μm is obtained in steel, increasing to ~5 μm in soft biological materials.

3.2. *Challenge 2: Deep drilling of metals*

The drilling of deep narrow holes in metals is a challenge. It is, however, also a core technology for many important industrial processes as, e.g., the formation of high-performance fuel-injection nozzles. Consequently much research has been dedicated to this endeavor (see Ref. 12 for a recent review).

One problem is associated with the use of linearly polarized laser light, which has been shown to lead to elongated exit holes, presumably due to the different reflectivity of the two polarizations under grazing incidence inside the laser-drilled channel. This problem can be solved through the application of circular polarization or simply by rotating the polarization during the laser drilling [13].

Another difficulty is to produce truly cylindrical holes, since the exit hole always tends to be smaller than the entrance hole. This problem is minimized by application of a so-called trepanning technique, which moves the laser spot constantly around the perimeter of the hole during its drilling [12].

A specific problem related with deep drilling is also the formation of particle-induced plasma break down. In all drilling processes, the debris from previous pulses in the form of small metal particles or clusters can lead to a reduced threshold for laser plasma break down. This is a potential problem, since it redistributes part of the laser energy from the area under direct irradiation to a more diffuse irradiation by the plasma emission. For shallow holes this effect is mainly a loss mechanism and it will often lead to a reduced transverse precision (larger holes) [12].

The formation of a particle-induced plasma will depend on the exact cutting geometry and the flow rate of any cutting gas. Again working in vacuum is a solution, since this will normally allow the escape of debris much further away from the ablation region. However, the alternative of using helium could also be advantageous, this time because of its low atomic mass, which gives minimal hindrance to escaping debris. The problem of course also depends on the time between laser pulses, i.e. the laser repetition rate, and will eventually put an upper limit on the range of repetition rates where the ablation per second simply increases linearly with the repetition rate.

When performing depth gauging (see Sec. 3.1) during the drilling of deep holes, the formation of a particle-induced plasma can be seen as the appearance of a signal corresponding to a depth between the original surface level and the bottom of the hole that has already been drilled. Preliminary investigations in our group show that for deep holes, the formation of such a particle-induced plasma can only be avoided at laser repetition rates much below 1 kHz. At parameters realistic for production (high pulse energies and kilohertz repetition rates), however, the plasma seems to be an inherent property of the drilling process. There are in fact indications that the plasma inside the channel is an important factor for the generation of cylindrical holes at least during percussion drilling. According to our recent observations, the propagation of the plasma through the already formed laser drilled channel seems to serve as a mechanism for broadening the hole to its final diameter. The completed migration of the plasma through samples of a few hundred micrometers thickness (observed by the depth-profiling technique) correlates closely with the time of complete laser breakthrough (i.e. breakthrough over the entire laser beam diameter) observed on a photodiode behind the sample.

3.3. *Challenge 3: "Non-constant" ablation rates*

A critical parameter for the ablation efficiency of metals is the reflectivity of the sample, see Eq. 4. The feedback technique described in the Sec. 3.1 can also be used to investigate the ablation rate as a function of the number of laser pulses or the depth. Experiments investigating the ablation rate of a metal surface for the first pulses at fluences just above the threshold fluence show that the ablation rate is increasing over the first hundreds of laser pulses [4].

The reason for this seems to be changes in the energy coupling efficiency to the sample: scanning-electron-microscope images taken after an increasing number of pulses show that the surface morphology changes from perfectly flat to strongly corrugated. The development of ripples seemingly reduces the effective reflectivity of the sample leading to a more efficient ablation.

This indicates that even for samples that are quite homogeneous, it is not straightforward to extrapolate from single-pulse ablation rate measurements to the correct parameters for laser machining to a specific geometry. Again the application of an on-line feedback technique as described in Sec. 3.1 is desirable.

3.4. *Challenge 4: Optimizing throughput by parallel processing*

As mentioned in Secs. 2.6 and 2.7 above, throughput is a serious limitation of ultrashort-pulse machining. This becomes particularly relevant in the case of nanomachining, where a process based on sub-diffraction-limited ablation demands the application of very small fluences (see Sec. 2.1). Of course it is

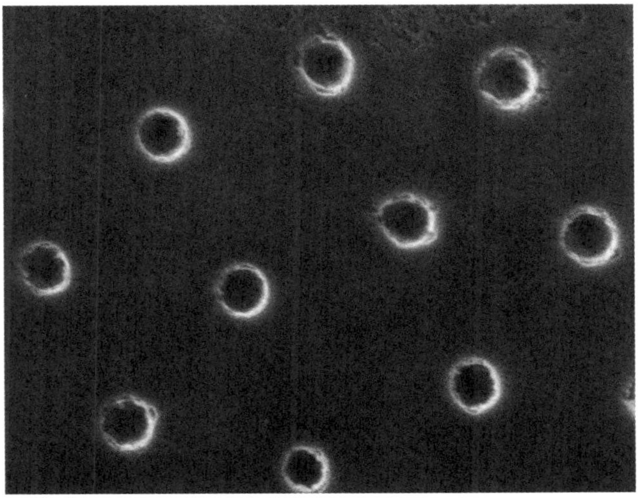

Fig. 2: A scanning-electron-microscope image of sub-micron holes formed in a copper surface by the method of a lens array formed directly on the surface. The distance between holes is ~7 μm.

tempting to use the procedure outlined in Sec. 2.7 and just increase the laser repetition rate while reducing the energy per laser pulse, and in some applications this may in fact be an option. In other applications it is, however, not possible to move the sample (or the laser beam) fast enough to take advantage of high laser repetition rates.

Another way to proceed is to apply parallel processing: The pulse energy from a typical laser system is large enough to perform ablation of more than 1000 holes at the same time. Unfortunately, a diffractive optic can in general not be used, cf. Sec. 2.5 above. Instead one can think of imaging a mask with many holes onto the sample, but this has several disadvantages: a mask for nanometer sized holes will have a poor transmission, and the procedure will require optically flat samples since the imaging system needs to have a short depth of focus. Alternatively, one can use a microlens array near the sample [14]. This is an attractive solution, but again only if the sample is optically flat and if the debris from the ablation process does not damage the lens array.

An alternative solution was recently demonstrated in our group [5]: A self-assembled layer of micrometer-sized quartz spheres works as an efficient lens array under proper conditions. The lens array is formed by direct deposition on the sample, which eliminates the complications associated with uneven samples. The use of a transparent spacer quartz layer under the quartz spheres ensures that the sample is struck by an array of well-focused spots.

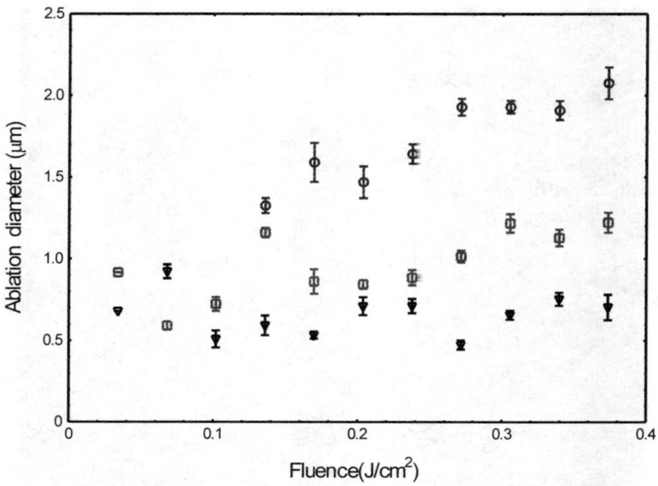

Fig. 3: The diameter of holes generated by the lens array as a function of the average fluence on the sample. Data are shown without a spacer layer (circles), and with spacer layers of 1.6 μm (squares) and 3.1 μm (triangles).

An example of a laser-produced patterns is shown in Fig. 2 on the previous page. The hole size was investigated as a function of the applied laser fluence and the thickness of the transparent spacer layer. At the optimum thickness of the spacer layer, the hole size was found to be a weakly increasing function of the laser fluence with hole sizes close to 500 nm over a large fluence range, as shown in Fig. 3 above. This shows that the method is applicable for the parallel production of sub-micron holes without significant sensitivity to small variations in intensity over the laser spot size.

4. Conclusion

An ultrashort-pulse laser *is* a unique tool for laser machining. The short pulse duration is the basis for very high precision and maximum strength of the surrounding material. However, the present paper demonstrates that it is a tool with certain limitations: Throughput is modest, so applications are most likely to be found in nano- and micromachining. The use of ultrashort pulses requires slightly more attention than longer-pulse laser light due to linear and non-linear propagation effects.

The paper also outlines different ways to optimize short-pulse laser machining. For nano-structuring, parallel processing is an attractive alternative, and depositing a lens array directly on a sample broadens the application to samples that are not necessarily optically flat, Sec. 3.4. In order to deal with the reality of non-uniform samples, advanced on-line process control can be implemented. In fact, the depth-profiling method of Sec. 3.2 will only work for ultrashort pulses, since it is based on the temporal information of the backscattered light.

Following the invention of the femtosecond laser, some years passed where operating the laser systems was a difficult task in itself, and their use was strictly limited to research. Now, due to the rapid engineering progress, new systems have been developed, based on a solid state gain medium, that are much better suited for industrial purposes. In the future, with the further development of compact, rugged and user-friendly laser systems, it seems likely that ultrashort-pulse machining will be an industrially relevant option for specific applications.

Acknowledgments

The results outlined in this paper represent a cumulative effort of the ultrashort-pulse-machining research group at the University of Aarhus with several generations of students: Rune Lausten, Jakob A. Olesen, Kasper Vestentoft, Bjarke H. Christensen, and Charlotte S. Nielsen. The work was financed by the Danish Natural Science Research Council (SNF).

References

1. P. B. Corkum, Phys. Rev. Lett. **71**, 1994 (1993).
2. *Springer Series in chemical physics, Vol. 71: Ultrafast Phenomena XIII* (Springer-Verlag Berlin Heidelberg 2002).
3. R. Lausten and P. Balling, Appl. Phys. Lett. **79**, 884 (2001).
4. R. Lausten, J. A. Olesen, K. Vestentoft, and P. Balling, in *Springer Series in chemical physics, Vol. 71: Ultrafast Phenomena XIII* (Springer-Verlag Berlin Heidelberg 2002).
5. K. Vestentoft, J. A. Olesen, B. H. Christensen, and P. Balling, Appl. Phys. A, **80,** 493, Rapid Communication (2004).
6. D. Bäuerle, *Laser Processing and Chemistry*, 3rd edition (Springer-Verlag Berlin Heidelberg 2000).
7. P. P. Pronko, S. K. Dutta, J. Squier, J. V. Rudd, D. Du, and G. Mourou, Opt. Comm. **114**, 106 (1995).
8. A. E. Siegman, *Lasers*, University Science Books, California (1986).
9. E. T. Nibbering *et al.*, J. Opt. Soc. Am. B **14**, 650 (1997).
10. S. Nolte, C. Momma, H. Jacobs, A. Tünnermann, B. N. Chichkov, B. Wellegehausen, and H. Welling, J. Opt. Soc. Am. B **14**, 2716 (1997).

11. K. Furusawa, K. Takahashi, H. Kumagai, K. Midorikawa, and M. Obara, Appl. Phys. A **69**, S359 (1999).
12. D. Breitling, C. Föhl, F. Dausinger, T. Kononenko, and V. Konov, in *Femtosecond Technology for Technical and Medical Applications* (Springer-Verlag Berlin Heidelberg 2004).
13. S. Nolte *et al.*, Appl. Phys. A **68**, 563 (1999).
14. K. Piglmayer, R. Denk, D. Bäuerle, Appl. Phys. Lett. **80**, 4693 (2002).

Chapter 15

OPTICAL SURFACE PROFILOMETRY OF LOW REFLECTANCE MATERIALS – EVALUATION AS A LASER PROCESSING DIAGNOSTIC

D M KANE, A M JOYCE

Department of Physics, Macquarie University ~ Sydney, NSW 2109, Australia

R J CHATER

Department of Materials, Exhibition Road, Imperial College, London SW7 2AZ, UK

Optical surface profilometry is a technique that has advantages over other profilometry techniques (stylus profilometry, AFM) of being non-contact and being able to profile comparatively large areas in a single "z-scan". Thus, it is employed in monitoring surface quality and measuring surface form in high technology manufacturing processes and quality assurance, as well as being applied as a diagnostic in research and development contexts. Its application to optical materials has been limited due to issues relating to the low reflectance of the surfaces. A feasibility study for profiling laser induced optical damage and "loose" microscopic sized pieces of optical material (particles introduced by design) on optical substrates is reported. Progress on profiling these difficult samples has been achieved.

1. Introduction

In laser micro-machining and nano-patterning, laser cleaning and other laser processing applications it is important to be able to measure the surface profile, both before and after processing. This is an integral part of characterizing the structures that have been machined or patterned onto the surface, and of identifying any unwanted surface or bulk damage to the sample. The latter can be easily missed in normal optical-microscopy-monitoring of highly transmitting surfaces or low reflection contrast samples (especially prevalent for normal incidence illumination of the sample). The three main methods used for surface profiling are stylus instruments, scanning probe instruments (such as AFM) and optical surface profilers. All these sophisticated instruments are subject to a range of potential measurement aberrations which require specialist training to avoid possible misinterpretation of the measured surface profiles. Prior research looking at the topology of laser induced optical damage associated with laser irradiated particle-on-glass-surface samples [1] has shown that the predominant type of high-laser-pulse-energy damage can be approximated by a "spherical cap" shaped pit which is several tens of microns in diameter – a size consistent with the laser beam size rather than the size of individual particles (which is micron or sub-micron). Thus, to optimally surface profile such damage sites requires a method that can quickly scan relatively large sample areas, with accurate spatial registration. Optical surface profilers are the superior instrument for such applications. Additionally, for "dirty samples", such as those that may have residual particle contamination, stylus and scanning probe profilers can be damaged and/or incapable of completing a spatially calibrated and accurately registered profile due to stylus/scanning probe contact with the contaminants. A non-contact method like optical surface profiling is required. Most managers of stylus or scanning probe surface profilers will not permit "dirty samples" to be used with the instruments. Thus, the optical surface profiler has advantages for many applications of being non-contact, able to scan comparatively large areas quickly, at vertical resolutions as good as 0.1 nm for smooth, flat surfaces. Its disadvantages are relatively poor lateral resolution (determined by the optics in the profiler) and difficulty in registering large step changes accurately.

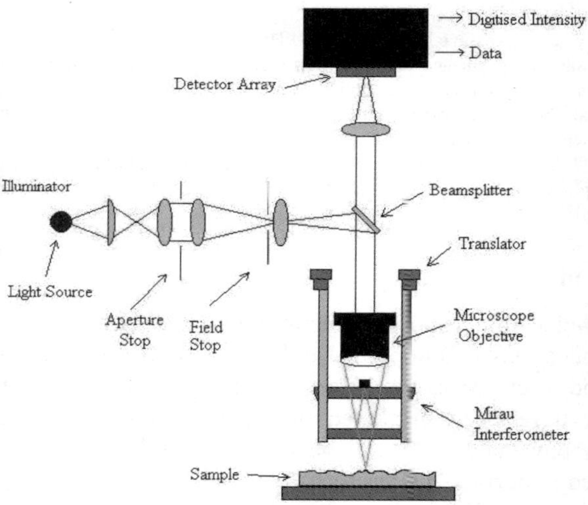

Figure 1: Interference Microscope - A schematic of the basic set-up of the OSP.

The Veeco (formerly Wyko) Optical Surface Profiler (OSP) [2, 3], used for most of this study, is shown in figure 1. There are several other manufacturers of optical surface profilers and the technology of the optical surface profiler is experiencing active research and development to effect improvements that will impact on applications such as those described here (improved results for low reflectance samples, measurement of a thin film on a surface) at this time. We have carried out comparative studies using a Zygo OSP (at Imperial College) and found no differences in the results that are attributable to instrumentation differences.

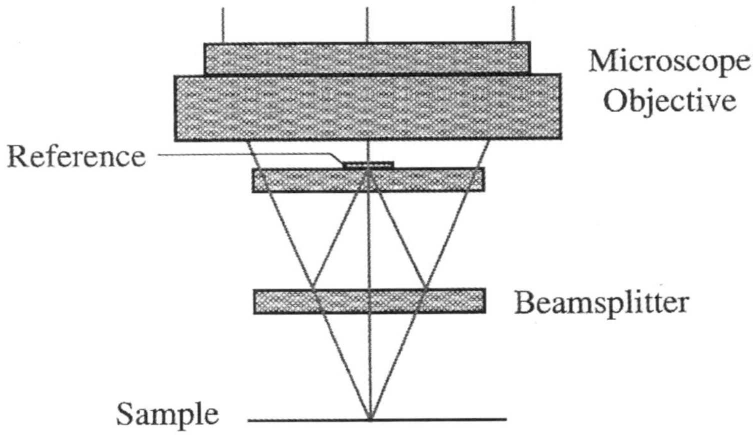

Figure 2: Detail of Mirau interferometer.

The OSP is a semi-custom WYKO NT 3300. The surface profiler has two different modes of operation: Phase Shifting Interferometry (PSI) mode and Vertical Scanning Interferometry (VSI) mode. PSI is the mode used exclusively for the measurement of very smooth surfaces, in particular the surface roughness of optics and semiconductor materials. The PSI mode has the best vertical resolution (~0.1 nm) but can not image height changes larger than $\lambda/4$ (~ 160 nm at the ~640 nm centre wavelength used in the instrument) and is not suitable for the topics covered in this paper. The OSP is used in white light vertical scanning interferometry (VSI) mode. The resolution in VSI mode is 3 nm vertically, and is 150 nm horizontally at the highest magnification (50x) and field of view (FOV) (2.0x). The lateral resolution is a larger number at lower overall magnification. The light source is a white light halogen lamp which is conditioned into a collimated beam, by input optics, and then directed by a 45° beamsplitter to a microscope objective, which in turn focuses the light onto the sample (illuminating the area of the field of view for the particular optics used). The two beam interference is produced by a Mirau interferometer (detail shown in figure 2). Light is split into two beams at the beamsplitter. The "reference beam" is directed to, and then reflected from, the plane reference mirror, while the sample beam gets directed to and then reflected from the sample. These two reflected beams interfere in the plane of the beamsplitter (figure 2) when the optical path length of each beam is the same.

The two beam interferogram is then imaged onto the detector array by the microscope objective and the upper lens in-line with the detector array (in figure 1). In white light interferometry, high contrast interference fringes are only obtained when the path difference between the reference and sample beams is within the short coherence length of the source (typically microns, to tens of microns using filters). As the mount holding the reference mirror is precision translated (in the vertical, z-direction) relative to the beamsplitter and sample, different height features on the sample pass in and out of satisfying the "within the coherence length" optical path difference requirement, to generate high contrast interference fringes, at different positions in the scan. The detection process monitors the fringe visibility as the translation proceeds. A description which gives the essence of generating the surface profile follows. The maximum fringe contrast (visibility) is recorded for each "pixel" in the detector array by sequentially updating a look-up table that records the greatest value of the fringe visibility for each pixel, as the scan progresses. An image frame is grabbed at each scan step. A higher fringe visibility value will replace the current value in the table, while a lower value will lead to no update. At the end of the scan these scan positions (frame numbers) of maximum fringe contrast are analysed to produce a surface height map (ie a surface profile). A more complete description would address the mathematical algorithms which are used and the need to consider nearest neighbour pixels in determining fringe visibility at all.

The maximum possible visibility of the interference fringes is 1 if the sample and reference beams have the same irradiance and both beams are generated from optically flat surfaces. In this case visibilities in the range from 0-1 will be possible giving the greatest achievable measurement sensitivity in the variation in fringe visibility. This will only occur when the reflectivity of the reference mirror is a close match to that of the sample. Where there is a strong mismatch in these reflectance values this will limit the range of fringe visibilities. For a 4% reflectance glass surface and a 100% reflecting reference mirror this fringe visibility range will be reduced to 0 - 0.20 for specular reflection (a perfectly flat sample) and less for rough samples. Thus, the nature of the sample being profiled has a significant impact on the dynamic range for fringe visibility measurement. Optical material surfaces with low reflectance and high transmission at the wavelengths used are a challenge for successful profiling. If insufficient signal is detected this shows up as black (areas of no information) in the surface profile. In VSI mode a neutral density filter allows control of the light level so that some optimization of fringe visibility for different samples can be affected.

In addition to the low light level and low interference fringe contrast issues for high transmission materials there is also the possibility of multiple reflections from the sample surface. This can arise in optical damage samples that have cracking in planes parallel or near parallel to the surface, in addition to surface damage. Another key sample of interest is that of a spherical silica particle sitting on a surface. We wanted to know if the upper half of such a sphere could be successfully profiled sitting on a flat glass or silica surface. A series of systematic OSP studies of this sample type have been carried out and are reported. But, firstly, characterisation of laser induced optical damage caused by high laser pulse energy irradiation of silica spherical particles on microscope slide glass or fused silica samples is reported. Near field focusing and backscattering of laser light between the surface and the underside of the particle are found to effect the topology of the damage sites significantly.

2. Surface Damage Sites Resulting from High Laser Pulse Energy Irradiation of a Silica Microsphere on a Transparent Surface

2.1 *Experiment*

The silica microspheres being used in this experiment are commonly used in laser cleaning experiments, which is one reason imaging for both pre- and post-cleaning is critical. The SMs are provided by Kromasil [4] who specialise in creating the spheres for chromatographic applications. The diameters used in these experiments are 3.5 μm and 5.0 μm nominally, and are both high purity silica microspheres with a small amount of metal impurities <10 parts per million (ppm) and are assumed to be transparent, isotropic and spherical. The substrates used were standard microscope slides (dimension 75 x 25 x 1 mm^3), and fused silica (UV7980) with dimensions 55 x 20 x 2 mm^3 and flatness better that $\lambda/2$. The silica microspheres were applied to the substrate by the "dip and tap method" [5, chapter 4]. The laser system used for the experiments is a GSI Lumonics PulseMaster PM-848 KrF excimer laser with a wavelength of 248 nm, pulse energy up to 450 mJ at an average power of up to 80 W. The pulse length is 8 ns. The experiments use a single laser pulse of irradiation. Images of the particle-on-surface samples are recorded by standard optical microscopy, both before and after laser irradiation, for comparison with optical surface profile images. A computer controlled translation stage is used to precisely move the sample from in front of the microscope, where a before image is taken, to the path of the KrF laser beam for irradiation, and back to the microscope where an after image is captured. Optical surface profilometry is carried out off-line. The evaluation of how the optical surface profiler "sees" microspheres on the surface is reported in section 3.

2.2 *Results & discussion*

Microscope slide glass samples required fluences between 1.5 and 3.0 J/cm^2 to be confident that damage was likely to occur to the substrate (in contrast to fluences of 200-300 mJ/cm^2 used for successful laser cleaning of such microspheres from the surface). Some before and after images, taken by the optical microscope, are shown in Figure 3. From the images we can see there is significant damage caused to the substrate. At almost every point where there was previously a microsphere, there is now a damage site. In the bottom row, after image there is an apparent rectangular shape to the damage sites that is similar to the shape of the laser beam, with the long axis of the beam in the horizontal direction, but generally little detail of the damage site is learned from the optical microscopy images. Surface profiling of the damage sites used the highest magnification and field of view (FOV) available, resulting in an image of an area 59 μm x 45 μm. The scan length and angle of the sample platform led to no change in the resulting profiles. This indicates an aberration free, reliable optical surface profile result has been obtained. Figure 4 shows the different topologies of damage site as laser pulse fluence is increased. Fig. 4(a) shows part of the particle being left behind on the surface of the substrate in the shape of the rectangular beam. Fig. 4(b) shows a small amount of the particle being left

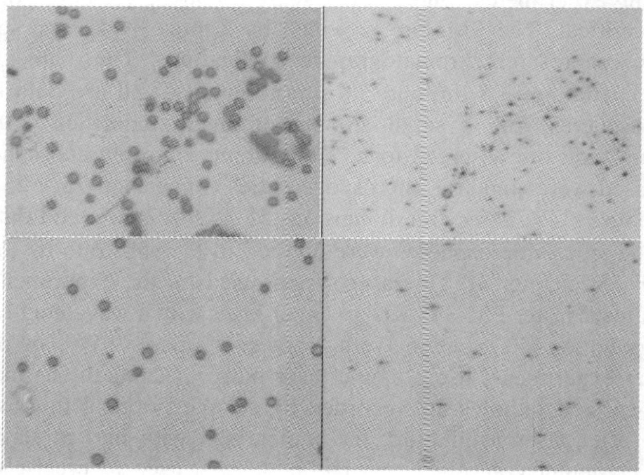

Figure 3: Before (left) and after (right) images of the laser cleaning process taken with an optical microscope, both images are taken using a 2500x magnification.

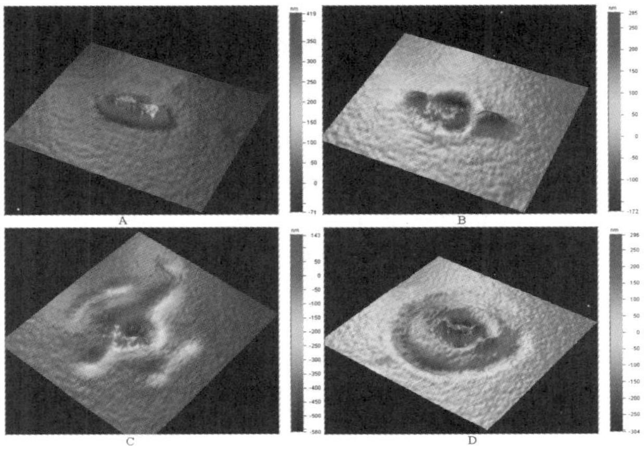

Figure 4: Damage caused to the substrate under high fluence irradiation, each image shows a different degree of damage; (a) showing a small amount of damage, (d) showing the most amount of damage.

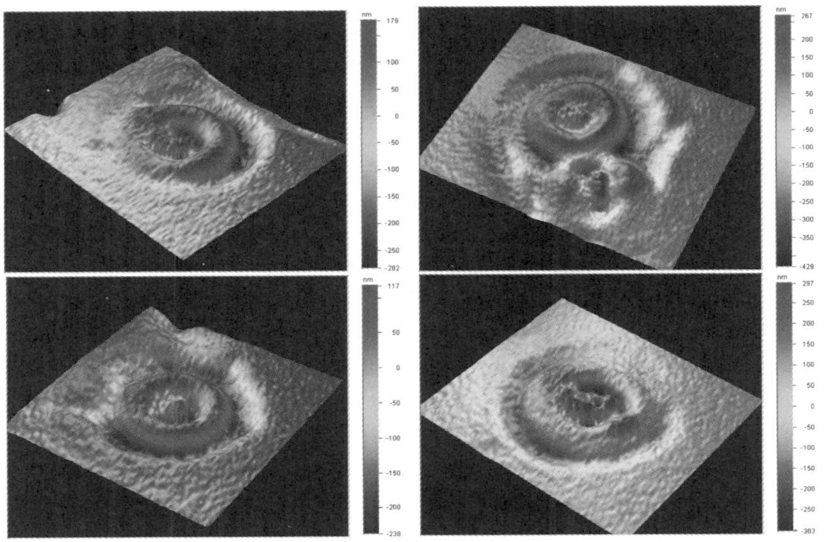

Figure 5. Optical surface profiles of the surface under a 5 μm diameter silica spherical particle on microscope slide glass, after irradiation of the particle with a laser pulse of fluence ~2 J/cm^2. The microsphere is removed as part of the laser/particle/surface interaction. The images shown vary slightly in their calibrated size, but all are of the order of 7 μm x 8 μm. The valley to peak heights of the "rings" range from 0.35 – 0.7 μm.

behind and the forward transmission by the particle being projected onto the surface in the form of circular rings, resulting in circular peak and trough contours.

Fig 4(c) is the next level of damage with no remnant of the particle but a much larger contoured ring site, starting to resemble both the shape and the dimension of the sphere that was previously present. Finally Fig. 4(d) shows the maximum damage with no debris from the particle present and the circular contoured damage site of size of order of the former particle.

Several of the damage sites produced at a laser pulse fluence of the order of 2 J/cm^2 are shown in fig. 5. These take the form of rings consistent with a sequence of a melt, photoelastic reforming, followed by resolidification of the glass surface, determined by the forward transmission irradiance of the laser beam by the microsphere. The radius of the rings scaled with the radius of the microsphere. A typical sequence was 0.50 ∓ 0.05 μm for the first "bright" ring, 1.15 ∓ 0.05 μm for the first "dark" ring, 2.0 ∓ 0.2 μm for the second "bright" ring, and 2.50 ∓ 0.05 μm for the microsphere radius. Values of these radii for ten samples were in the range 0.5 - 0.7 μm for the first "bright" ring, 1.15 -1.70 μm for the first "dark" ring, 2.00 – 2.55 μm for the second "bright" ring, and 2.50 – 3.25 μm for the microsphere radius. The existence of the second bright ring at ~2 μm is strong evidence that enhancement of the subsidiary rings as predicted by the particle on a surface, Bobbert and Vliger (POS-BV) theory [6] is appropriate for the particle/laser/ surface interaction being observed.

When the substrate is fused silica the topology of the optical damage is significantly different. The low absorbing substrate in this case prevents strong coupling of the irradiance pattern under the particle to the surface and no optical damage of the type shown in figs. 4 & 5 is observed. Instead at fluences of the order of 2 J/cm^2 the focussing of the laser pulse under the particle initiates a cracking phenomenon. Two new surfaces are created, initiated from an apex at the centre of the particle. Figure 6 shows the trigonal, bivalve shell-shaped pit left on the surface after, two new surfaces have been created by cracking and the "chip" of silica has been ejected from the site. It is possible to flow these chips away from the surface in the laminar flow of gas that is used to remove the ejected particles. It is unknown at this time whether the cracking is initiated by nano-indentation, by virtue of the spherical particle being pressed into the surface, or whether it is the direct result of absorption of the focused laser pulse in the substrate.

The results in figs. 4 - 6 clearly demonstrate that the optical surface profiler is an excellent tool for characterising laser induced optical damage of the types generated by irradiating microsphere-on-a-glass-surface samples.

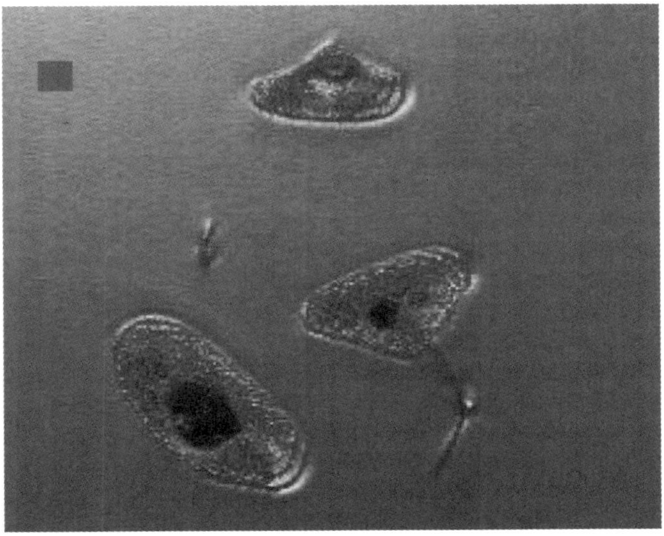

Figure 6. Optical surface profiles of the surface under a 5 μm diameter silica spherical particle on fused silica, after irradiation of the particle with a laser pulse of fluence ~2 J/cm^2. The microsphere is removed as part of the laser/particle/surface interaction. The two smaller bivalve shaped pits shown are of the order of 40 μm x 20 μm. The depth of the pit valley is ~0.5 μm.

3. Optical Surface Profiler Investigation of Transparent Silica Microspheres on a Transparent Substrate

3.1 Issues

The physical optics principles of the OSP determine that it is not possible to get a complete three-dimensional image of the sphere. The OSP is primarily a profilometric instrument and is expected to 'profile' the surface of the sample, sphere and substrate included. Thus, even if the sphere and the surface is perfectly reflective we would predict a surface profile that shows the flat substrate, a discontinuity at the equator of the spherical particle, which would lead to an inaccurate height step between the substrate surface and the profile of the top hemisphere of the particle. At best we would expect to see a profile of a half sphere, but all the angles of reflection from this hemispherical surface may not be collected within the numerical aperture of the microscope objective of the OSP. Additionally, the reflection coefficient is dependent on the angle of incidence being lowest for normal incidence (the "top" point of the microsphere). Given that the silica microspheres are largely transparent they will also act as a "lens" imaging the surface, and the effects of a secondary reflection from the substrate surface, imaged by the microsphere becomes an additional light field contributing to the interferogram. There has been previous work using White-Light Interferometry (WLI) to measure the thickness of thin-films [7], that has taken into account two reflected fields from the sample, one from the upper and one from the lower boundaries of the thin film. The microspheres are more complicated than this. In addition to the two contributions to the sample reflection described above there is also the possibility of a third component of light reflected from the lower surface of the microsphere. It is this multi-component field that combines with the reference beam to form the interferogram from which the profile is derived. Thus, there is the potential for there to be more than one maximum in the fringe visibility during a z-scan. The "look-up table" will record the largest such maximum, which may be primarily derived from reflection from the substrate surface rather than the surface of the microsphere. Best results of profiling the microsphere will be expected when measures are taken (eg. platform tilting or scan length tailoring) to avoid collecting light reflected from the substrate surface and the lower surface of the particle. We note that we included particle-on-surface samples, as samples of interest to be profiled, in the technical brief we submitted to the manufacturer, as part of the requisitioning process. We were not appraised of any of the issues being discussed here in their response. It has been necessary to independently research the physical optics of the OSP [2,8] to interpret the results of profiling these samples. An idealized optical surface profile of the nearly perfectly spherical particles would be the hemispherical surface convolved with an instrument spread function, characterised by a full width half maximum proportional to the lateral resolution of the instrument (for a given optics combination). Thus, the instrument is expected to predict a width of the microsphere that is larger than the actual width and an accurate half height for the hemisphere, as a first prediction, if reflection from the complete half sphere is collected and contributes to producing an interferogram. The difference between half the measured width and the half-height (radius) should be less than 160 nm, at maximum magnification.

Note the laser induced optical damage sites discussed in section 2 represent a simpler case for study with the OSP. Even though the topology of the damage sites is not necessarily simple or as smooth as the undamaged spheres, the samples are predominantly a single surface to be profiled and the OSP can image samples with high average roughness values.

3.2 *Experiments, results and discussion*

The microspheres, and the preparation of the microsphere-on-glass samples are discussed in section 2.1. Further processing of some samples, involving metal coating, is described in the experiments below. Four experiments have been completed to investigate the effect on the resulting optical surface profile, of multiple reflection fields contributing to the interferogram, for microsphere on a surface samples.

3.2.1 *Silica microsphere on a microscope slide*

This involved optical surface profiling 3.5 μm and 5.0 μm diameter silica microspheres on an ordinary microscope slide, of dimensions ~ 75 mm x 25 mm x 1 mm. A scan sequence for these samples has the reflected field first being derived from the top surface of the sphere, but as the scan proceeds it is replaced by the reflection from either the bottom surface of the sphere or the surface of the substrate, which being planar rather than spherical represents a much better reflector than the microsphere. Using the optical surface profiler with highest magnification (50x) and FOV(2x), a series of profiles were recorded using different scan length and different tilt angles for the sample platform. All scans sampled the surface (ie the scan length and the scan start and finish depths were such that the scan image plane traversed the surface and went some distance into the transparent microscope slide. All such scans produced a similar result. A typical three-dimensional topography and a two-dimensional cross sectional image obtained are shown in Figure 7.

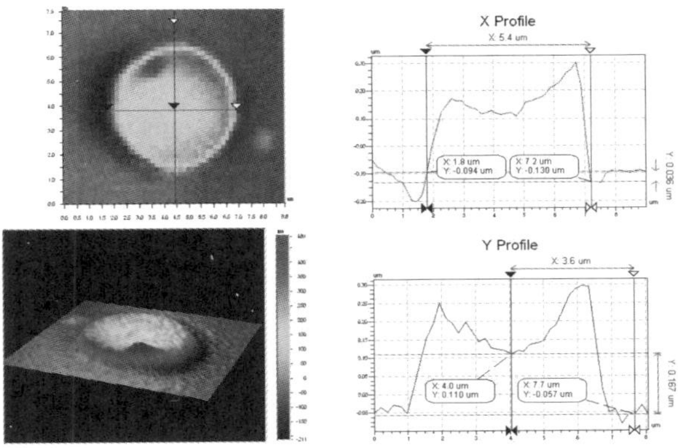

Figure 7: Image produced by the OSP of a "5 μm" transparent silica microsphere on a transparent glass microscope slide. The 2D image has an area of 7.9 μm x 8.8 μm.

Looking at the optical surface profile of the silica microsphere on a glass slide sample we note the horizontal dimension seems to be showing a spherical particle of diameter with reasonable accuracy - 5.4 μm. But, "errors" in the "calibration" are observed, associated with the vertical dimension, which is aberrantly small (relating to neither the radius nor diameter of the microsphere). The profile of the microsphere is not spherical. It actually shows a 'crater' effect, with a height of 0.330 μm for the highest point on the "microsphere", relative to the microscope slide surface. Other microspheres imaged on the same slide gave similar crater results, with a width of approximately 5 μm and a maximum height toward the edge of the crater of no larger than 0.6 μm. Another thing to notice is the channel into the substrate at the sides of the microsphere. This 'shadowing' effect is apparent on all the imaged microspheres and has an apparently random range of values up to ~0.3 μm. Finally the surface of the sphere is measured as being quite rough, not smooth as expected. The image obtained is clearly compromised by the reflection from the microscope slide surface, as seen through the microsphere, being the strongest interfering field with that from the reference mirror, especially near to the vertical axis through the centre of the microsphere.

3.2.2 *Metal coated "silica microsphere on a microscope slide" samples*

The next experiment aims to produce a "single surface" microsphere on a microscope slide surface sample by metal coating the sample prior to optical surface profiling. This should avoid a significant contribution to the interferogram, captured by the z-scan process, from the reflection from the surface of the substrate viewed through the microsphere. The microspheres were applied to the slide as before, but a highly reflective layer was added on top of both the spheres and the substrate. This was achieved using the Edwards E306A evaporative coating system. First a layer of aluminium was applied and then a layer of silver was added. Both of the layers were approximately 300 nm in thickness and served to have a sample for which both the microspheres and the substrate had the same reflectance. No transmission through the microsphere occurred for this metal coating thickness. A typical result is shown in Figure 8.

The scan for this sample shows a circular, flattened shape for the top of the microsphere, with no crater effect in the middle, and no channel/shadowing at the edges of the sphere. Most importantly from a measurement point of view is the difference in the values of the height and width as compared to the first experiment. The measured width is 4.3 μm and the measured height is 4.5 μm. The difference comes from an ambiguity in choosing where the sides of the particles start. The two-dimensional cross section of the sphere also shows that the form is much smoother than the rough surface obtained for the experiment in 3.2.1. The optical surface profiles of the coated microspheres have close to the expected dimensions, and exhibit a smooth surface. However the shape of the microsphere is significantly flattened compared to expectation, even taking the lateral resolution of 160 nm into account.

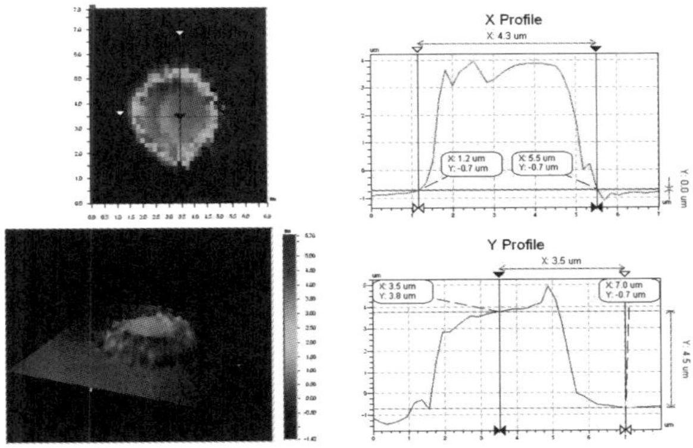

Figure 8: Image produced by the OSP of a "5 μm" transparent Silica Microsphere on a transparent glass microscope slide, all coated with a highly reflective layer of aluminium followed by silver. The 2D image has an area of 6.9 μm x 7.8 μm.

3.2.3 *Silica microsphere on metal coated microscope slide*

The results in 3.2.2 compared to those in 3.2.1 show that optical surface profiling of coated samples produces a more "correct" result than optical surface profiling of uncoated samples. This in turn shows the negative impact on the resulting surface profile of the competition of the reflectance from the surface of the microsphere with the reflection of the substrate as imaged through the microspheres, in producing the highest contrast interference fringe that is mapped as the location of the surface. The effect of the "unwanted" glass substrate surface reflection contribution can be enhanced, relative to the reflection from the microsphere, by metal coating the microscope slide with the same highly reflective layer (Aluminium and Silver), as in 3.2.2, and then introducing the microspheres. A typical optical surface profile result of such a sample is shown in Figure 9. Looking at other spheres on this slide gave similar results where the height (depth) seemed to be half to a third of the measured width.

The image from this case contrasts with both of the previous experiments. It shows a more spherical shape once allowance for the convolution with the lateral instrumentation spread function has been made. The app7arent height (or depth) in fig.9 of 2.6 μm is consistent with the radius of the microsphere and the apparent width of 6.0 μm is consistent with the diameter. It is somewhat larger

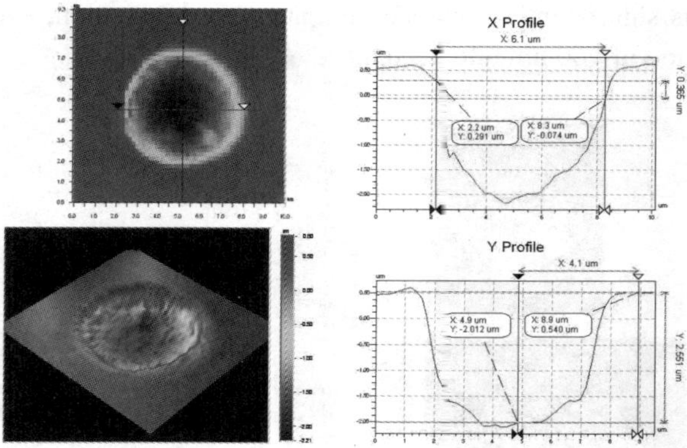

Figure 9: Optical surface profile of a transparent silica microsphere on a metal coated glass substrate (highly reflective). The 2D image has an area of 10.0 μm x 9.3 μm.

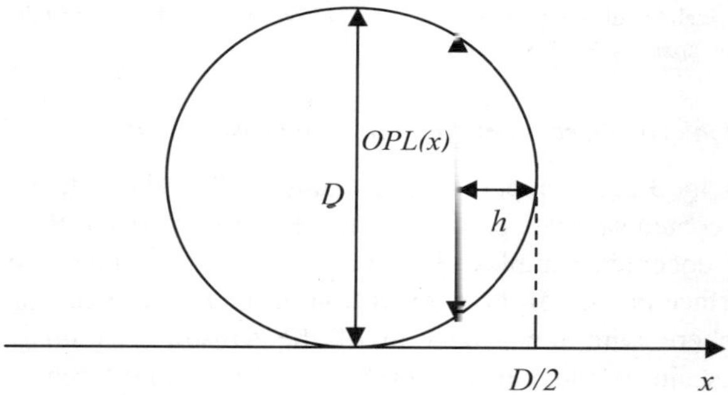

Figure 10: Schematic of the microsphere sitting on a flat substrate showing the quantities used to calculate the optical pathlength increase for paths transmitted vertically through the microsphere.

(~10-20%) than the diameter as is consistent with the result of convolution with the lateral resolution spread function of the OSP Analysis of the increase in the optical path length, OPL(x), for white light incident on the highly reflective substrate surface after transmission through the microsphere (see fig. 10) results in equation 1, which gives this quantity as a function of surface position, x, measured from an origin at the centre, contact point of the microsphere. D is the microsphere diameter and n is the refractive index of the silica microsphere.

$$OPL(x) = 2(n-1)\left[hD - h^2\right]^{1/2}, \quad h = (D/2) - x \tag{1}$$

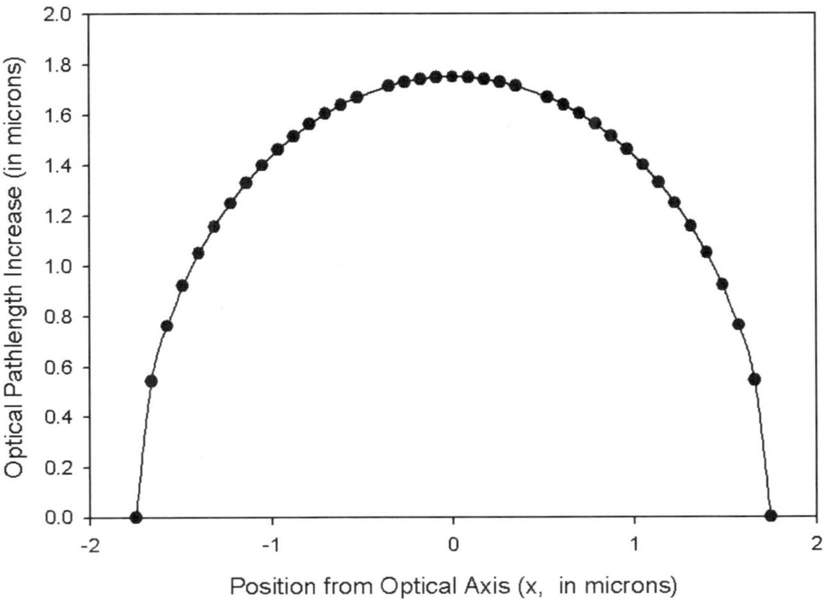

Figure 11: Optical pathlength increase due to a single pass vertically across a 3.5 μm microsphere as a function of the position of that path relative to the point of contact of the microsphere with the surface.

Calculations of the OPL(x) for a 3.5 μm (single pass through the microsphere) are shown in fig. 11. There is qualitative agreement between the shape of this curve, and the surface profile of the microsphere on the coated surface, suggesting the surface profile is consistent with viewing the highly reflective surface through the transparent microsphere. The microsphere acts to increase the OPL to the surface, thus making the surface appear further away than it is. However, there is no quantitative agreement as the "depth" of the depression in fig. 9 is consistent with the radius of the microsphere, not its diameter. It is the latter that is expected for a double pass through a transparent microsphere with a refractive index of about 1.5. The depression in the surface profile is about half (or somewhat less) than the expected depth and is consistent with a single rather than a double pass through the microsphere. It is possible that edge effects due to the radius-high discontinuity between the equator of the microsphere and surface below it may effect the apparent depth. The qualitative agreement and the logic of the argument of viewing the surface through the microsphere lends confidence to this being of importance in the complete explanation. It also allows further insight to be gained into the result for a silica microsphere on a microscope slide shown in fig. 7. The central dip in this profile is likely due to viewing the surface through the microsphere while the outer rim is likely due to the microsphere itself.

The results shown in 3.2.1-3.2.3 are not unique to the WYKO/VEECO profiler. A series of similar experiments were performed using a ZYGO profiler at Imperial College in London and showed the same results with respect to the measurement and shape ambiguities and aberrations of the various microsphere-on-a-surface samples.

3.2.4 *Silica microsphere on a microscope slide with scan length set to avoid the slide surface*

The experiments above show that the reflected field from the surface is a complicating and competing field which contributes in an unwanted way to producing the maximum contrast interference fringes within a z-scan that are ultimately used to map the surface profile. To avoid generating this field altogether it is possible to tailor the z-scan length and the scan start and stop positions so that the scan does not "see" the surface of the substrate. In this way an optical surface profile of the microsphere alone can be formed. Figure 12 shows such a surface profile of a silica microsphere on a transparent, microscope slide substrate. The scan length was 6 μm and tailored to stop just above the surface. The surface profile gave a height of 5.4 μm from the centre position of the particle to the substrate, and a width of 5.3 μm from the outer roughened surface. However, it is only approximately the middle third of the particle that is profiled with good shape and smooth form. Monitoring the interference fringes on the OSPs screen, as the scan progresses, shows that high contrast interference fringes are generated in the top-of-the-microsphere section of the scan but then fade to very low contrast fringes. This indicates that it is only the top cap of the microsphere that generates a reflection field of sufficient irradiance to generate interference fringes of sufficient contrast, when mixed with the reflected field from the reference mirror, to be mapped to a profile with reasonable accuracy of the top half of the microsphere. It is important to note that avoiding a reflection field from the underlying surface does allow an accurate measure of the particle height to be generated, for the central section of the microsphere, and the phase perturbations that occur within the outer thirds of the microsphere do show its width. Monitoring the interference fringes during the z-scan is a diagnostic of whether a reflected field of sufficient irradiance (compared to that from the reference mirror), within the collection numerical aperture of the instrument, is achieved, for a surface of given topology to be successfully profiled. Doing these types of measurements requires great care and is time-consuming. Scans of individual microspheres that happened to be sitting on top of several others also generate profiles of the type shown in fig 12. This is another instance where the effect of a competing reflected field from the surface is avoided. The tilt of the sample stage can also be increased to avoid this reflected field from the surface, but the angles required are very large (10-20°).

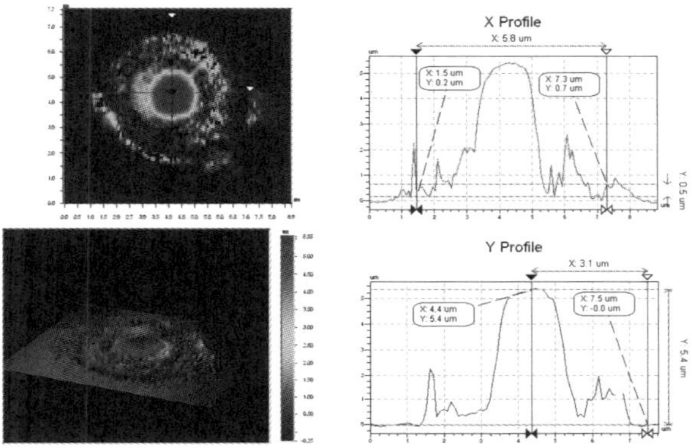

Figure 12: Tailored scan length and stop position to avoid reflection from the surface. The optical surface profile gives the a measured height as expected, for a 5 μm diameter microsphere. The 2D image has an area of 8.8 x 7.7 μm^2.

4. Conclusion

The results presented in section 2 clearly demonstrate the effectiveness of the optical surface profiler in characterization of contouring; patterning; and laser induced optical damage sites, primarily caused by fracture phenomena; of glass and silica surfaces. These samples present the challenge of low reflectance surfaces but successful optical surface profiling has been achieved. Confirmation that the optical surface profiles give a true representation of the sample is obtained from observing that tailoring the scan length to preclude any multiple reflections or using different scan lengths does not result in any measurable changes in the image produced. The technique allows quick and easy observation of relatively large areas of micro-patterning and contouring as compared to stylus and scanning probe profilometry.

For microsphere-on-a-surface samples discussed in section 3, none of the resulting optical surface profiles show an accurate topology for the top surface shape of the microsphere. However, if care is taken to avoid any light reflected from the surface on which the microsphere is contacted, while also scanning to within a few tens of nanometres of the surface, then values for the height and width of the microsphere that are accurate to of the order of 10% are obtained. Such a measurement requires care and time to complete reliably.

All the different samples investigated: microsphere-on-a-glass-surface; microsphere-on-a-metal-coated-surface; and metal-coated-microsphere-on-a-surface; can be used to clearly show the positions of microspheres on a surface. All the samples give reasonable values for the microsphere width and metal-coated-microsphere-on-a-surface samples also give reasonable values for the height. However, it is a significant disadvantage to have to coat samples for measurement. Microsphere-on-a-metal-coated surface samples give values for the height which are linked to the actual sphere diameter, being of the order of the radius. Microsphere-on-a-glass-surface samples give heights which are small (by a factor of the order of 10) compared to the microsphere radius, and which vary considerably depending on sample stage tilt angle and scan length. The results of the microsphere-on-a-glass-surface and microsphere-on-a-metal-coated-surface samples give insight into how multiple reflection fields compete to give different aberrations in the shape of the optical surface profile recorded as compared to the known form of the microsphere on the surface. Application of optical surface profilometry to characterising transparent micorspheres requires very careful methodology, analysis and interpretation. When appropriate methodology is used quantitative information on the diameter of the microsphere can be obtained and the presence and distribution of microspheres is measurable.

Acknowledgements

The Australian Research Council (through an ARC Large Grant) and Macquarie University are thanked for funding this research. DMK would like to thank Dr David McPhail for hosting her visit of The Department of Materials, Imperial College, during which comparative optical surface profiling studies using the Zygo Optical Surface Profiler were completed.

References

1. Kane, D. M., Halfpenny, D. R., (2000) Reduced threshold ultraviolet laser ablation of glass substrates with surface particle coverage: A mechanism for systematic surface laser damage, Journal of Applied Physics, Vol. 87, No. 9, pp 4548-4552.
2. Wyant, J. C., Koliopoulos, C. L., Bhushan, B. and Basila, D., (1986) development of a three-dimensional noncontact digital optical profiler, J Tribology, Vol 108, pp 1-8.
3. Lamb, C., Zecchino, M., (1999) WYKO Surface Profilers Technical Reference Manual, Veeco Metrology Group, Version 2.1.1.
4. Kromasil website with the details of the SM: http://www.kromasil.com/products/products/kromasil/index.htm

5. B Luk'yanchuk (Ed.), Laser Cleaning, World Scientific Publishing (2002). Ibid. D.M. Kane, A.J. Fernandes and D.R. Halfpenny, Pulsed laser cleaning of particles from surfaces and optical materials, ch. 4. pp 181-228.
6. B S Luk'yanchuk , Z B Wang, Y Zhou, M H Hong, W D Song and T C Chong, Particle on asurface: about possible acoustic and plasmonic effects in dry laser cleaning, ch 3, this book.
7. Kim, S., Kim, G., (1999) Thickness-profile measurement of transparent thin-film layers by white-light scanning interferometry, Applied Optics, Vol. 38, No. 28, pp 5968-5973.
8. Deck, L., de Groot, P., (1994) High-speed non-contact profiler based on scanning white-light interferometry, Applied Optics, Vol. 33, No. 31, pp 7334-7338.